The Calculus of Variations and Functional Analysis

With Optimal Control and Applications in Mechanics

SERIES ON STABILITY, VIBRATION AND CONTROL OF SYSTEMS

About the Series

Rapid developments in system dynamics and control, areas related to many other topics in applied mathematics, call for comprehensive presentations of current topics. This series contains textbooks, monographs, treatises, conference proceedings and a collection of thematically organized research or pedagogical articles addressing dynamical systems and control.

The material is ideal for a general scientific and engineering readership, and is also mathematically precise enough to be a useful reference for research specialists in mechanics and control, nonlinear dynamics, and in applied mathematics and physics.

Selected Volumes in Series B

Proceedings of the First International Congress on Dynamics and Control of Systems, Chateau Laurier, Ottawa, Canada, 5–7 August 1999
Editors: A. Guran, S. Biswas, L. Cacetta, C. Robach, K. Teo, and T. Vincent

Selected Volumes in Series A

Vol. 2 Stability of Gyroscopic Systems
Authors: A. Guran, A. Bajaj, Y. Ishida, G. D'Eleuterio, N. Perkins, and C. Pierre

Vol. 3 Vibration Analysis of Plates by the Superposition Method
Author: Daniel J. Gorman

Vol. 4 Asymptotic Methods in Buckling Theory of Elastic Shells
Authors: P. E. Tovstik and A. L. Smirinov

Vol. 5 Generalized Point Models in Structural Mechanics
Author: I. V. Andronov

Vol. 6 Mathematical Problems of Control Theory: An Introduction
Author: G. A. Leonov

Vol. 7 Analytical and Numerical Methods for Wave Propagation in Fluid Media
Author: K. Murawski

Vol. 8 Wave Processes in Solids with Microstructure
Author: V. I. Erofeyev

Vol. 9 Amplification of Nonlinear Strain Waves in Solids
Author: A. V. Porubov

Vol. 10 Spatial Control of Vibration: Theory and Experiments
Authors: S. O. Reza Moheimani, D. Halim, and A. J. Fleming

Vol. 11 Selected Topics in Vibrational Mechanics
Editor: I. Blekhman

SERIES ON STABILITY, VIBRATION AND CONTROL OF SYSTEMS

Series A Volume 12

The Calculus of Variations and Functional Analysis

With Optimal Control and Applications in Mechanics

Leonid P. Lebedev
National University of Colombia, Colombia &
Rostov State University, Russia

Michael J. Cloud
Lawrence Technological University, USA

World Scientific

NEW JERSEY • LONDON • SINGAPORE • SHANGHAI • HONG KONG • TAIPEI • BANGALORE

Published by

World Scientific Publishing Co. Pte. Ltd.
5 Toh Tuck Link, Singapore 596224
USA office: Suite 202, 1060 Main Street, River Edge, NJ 07661
UK office: 57 Shelton Street, Covent Garden, London WC2H 9HE

British Library Cataloguing-in-Publication Data
A catalogue record for this book is available from the British Library.

THE CALCULUS OF VARIATIONS AND FUNCTIONAL ANALYSIS: WITH OPTIMAL CONTROL AND APPLICATIONS IN MECHANICS

ISBN 981-238-581-9

Printed in Singapore by World Scientific Printers (S) Pte Ltd

Foreword

A foreword is essentially an introductory note penned by an invited writer, scholar, or public figure. As a new textbook does represent a pedagogical experiment, a foreword can serve to illuminate the author's intentions and provide a bit of insight regarding the potential impact of the book.

Alfred James Lotka — the famous chemist, demographer, ecologist, and mathematician — once stated that "The preface is that part of a book which is written last, placed first, and read least." Although the following paragraphs do satisfy Lotka's first two conditions, I hope they will not satisfy the third. For here we have a legitimate chance to adopt the sort of philosophical viewpoint so often avoided in modern scientific treatises. This is partly because the present authors, Lebedev and Cloud, have accepted the challenge of unifying three fundamental subjects that were all rooted in a philosophically-oriented century, and partly because the variational method itself has been the focus of controversy over its philosophical interpretation. The mathematical and philosophical value of the method is anchored in its coordinate-free formulation and easy transformation of parameters. In mechanics it greatly facilitates both the formulation and solution of the differential equations of motion. It also serves as a rigorous foundation for modern numerical approaches such as the finite element method. Through some portion of its history, the calculus of variations was regarded as a simple collection of recipes capable of yielding necessary conditions of minimum for interesting yet very particular functionals. But simple application of such formulas will not suffice for reliable solution of modern engineering problems — we must also understand various convergence-related issues for the popular numerical methods used, say, in elasticity. The basis for this understanding is functional analysis: a relatively young branch of mathematics pioneered by Hilbert, Wiener, von

Neumann, Riesz, and many others. It is worth noting that Stefan Banach, who introduced what we might regard as the core of modern functional analysis, lectured extensively on theoretical mechanics; it is therefore not surprising that he knew exactly what sort of mathematics was most needed by engineers.

For a number of years I have delivered lecture courses on system dynamics and control to students and researchers interested in Mechatronics at Johannes Kepler University of Linz, the Technical University of Vienna, and the Technical University of Graz. Mechatronics is an emerging discipline, frequently described as a mixture of mechanics, electronics, and computing; its principal applications are to controlled mechanical devices. Some engineers hold the mistaken view that mechatronics contains nothing new, since both automatic control and computing have existed for a long time. But I believe that mechatronics is a philosophy which happens to overlap portions of the above-mentioned fields without belonging to any of them exclusively. Mechanics, of course, rests heavily on the calculus of variations, and has a long history dating from the works of Bernoulli, Leibniz, Euler, Lagrange, Fermat, Gauss, Hamilton, Routh, and the other pioneers. The remaining disciplines — electronics and computing — are relatively young. Optimal control theory has become involved in mechatronics for obvious reasons: it extends the idea of optimization embodied in the calculus of variations. This involves a significant extension of the class of problems to which optimization can be applied. It also involves an extension of traditional "smooth" analysis tools to the kinds of "non-smooth" tools needed for high-powered computer applications. So again we see how the tools of modern mathematics come into contact with those of computing, and are therefore of concern to mechatronics.

Teaching a combination of the calculus of variations and functional analysis to students in engineering and applied mathematics is a real challenge. These subjects require time, dedication, and creativity from an instructor. They also take special care if the audience wishes to understand the rigorous mathematics used at the frontier of contemporary research. A principal hindrance has been the lack of a suitable textbook covering all necessary topics in a unified and sensible fashion. The present book by Professors Lebedev and Cloud is therefore a welcome addition to the literature. It is lucid, well-connected, and concise. The material has been carefully chosen. Throughout the book, the authors lay stress on central ideas as they present one powerful mathematical tool after another. The reader is thus prepared not only to apply the material to his or her own work, but also to delve further into the literature if desired.

An interesting feature of the book is that optimal control theory arises as a natural extension of the calculus of variations, having a more extensive set of problems and different methods for their solution. Functional analysis, of course, is the basis for justifying the methods of both the calculus of variations and optimal control theory; it also permits us to qualitatively describe the properties of complete physical problems. Optimization and extreme principles run through the entire book as a unifying thread.

The book could function as both (i) an attractive textbook for a course on engineering mathematics at the graduate level, and (ii) a useful reference for researchers in mechanics, electrical engineering, computer science, mechatronics, or related fields such as mechanical, civil, or aerospace engineering, physics, etc. It may also appeal to those mathematicians who lean toward applications in their work. The presence of homework problems at the end of each chapter will facilitate its use as a textbook.

As Poincaré once said, mathematicians do not destroy the obstacles with which their science is spiked, but simply push them toward its boundary. I hope that some particular obstacles in the unification of these three branches of science (the calculus of variations, optimal control, and functional analysis) and technology (mechanics, control, and computing) will continue to be pushed out as far as possible. Professors Lebedev and Cloud have made a significant contribution to this process by writing the present book.

Ardeshir Guran
Wien, Austria
March, 2003

Preface

The successful preparation of engineering students, regardless of specialty, depends heavily upon the basics taught in the junior year. The general mathematical ability of students at this level, however, often forces instructors to simplify the presentation. Requiring mathematical content higher than simple calculus, engineering lecturers must present this content in a rapid, often cursory fashion. A student may see several different lecturers present essentially the same material but in very different guises. As a result "engineering mathematics" often comes to be perceived as a succession of procedures and conventions, or worse, as a mere bag of tricks. A student having this preparation is easily confounded at the slightest twist of a problem. Next, the introduction of computers has brought various approximate methods into engineering practice. As a result the standard mathematical background of a modern engineer should contain tools that belonged to the repertoire of a scientific researcher 30–40 years ago. Computers have taken on many functions that were once considered necessary skills for the engineer; no longer is it essential for the practitioner to be able to carry out extensive calculations manually. Instead, it has become important to understand the background behind the various methods in use: how they arrive at approximations, in what situations they are applicable, and how much accuracy they can provide. In large part, for solving the boundary value problems of mathematical physics, the answers to such questions require knowledge of the calculus of variations and functional analysis. The calculus of variations is the background for the widely applicable method of finite elements; in addition, it can be considered as the first part of the theory of optimal control. Functional analysis allows us to deal with solutions of problems in more or less the same way we deal with vectors in space. A unified treatment of these portions of mathematics, together with examples of how to exploit them in mechanics, is the objective of this book. In this way we hope to contribute in some small way to the preparation of the current and next generations of engineering analysts. The book is introductory in nature, but should provide the reader with a fairly complete picture of the area. Our choice of material is centered around various minimum and optimization problems that play extremely important roles in physics and engineering. Some of the tools presented are absolutely classical, some are quite recent. We collected this material to demonstrate the unity of classical and modern methods, and to enable the reader to understand modern work in this important area.

We would like to thank the World Scientific editorial staff — in particular, Mr. Yeow-Hwa Quek — for assistance in the production of this book. The book appears in the Series on Stability, Vibration and Control of Systems. We owe special thanks to Professors Ardeshir Guran (series Editor-in-Chief, Institute of Structronics in Canada and Johannes Kepler University of Linz in Austria) and Georgios E. Stavroulakis (series Editor, University of Ioannina and Technical University of Braunschweig) for their valuable comments and encouragement. Finally, we are grateful to Natasha Lebedeva and Beth Lannon-Cloud for their patience and support.

Department of Mechanics and Mathematics — L.P. Lebedev
Rostov State University, Russia

&

Department of Mathematics
National University of Colombia, Colombia

Department of Electrical and Computer Engineering — M.J. Cloud
Lawrence Technological University, USA

Contents

Chapter 1

Basic Calculus of Variations

1.1 Introduction

Optimization is a universal human goal. Students would like to learn more, receive better grades, and have more free time; professors (at least some of them!) would like to give better lectures, see students learn more, receive higher pay, and have more free time. These are the optimization problems of real life. In mathematics, optimization makes sense only when formulated in terms of a function $f(x)$ or other expression. We then seek to minimize the value of the expression.[1]

In this book we consider the minimization of *functionals*. The notion of functional generalizes that of function. Although generalization does yield results of greater generality, as a rule we cannot expect these to be sharper in particular cases. So to understand what we can expect of the calculus of variations, we should review the minimization of ordinary functions. We assume everything to be sufficiently differentiable for our purposes.

Let us begin with the one-variable case $y = f(x)$. First we recall some terminology.

Definition 1.1.1 The function $f(x)$ is said to have a *local minimum* at a point x_0 if there is a neighborhood $(x_0 - d, x_0 + d)$ in which $f(x) \geq f(x_0)$. We call x_0 the *global minimum* of $f(x)$ on $[a, b]$ if $f(x) \geq f(x_0)$ holds for all $x \in [a, b]$.

The necessary condition for a differentiable function $f(x)$ to have a local minimum at x_0 is

$$f'(x_0) = 0. \tag{1.1.1}$$

[1] Since the problem of *maximum* of f is equivalent to the problem of minimum of $-f$, it suffices to discuss only the latter type of problem.

A simple and convenient sufficient condition is

$$f''(x_0) > 0. \tag{1.1.2}$$

Unfortunately, no available criterion for a local minimum is both sufficient and necessary. Our approach, then, is to solve (1.1.1) for possible points of local minimum of $f(x)$, and then to test these using one of the available sufficient conditions.

The global minimum on $[a, b]$ can be attained at a point of local minimum. However there are two points, a and b, where (1.1.1) may not be fulfilled (because the corresponding neighborhoods are one-sided) but where the global minimum may still occur. Hence given a differentiable function $f(x)$ on $[a, b]$, we first find all x_k at which $f'(x_k) = 0$. We then calculate $f(a)$, $f(b)$, and $f(x_k)$ at the x_k, and choose the minimal one. This gives us the global minimum. We see that although this method can be formulated as an algorithm suitable for machine computation, it still cannot be reduced to the solution of an equation or system of equations.

These tools are extended to multivariable functions and to more complex objects called functionals. A simple example of a functional is an integral whose integrand depends on an unknown function and its derivative. Since the extension of ordinary minimization methods to functionals is not straightforward, we continue to examine some notions that come to us from calculus.

For a continuously differentiable function $y = f(x)$ we have Lagrange's formula

$$f(x+h) - f(x) = f'(x+\theta h)h \qquad (0 \le \theta \le 1).$$

Since continuity of f' means that

$$f'(x+\theta h) - f'(x) = r_1(x, \theta, h) \to 0 \quad \text{as } h \to 0,$$

we have

$$f(x+h) = f(x) + f'(x)h + r_1(x, \theta, h)\, h$$

where $r_1(x, \theta, h) \to 0$ as $h \to 0$. The term $r_1(x, \theta, h)\, h$ is Lagrange's form of the remainder. There is also Peano's form

$$f(x+h) = f(x) + f'(x)h + o(h),$$

which means that[2]

$$\lim_{h\to 0} \frac{f(x+h) - f(x) - f'(x)h}{h} = 0.$$

The principal (linear in h) part of the increment of f is the *first differential* of f at x. Writing $dx = h$ we have

$$df = f'(x)\,dx.$$

"Infinitely small" quantities are *not* implied by this notation; here dx is a finite increment of x (when used for approximation it should be sufficiently small). The first differential is invariant under the change of variable $x = \varphi(s)$:

$$df = f'(x)\,dx = \frac{df(\varphi(s))}{ds}\,ds,$$

where $dx = \varphi'(s)\,ds$.

Lagrange's formula extends to functions having m continuous derivatives in some neighborhood of x. The extension for $x + h$ lying in the neighborhood is Taylor's formula:

$$f(x+h) = f(x) + \frac{1}{1!}f'(x)h + \frac{1}{2!}f''(x)h^2 + \cdots + \frac{1}{(m-1)!}f^{(m-1)}(x)h^{m-1}$$
$$+ \frac{1}{m!}f^{(m)}(x+\theta h)h^m \qquad (0 \le \theta \le 1).$$

Continuity of $f^{(m)}$ at x yields

$$f^{(m)}(x+\theta h) - f^{(m)}(x) = r_m(x,\theta,h) \to 0 \quad \text{as } h \to 0,$$

hence Taylor's formula becomes

$$f(x+h) = f(x) + \frac{1}{1!}f'(x)h + \frac{1}{2!}f''(x)h^2 + \cdots + \frac{1}{m!}f^{(m)}(x)h^m$$
$$+ \frac{1}{m!}r_m(x,\theta,h)h^m$$

with remainder in Lagrange form. When we do not wish to carefully display the dependence of the remainder on the parameters in Taylor's formula, we

---

[2]We write $g(x) = o(r(x))$ as $x \to x_0$ if $g(x)/r(x) \to 0$ as $x \to x_0$. See § 1.9 for further discussion of this notation.

use Peano's form

$$f(x+h) = f(x) + \frac{1}{1!}f'(x)h + \frac{1}{2!}f''(x)h^2 + \cdots + \frac{1}{m!}f^{(m)}(x)h^m + o(h^m). \tag{1.1.3}$$

The conditions of minimum (1.1.1)–(1.1.2) can be derived via Taylor's formula for a twice continuously differentiable function having

$$f(x+h) - f(x) = f'(x)h + \frac{1}{2}f''(x)h^2 + o(h^2).$$

Indeed $f(x+h) - f(x) \geq 0$ if x is a local minimum. The right-hand side has the form $ah + bh^2 + o(h^2)$. If $a = f'(x) \neq 0$, for example when $a < 0$, it is clear that for $h < h_0$ with sufficiently small h_0 the sign of $f(x+h) - f(x)$ is determined by that of ah; hence for $0 < h < h_0$ we have $f(x+h) - f(x) < 0$, which contradicts the assertion that x minimizes f. The case $a > 0$ is similar, and we arrive at the necessary condition (1.1.1). Returning to the increment formula we now get

$$f(x+h) - f(x) = \frac{1}{2}f''(x)h^2 + o(h^2).$$

The term $f''(x)h^2$ defines the value of the right-hand side when h is sufficiently close to 0, hence when $f''(x) > 0$ we see that for sufficiently small $|h| \neq 0$

$$f(x+h) - f(x) > 0.$$

So (1.1.2) is sufficient for x to be a minimum point of f.

A function in n variables

We cannot expect more from the theory of minimum of a function $y = f(\mathbf{x})$ with $\mathbf{x} = (x_1, \ldots, x_n)$.[3]

We say that $f(\mathbf{x})$ has a *global minimum* at the point $\mathbf{x}^*$ if the inequality

$$f(\mathbf{x}^*) \leq f(\mathbf{x}^* + \mathbf{h}) \tag{1.1.4}$$

holds for all nonzero $\mathbf{h} = (h_1, \ldots, h_n) \in \mathbb{R}^n$. We call $\mathbf{x}^*$ a *local minimum* if there exists $\rho > 0$ such that (1.1.4) holds whenever

$$\|\mathbf{h}\| = (h_1^2 + \cdots + h_n^2)^{1/2} < \rho.$$

[3] We will use the notations $f(\mathbf{x})$ and $f(x_1, \ldots, x_n)$ interchangeably.

Let $\mathbf{x}^*$ be a minimum point of a continuously differentiable function $f(\mathbf{x})$. Then $f(x_1, x_2^*, \ldots, x_n^*)$ is a function in one variable x_1 and takes its minimum at x_1^*. It follows that $\partial f/\partial x_1 = 0$ at $x_1 = x_1^*$. Similarly we see that the rest of the partial derivatives of f are zero at $\mathbf{x}^*$:

$$\left.\frac{\partial f}{\partial x_i}\right|_{\mathbf{x}=\mathbf{x}^*} = 0, \qquad i = 1, \ldots, n. \tag{1.1.5}$$

This is a necessary condition of minimum for a continuously differentiable function in n variables at the point $\mathbf{x}^*$.

To get sufficient conditions we must extend Taylor's formula. Let $f(\mathbf{x})$ possess all continuous derivatives up to order $m \geq 2$ in some neighborhood of a point $\mathbf{x}$, and suppose $\mathbf{x} + \mathbf{h}$ lies in this neighborhood. Fixing these, we apply (1.1.3) to $f(\mathbf{x} + t\mathbf{h})$ and get Taylor's formula in the variable t:

$$f(\mathbf{x}+t\mathbf{h}) = f(\mathbf{x}) + \frac{1}{1!}\left.\frac{df(\mathbf{x}+t\mathbf{h})}{dt}\right|_{t=0} t + \frac{1}{2!}\left.\frac{d^2 f(\mathbf{x}+t\mathbf{h})}{dt^2}\right|_{t=0} t^2 + \cdots + \frac{1}{m!}\left.\frac{d^m f(\mathbf{x}+t\mathbf{h})}{dt^m}\right|_{t=0} t^m + o(t^m).$$

The remainder term is for the case when $t \to 0$. We underline that this is an equality for sufficiently small t. From this, the general Taylor formula can be derived.

To study the problem of minimum of $f(\mathbf{x})$, we need consider only the first two terms of this formula:

$$f(\mathbf{x}+t\mathbf{h}) = f(\mathbf{x}) + \frac{1}{1!}\left.\frac{df(\mathbf{x}+t\mathbf{h})}{dt}\right|_{t=0} t + \frac{1}{2!}\left.\frac{d^2 f(\mathbf{x}+t\mathbf{h})}{dt^2}\right|_{t=0} t^2 + o(t^2). \tag{1.1.6}$$

Calculating $df(\mathbf{x}+t\mathbf{h})/dt$ as a derivative of a composite function, we have

$$\left.\frac{df(\mathbf{x}+t\mathbf{h})}{dt}\right|_{t=0} = \frac{\partial f(\mathbf{x})}{\partial x_1} h_1 + \frac{\partial f(\mathbf{x})}{\partial x_2} h_2 + \cdots + \frac{\partial f(\mathbf{x})}{\partial x_n} h_n.$$

Writing $dx_i = th_i$ we can define the first differential

$$df = \frac{\partial f(\mathbf{x})}{\partial x_1} dx_1 + \frac{\partial f(\mathbf{x})}{\partial x_2} dx_2 + \cdots + \frac{\partial f(\mathbf{x})}{\partial x_n} dx_n.$$

Similarly for the next term we have

$$\left.\frac{d^2 f(\mathbf{x}+t\mathbf{h})}{dt^2}\right|_{t=0} = \sum_{i,j=1}^{n} \frac{\partial^2 f(\mathbf{x})}{\partial x_i \, \partial x_j} h_i h_j.$$

This defines the second differential of f:

$$d^2 f = \sum_{i,j=1}^{n} \frac{\partial^2 f(\mathbf{x})}{\partial x_i\, \partial x_j}\, dx_i\, dx_j.$$

Taylor's formula of the second order can now be written as

$$f(\mathbf{x}+\mathbf{h}) = f(\mathbf{x}) + \frac{1}{1!}\sum_{i=1}^{n} \frac{\partial f(\mathbf{x})}{\partial x_i} h_i + \frac{1}{2!}\sum_{i,j=1}^{n} \frac{\partial^2 f(\mathbf{x})}{\partial x_i \partial x_j} h_i h_j + o(\|\mathbf{h}\|^2).$$

As with the one-variable case, from (1.1.6) we have the necessary condition $df = 0$ at a point of minimum which, besides, follows from (1.1.5). It also follows from (1.1.6) that

$$\left.\frac{d^2 f(\mathbf{x}+t\mathbf{h})}{dt^2}\right|_{t=0} > 0 \text{ for any sufficiently small } \|\mathbf{h}\|$$

suffices for $\mathbf{x}$ to minimize f. The corresponding quadratic form in the variables h_i is

$$\frac{1}{2}\begin{pmatrix} h_1 & \cdots & h_n \end{pmatrix} \begin{pmatrix} \dfrac{\partial^2 f(\mathbf{x})}{\partial x_1^2} & \cdots & \dfrac{\partial^2 f(\mathbf{x})}{\partial x_1 x_n} \\ \vdots & \ddots & \vdots \\ \dfrac{\partial^2 f(\mathbf{x})}{\partial x_n x_1} & \cdots & \dfrac{\partial^2 f(\mathbf{x})}{\partial x_n^2} \end{pmatrix} \begin{pmatrix} h_1 \\ \vdots \\ h_n \end{pmatrix}.$$

The $n \times n$ *Hessian matrix* is symmetric under our smoothness assumptions regarding f. Positive definiteness of the quadratic form can be verified with use of Sylvester's criterion.

The problem of global minimum for a function in many variables on a closed domain Ω is more complicated than the corresponding problem for a function in one variable. Indeed, the set of points satisfying (1.1.5) can be infinite for a function in many variables. Trouble also arises concerning the domain boundary $\partial\Omega$: since it is no longer a finite set (unlike $\{a, b\}$) we must also solve the problem of minimum on $\partial\Omega$, and the structure of such a set can be complicated. The algorithm for finding a point of global minimum of a function $f(\mathbf{x})$ cannot be described in several phrases; it depends on the structure of both the function and the domain.

To at least avoid the trouble connected with the boundary, we can consider the problem of global minimum of a function on an open domain. We shall do this same thing in our study of the calculus of variations: consider only open domains. Although analogous problems with closed domains arise in applications, the difficulties are so great that no general results are applicable to many problems. One must investigate each such problem separately.

When we have constraints

$$g_i(\mathbf{x}) = 0, \qquad i = 1, \ldots, m,$$

we can reduce the problem of constrained minimization to an unconstrained problem provided we can solve the above equations in the form

$$x_k = \psi_k(x_1, \ldots, x_{n-m}), \qquad k = n - m + 1, \ldots, n.$$

Substitution into $f(\mathbf{x})$ would yield an ordinary unconstrained minimization problem for a function in $n - m$ variables

$$f(x_1, \ldots, x_{n-m}, \ldots, \psi_n(x_1, \ldots, x_{n-m})).$$

The resulting system of equations is nonlinear in general. This situation can be circumvented by the use of Lagrange multipliers. The method proceeds with formation of the *Lagrangian function*

$$\mathcal{L}(x_1, \ldots, x_n, \lambda_1, \ldots, \lambda_m) = f(\mathbf{x}) + \sum_{j=1}^{m} \lambda_j g_j(\mathbf{x}),$$

by which the constraints g_j are adjoined to the function f. Then the x_i and λ_i are all treated as independent, unconstrained variables. The resulting necessary conditions form a system of $n + m$ equations

$$\frac{\partial f(\mathbf{x})}{\partial x_i} + \sum_{j=1}^{m} \lambda_j \frac{\partial g_j(\mathbf{x})}{\partial x_i} = 0, \qquad i = 1, \ldots, n,$$

$$g_j(\mathbf{x}) = 0, \qquad j = 1, \ldots, m,$$

in the $n + m$ unknowns x_i, λ_j.

Functionals

The kind of dependence in which one real number corresponds to another (or to a finite set) is not enough to describe many natural processes. Areas such as physics and biology spawn formulations not amenable to such simple description. Consider the deformations of an airplane in flight. At a certain point near an engine say, the deformation is not merely a function of the force produced by the engine — it also depends on the other engines, air resistance, and passenger positions and movements. (Hence the admonition that everyone remain seated during potentially dangerous parts of the flight.) In general, many real processes in a body are described by the dependence of the displacement field (e.g., the field of strains, stresses, heat, voltage) on other fields (e.g., loads, heat radiation) in the same body. Each field is described by one or more functions, so the dependence here is that of a function uniquely defined by a set of other functions acting as whole objects (arguments). A dependence of this type, provided we specify the classes to which all functions belong, is called an *operator* (or *map*, or sometimes just a "function" again). Problems of finding such dependences are usually formulated as boundary or initial-boundary value problems for partial differential equations. These and their analysis form the main content of any course in a particular science. Since a full description of any process is complex, we often work with simplified models that retain only essential features. However, even these can be quite challenging when we seek solutions.

As humans we often try to optimize our actions through an intuitive — not mathematical — approach to fuzzily-posed problems on minimization or maximization. This is because our nature reflects the laws of nature in total. In physics there are quantities, like energy and enthalpy, whose values in the state of equilibrium or real motion are minimal or maximal in comparison with other "nearby admissible" states. Younger sciences like mathematical biology attempt to follow suit: when possible they seek to describe system behavior through the states of certain fields of parameters, on which functions of energy type attain maxima or minima. The energy of a system (e.g., body or set of interacting bodies) is characterized by a number which depends on the fields of parameters inside the system. Thus the dependence described by quantities of energy type is such that *a numerical value E is uniquely defined by the distribution of fields of parameters characterizing the system.* We call this sort of dependence a *functional.* Of course, in mathematics we must also specify the classes to which the above fields may belong. The notion of functional generalizes that of function so that the minimization problem remains sensible. Hence we come to the object of investigation of our main subject: the calculus of variations. In actuality we shall consider a somewhat restricted class of functionals. (Optimization of general functionals belongs to *mathematical programming*, a younger science that contains the calculus of variations — a subject some 300 years old — as a special case.) In the calculus of variations we minimize functionals of integral type. A typical problem involves the total energy functional for an elastic membrane under load $F = F(x, y)$:

$$E(u) = \frac{1}{2}a \iint_S \left[\left(\frac{\partial u}{\partial x} \right)^2 + \left(\frac{\partial u}{\partial y} \right)^2 \right] dx\,dy - \iint_S Fu\,dx\,dy.$$

Here $u = u(x, y)$ is the deflection of a point (x, y) of the membrane, which occupies a domain S and has tension described by parameter a (we can put $a = 1$ without loss of generality). For a membrane with fixed edge, in equilibrium $E(u)$ takes its minimal value relative to all other *admissible* (or *virtual*) states. (An "admissible" function takes appointed boundary values and is sufficiently smooth, in this case having first and second continuous derivatives in S.) The equilibrium state is described by Poisson's equation

$$\Delta u = -F. \tag{1.1.7}$$

Let us also supply the boundary condition

$$u\big|_{\partial S} = \phi. \tag{1.1.8}$$

The problem of minimum of $E(u)$ over the set of smooth functions satisfying (1.1.8) is equivalent to the boundary value problem (1.1.7)–(1.1.8). Analogous situations arise in electrodynamics, geology, biology, and hydromechanics. Eigenfrequency problems can also be formulated within the calculus of variations.

Other interesting problems come from geometry. Consider the following *isoperimetric problem*:

> *Of all possible smooth closed curves of unit length in the plane, find the equation of that curve L which encloses the greatest area.*

With $r = r(\phi)$ the polar equation of a curve, we seek to have

$$\int_0^{2\pi} \sqrt{r^2 + \left(\frac{dr}{d\phi} \right)^2}\, d\phi = 1, \qquad \frac{1}{2} \int_0^{2\pi} r^2\, d\phi \to \max.$$

Observe the way in which we have denoted the problem of maximization. Every high school student knows the answer, but certainly not the method of solution.

We cannot enumerate all problems solvable by the calculus of variations. It is safe to say only that the relevant functionals possess an integral form, and that the integrands depend upon unknown functions and their derivatives.

Minimization of a simple functional using calculus

Consider a general functional of the form

$$F(y) = \int_a^b f(x, y, y')\, dx, \tag{1.1.9}$$

where $y = y(x)$ is smooth. (At this stage we do not stop to formulate strict conditions on the functions involved; we simply assume they have as many continuous derivatives as needed. Nor do we clearly specify the neighborhood of a function for which it is a local minimizer of a functional.)

From the time of Newton's *Principia*, mathematical physics has formulated and considered each problem so that it has a solution which, at least under certain conditions, is unique. Although the idea of determinism in nature was buried by quantum mechanics, it remained an important part of the older subject of the calculus of variations. We know that for a membrane we must impose boundary conditions. So let us first understand whether the problem of minimum for (1.1.9) is well-posed; i.e., whether (at least for simple particular cases) a solution exists and is unique.

The particular form

$$\int_a^b \sqrt{1 + (y')^2}\, dx$$

yields the length of the plane curve $y = y(x)$ from $(a, y(a))$ to $(b, y(b))$. The obvious minimizer is a straight line $y = kx + d$. Without boundary conditions (i.e., with $y(a)$ or $y(b)$ unspecified), k and d are arbitrary and the solution is not unique. We can clearly impose no more than two restrictions on $y(x)$ at the ends a and b, because $y = kx + d$ has only two indefinite constants. However, the problem without boundary conditions is also sensible.

Problem setup is a tough yet important issue in mathematics. We shall eventually face the question of how to pose the main problems of the calculus of variations in a sensible way.

Let us consider the problem of minimum of (1.1.9) without additional restrictions, and attempt to solve it using calculus. Discretization will reduce the functional to a function in many variables. In the calculus of variations other methods of investigation are customary; however, the current approach is instructive because it leads to some central results of the calculus of variations and shows that certain important ideas are extensions of ordinary calculus.

We begin by subdividing $[a, b]$ into n partitions each of length $h = (b-a)/n$. Denote $x_i = a + ih$ and $y_i = y(x_i)$, so $y_0 = y(a)$ and $y_n = y(b)$.

Take an approximate value of $y'(x_i)$ as $(y_{i+1} - y_i)/h$. Approximating (1.1.9) by the Riemann sum

$$\int_a^b f(x, y, y')\,dx \approx h \sum_{k=0}^{n-1} f(x_k, y_k, y'(x_k)),$$

we get

$$\begin{aligned}\int_a^b f(x, y, y')\,dx &\approx h \sum_{k=0}^{n-1} f(x_k, y_k, (y_{k+1} - y_k)/h) \\ &= \Phi(y_0, \dots, y_n).\end{aligned} \tag{1.1.10}$$

Since $\Phi(y_0, \dots, y_n)$ is an ordinary function in $n+1$ independent variables, we set

$$\frac{\partial \Phi(y_0, y_1, \dots, y_n)}{\partial y_i} = 0, \qquad i = 0, \dots, n. \tag{1.1.11}$$

Again, any function f encountered is assumed to possess all needed derivatives. Henceforth we denote partial derivatives using

$$f_y = \frac{\partial f}{\partial y}, \qquad f_{y'} = \frac{\partial f}{\partial y'}, \qquad f_x = \frac{\partial f}{\partial x},$$

and the total derivative using

$$\begin{aligned}\frac{df(x, y(x), y'(x))}{dx} &= f_x(x, y(x), y'(x)) \\ &\quad + f_y(x, y(x), y'(x))y'(x) \\ &\quad + f_{y'}(x, y(x), y'(x))y''(x).\end{aligned}$$

Observe that in the notation $f_{y'}$ we regard y' as the name of a simple variable; we temporarily ignore its relation to y and even its status as a function in its own right.

Consider the structure of (1.1.11). The variable y_i appears in the sum (1.1.10) only once when $i = 0$ or $i = n$, twice otherwise. In the latter case (1.1.11) gives, using the chain rule and omitting the factor h,

$$\begin{aligned}&\frac{f_{y'}(x_{i-1}, y_{i-1}, (y_i - y_{i-1})/h)}{h} - \frac{f_{y'}(x_i, y_i, (y_{i+1} - y_i)/h)}{h} \\ &\quad + f_y(x_i, y_i, (y_{i+1} - y_i)/h) = 0.\end{aligned} \tag{1.1.12}$$

For $i = 0$ the result is

$$h\left[f_y(x_0, y_0, (y_1 - y_0)/h) - \frac{f_{y'}(x_0, y_0, (y_1 - y_0)/h)}{h}\right] = 0$$

or

$$f_{y'}(x_0, y_0, (y_1 - y_0)/h) - h\, f_y(x_0, y_0, (y_1 - y_0)/h) = 0. \tag{1.1.13}$$

For $i = n$ we obtain

$$f_{y'}(x_{n-1}, y_{n-1}, (y_n - y_{n-1})/h) = 0. \tag{1.1.14}$$

In the limit as $h \to 0$, (1.1.14) gives

$$f_{y'}(x, y(x), y'(x))\big|_{x=b} = 0$$

while (1.1.13) gives

$$f_{y'}(x, y(x), y'(x))\big|_{x=a} = 0.$$

Finally, considering the first two terms in (1.1.12),

$$\frac{f_{y'}(x_{i-1}, y_{i-1}, (y_i - y_{i-1})/h)}{h} - \frac{f_{y'}(x_i, y_i, (y_{i+1} - y_i)/h)}{h} = \\ -\frac{f_{y'}(x_i, y_i, (y_{i+1} - y_i)/h) - f_{y'}(x_{i-1}, y_{i-1}, (y_i - y_{i-1})/h)}{h},$$

we recognize an approximation for the total derivative $-df_{y'}/dx$ at y_{i-1}. Hence (1.1.12), after $h \to 0$ in such a way that $x_{i-1} = c$, reduces to the equation

$$f_y - \frac{d}{dx} f_{y'} = 0 \tag{1.1.15}$$

at any $x = c \in (a, b)$.

In expanded form (1.1.15) is

$$f_y - f_{y'x} - f_{y'y}y' - f_{y'y'}y'' = 0, \qquad x \in (a, b). \tag{1.1.16}$$

The limit passage has given us this second-order ordinary differential equation and two point conditions

$$f_{y'}\big|_{x=a} = 0, \qquad\qquad f_{y'}\big|_{x=b} = 0. \tag{1.1.17}$$

Equations (1.1.15) and (1.1.17) play the same role for the functional (1.1.9) as do equations (1.1.5) for a function in many variables. Hence if we impose no boundary conditions on $y(x)$, we get necessarily two boundary conditions for a function on which (1.1.9) attains a minimum.

Since the resulting equation is of second order, we can impose no more than two boundary conditions on its solution (see, however, Remark 1.5.1). We could, say, fix the ends of the curve $y = y(x)$ by putting

$$y(a) = c_0, \qquad y(b) = c_1. \tag{1.1.18}$$

If we repeat the above process under this restriction we get (1.1.12) and correspondingly (1.1.15), whereas (1.1.17) is replaced by (1.1.18). We can consider the problem of minimum of this functional on the set of functions satisfying (1.1.18). Then the necessary condition which a minimizer should satisfy is the boundary value problem consisting of (1.1.15) and (1.1.18).

We may wonder what happens if we require

$$y(a) = 0, \qquad y'(a) = 0.$$

After all, these are normally posed for a Cauchy problem involving a second-order differential equation. In the present case, however, a repetition of the above steps implies the *additional* restriction

$$f_{y'}\Big|_{x=b} = 0.$$

A problem for (1.1.15) with three boundary conditions is, in general, inconsistent.

So we now have some possible forms of the setup for the problem of minimum of the functional (1.1.9).

Brief summary of important terms

A *functional* is a correspondence assigning a real number to each function in some class of functions. The calculus of variations is concerned with *variational problems*: i.e., those in which we seek the *extrema* (maxima or minima) of functionals.

An *admissible function* for a given variational problem is a function that satisfies all the constraints of that problem.

We say that a function is "sufficiently smooth" for a particular development if all required actions (e.g., differentiation, integration by parts) are possible and yield results having the properties needed for that development.

1.2 Euler's Equation for the Simplest Problem

We begin with the problem of local minimum of the functional

$$F(y) = \int_a^b f(x, y, y')\,dx \tag{1.2.1}$$

on the set of functions $y = y(x)$ that satisfy the boundary conditions

$$y(a) = c_0, \qquad y(b) = c_1. \tag{1.2.2}$$

We now become explicit about this set, since on its properties the very existence of a solution can depend. In the present problem we must compare the values of $F(y)$ on all functions y satisfying (1.2.2). In view of (1.1.15) it is reasonable to seek minimizers that have continuous first and second derivatives on $[a, b]$.[4] Next, how do we specify a neighborhood of a function $y = y(x)$? Since all admissible functions must satisfy (1.2.2), we can consider the set of functions of the form $y(x) + \varphi(x)$ where

$$\varphi(a) = \varphi(b) = 0. \tag{1.2.3}$$

Since we wish to employ tools close to those of classical calculus, we first introduce the idea of continuity of a functional with respect to an argument which, in turn, is a function on $[a, b]$. A suitably modified version of the classical definition of function continuity is as follows: given any small $\varepsilon > 0$, there exists a δ-neighborhood of $y(x)$ such that when $y(x) + \varphi(x)$ belongs to this neighborhood we have

$$|F(y + \varphi) - F(y)| < \varepsilon.$$

It is seen that if the neighborhood of the zero function is specified by the inequality

$$\max_{x\in[a,b]} |\varphi(x)| + \max_{x\in[a,b]} |\varphi'(x)| < \delta, \tag{1.2.4}$$

the definition can become workable when $f(x, y, y')$ is continuous in the three independent variables x, y, y'. Of course, this is not the only possible

[4] It is good to prove statements under minimally restrictive conditions. However, new techniques are often developed without worrying too much about the degree of function smoothness required at each step: it is okay to suppose whatever degree of smoothness is needed and go ahead. When the desired result is obtained, then one can begin to consider which hypotheses could be weakened. Such refinement is important but should not be attempted at the outset, lest one become overwhelmed by details and never reach any valuable results.

definition of a neighborhood, and later we shall discuss other possibilities. But one benefit is that the left side of (1.2.4) contains the expression usually used to define the norm on the set of all functions continuously differentiable on $[a, b]$:

$$\|\varphi(x)\| = \max_{x\in[a,b]} |\varphi(x)| + \max_{x\in[a,b]} |\varphi'(x)|. \tag{1.2.5}$$

This set, supplied with the norm (1.2.5), is called the normed space $C^{(1)}(a, b)$. Its subspace of functions satisfying (1.2.3) we shall denote by $C_0^{(1)}(a, b)$. The space $C^{(1)}(a, b)$ is considered in functional analysis; it has many important properties, but in the first part of this book we shall need nothing further than the convenient notation. We denote by $C^{(k)}(a, b)$ the set of all functions having k continuous derivatives on $[a, b]$.

Thus a δ-neighborhood of $y(x)$ is the set of all functions of the form $y(x) + \varphi(x)$ where $\varphi(x)$ is such that $\varphi(x) \in C_0^{(1)}(a, b)$ and $\|\varphi(x)\| < \delta$.

Definition 1.2.1 We say that $y(x)$ is a *point of local minimum* of $F(y)$ on the set of functions satisfying (1.2.2) if there is a δ-neighborhood of $y(x)$, i.e., a set of functions $z(x)$ such that $z(x) - y(x) \in C_0^{(1)}(a, b)$ and $\|z(x) - y(x)\| < \delta$, in which

$$F(z) - F(y) \geq 0.$$

If in a δ-neighborhood we have $F(z) - F(y) > 0$ for all $z(x) \neq y(x)$, then $y(x)$ is a *point of strict local minimum.*

It is possible to speak of more than one type of local minimum. According to Definition 1.2.1, a function y is a minimum if there is a δ such that

$$F(y + \varphi) - F(y) \geq 0 \text{ whenever } \|\varphi\|_{C_0^{(1)}(a,b)} < \delta.$$

Historically this type of minimum is called "weak" and in what follows we will use only this type and refer to it simply as a minimum. But those who pioneered the calculus of variations also considered so-called strong local minima, defining these as values of y for which there is a δ such that $F(y + \varphi) \geq F(y)$ whenever $\max|\varphi| < \delta$ on $[a, b]$. Here the modified condition on φ permits "strong variations" into consideration: i.e., functions φ for which φ' may be large even though φ itself is small. Note that when we "weaken" the condition on φ by changing the norm from the norm of $C_0^{(1)}(a, b)$ to the norm of $C_0(a, b)$ which contains only φ and not φ', we simultaneously strengthen the statement we make regarding y when we assert the inequality $F(y+\varphi) \geq F(y)$.

Let us now turn to a rigorous justification of (1.1.15). We restrict the class of possible integrands $f(x,y,z)$ of (1.2.1) to the set of functions that are continuous in (x,y,z) when $x \in [a,b]$ and $|y - y(x)| + |z - y'(x)| < \delta$. Suppose the existence of a minimizer $y(x)$ for $F(y)$.[5] Consider $F(y+t\varphi)$ for an arbitrary but fixed $\varphi(x) \in C_0^{(1)}(a,b)$. It is a function in the single variable t, taking its minimum at $t = 0$. If it is differentiable then

$$\left.\frac{dF(y+t\varphi)}{dt}\right|_{t=0} = 0. \tag{1.2.6}$$

In order to justify differentiation under the integral sign, we assume $f(x,y,y')$ is continuously differentiable in the variables y and y'. In fact, (1.1.16) demonstrates that we shall need the existence of other derivatives of f as well. We shall end up assuming that $f(x,y,y')$ is twice continuously differentiable, in any combination of its arguments, in the domain of interest.

Let us carry out the derivative in (1.2.6) using the chain rule:

$$\begin{aligned} 0 &= \frac{d}{dt}\int_a^b f(x, y+t\varphi, y'+t\varphi')\,dx\bigg|_{t=0} \\ &= \int_a^b \left[f_y(x,y,y')\varphi + f_{y'}(x,y,y')\varphi'\right]dx. \end{aligned} \tag{1.2.7}$$

We denote the right member of (1.2.7) by $\delta F(y,\varphi)$ and call it the *first variation* of the functional (1.2.1). Integration by parts applied to the second term on the right in (1.2.7) gives

$$\int_a^b f_{y'}(x,y,y')\varphi'\,dx = -\int_a^b \varphi \frac{d}{dx} f_{y'}(x,y,y')\,dx$$

where the boundary terms vanish by (1.2.3). It follows that

$$\int_a^b \left[f_y(x,y,y') - \frac{d}{dx} f_{y'}(x,y,y')\right]\varphi\,dx = 0. \tag{1.2.8}$$

[5]This can lead to incorrect conclusions, and it is normally necessary to prove the existence of an object having needed properties. Perron's paradox illustrates the sort of consequences we may reach by supposing the existence of a non-existent object. Suppose there exists a greatest positive integer N. Since N^2 is also a positive integer we must have $N^2 \leq N$, from which it follows that $N = 1$. If we knew nothing about the integers we might believe this result and attempt to base an entire theory on it.

In the integrand we see the left-hand side of (1.1.15). To deduce (1.1.15) from (1.2.8) we need the "fundamental lemma" of the calculus of variations.

Lemma 1.2.1 *Let $g(x)$ be continuous on $[a,b]$, and let*

$$\int_a^b g(x)\varphi(x)\,dx = 0 \tag{1.2.9}$$

hold for any function $\varphi(x)$ that is differentiable on $[a,b]$ and vanishes in some neighborhoods of a and b. Then $g(x) \equiv 0$.

Proof. Suppose to the contrary that (1.2.9) holds while $g(x_0) \neq 0$ for some $x_0 \in (a,b)$. Without loss of generality we may assume $g(x_0) > 0$. By continuity, $g(x) > 0$ in a neighborhood $[x_0 - \varepsilon, x_0 + \varepsilon] \subset (a,b)$. It is easy to construct a nonnegative bell-shaped function $\varphi_0(x)$ such that $\varphi_0(x)$ is differentiable, $\varphi_0(x_0) > 0$, and $\varphi_0(x) = 0$ outside $(x_0 - \varepsilon, x_0 + \varepsilon)$. See Fig. 1.1. The product $g(x)\varphi_0(x)$ is nonnegative everywhere and positive near x_0. Hence $\int_a^b g(x)\varphi(x)\,dx > 0$, a contradiction. □

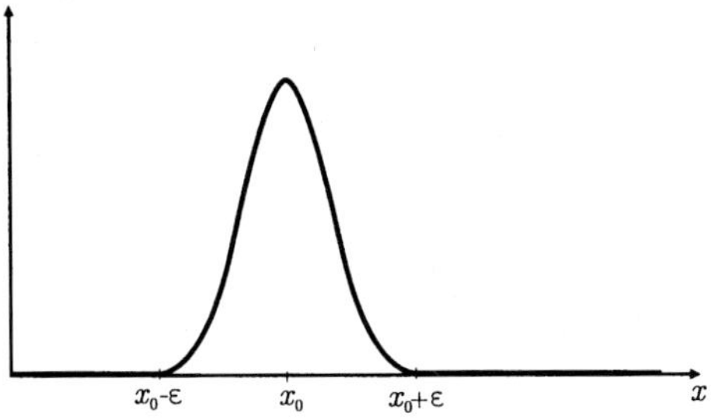

Fig. 1.1 Bell-shaped function for the proof of Lemma 1.2.1.

Note that in Lemma 1.2.1 it is possible to further restrict the class of functions $\varphi(x)$.

Lemma 1.2.2 *Let $g(x)$ be continuous on $[a,b]$, and let* (1.2.9) *hold for any function $\varphi(x)$ that is infinitely differentiable on $[a,b]$ and vanishes in some neighborhoods of a and b. Then $g(x) \equiv 0$.*

The proof is the same as that for Lemma 1.2.1: it is necessary to construct the same bell-shaped function $\varphi(x)$ that is infinitely differentiable. This form of the fundamental lemma provides a basis for the so-called theory of generalized functions or distributions. These are linear functionals on the sets of infinitely differentiable functions, and arise as elements of the Sobolev spaces to be discussed later.

Now we can formulate the main result of this section.

Theorem 1.2.1 *Suppose* $y = y(x) \in C^{(2)}(a,b)$ *locally minimizes the functional* (1.2.1) *on the subset of* $C^{(1)}(a,b)$ *consisting of those functions satisfying* (1.2.2). *Then* $y(x)$ *is a solution of the equation*

$$f_y - \frac{d}{dx} f_{y'} = 0. \tag{1.2.10}$$

Proof. Under the assumptions of this section (including that $f(x,y,y')$ is twice continuously differentiable in its arguments), the bracketed term in (1.2.8) is continuous on $[a,b]$. Since (1.2.8) holds for any $\varphi(x) \in C_0^{(1)}(a,b)$, Lemma 1.2.1 applies. □

Definition 1.2.2 Equation (1.2.10) is known as the *Euler equation*, and a solution $y = y(x)$ is called an *extremal* of (1.2.1). A functional is *stationary* if its first variation vanishes.

Observe that (1.2.10) and (1.2.2) taken together constitute a boundary value problem for the unknown $y(x)$.

Example 1.2.1 Find a function $\bar{y} = \bar{y}(x)$ that minimizes the functional

$$F(y) = \int_0^1 [y^2 + (y')^2 - 2y]\, dx$$

subject to the conditions $y(0) = 1$ and $y(1) = 0$.

Solution Here $f(x,y,y') = y^2 + (y')^2 - 2y$, so we obtain

$$f_y = 2y - 2, \qquad f_{y'} = 2y',$$

and the Euler equation is

$$y'' - y + 1 = 0.$$

Subject to the given boundary conditions, the solution is

$$\bar{y}(x) = 1 - \frac{e^x - e^{-x}}{e - e^{-1}}.$$

We stress that this is an extremal: only supplementary investigation can determine whether it is an actual minimizer of $F(y)$. Consider the difference

$F(\bar{y}+\varphi)-F(\bar{y})$ where $\varphi(x)$ vanishes at $x=0,1$. It is easily shown that

$$F(\bar{y}+\varphi)-F(\bar{y})=\int_0^1 [\varphi^2+(\varphi')^2]\,dx \geq 0,$$

so $\bar{y}(x)$ really is a global minimum of $F(y)$.

We should point out that such direct verification is not always straightforward. However, a large class of important problems in mechanics (e.g., problems of equilibrium for linearly elastic structures under conservative loads) can be solved by minimizing a total energy functional. In such cases we will always encounter a single extremal that minimizes the total energy. This happens because of the quadratic structure of the functional, just as in the present example.

Certain forms of f can lead to simplification of the Euler equation. The reader can easily show the following:

(1) If f does not depend explicitly on y, then $f_{y'}=\text{constant}$.
(2) If f does not depend explicitly on x, then $f-f_{y'}y'=\text{constant}$.
(3) If f depends explicitly on y' only and $f_{y'y'}\neq 0$, then $y(x)=c_1x+c_2$.

1.3 Some Properties of Extremals of the Simplest Functional

In our attempt to seek a minimizer on a subset of $C^{(1)}(a,b)$, we imposed the illogical restriction (f does not depend on y''!) that it must belong to $C^{(2)}(a,b)$. Let us consider how to circumvent this requirement.

Lemma 1.3.1 *Let $g(x)$ be a continuous function on $[a,b]$ for which the equality*

$$\int_a^b g(x)\varphi'(x)\,dx=0 \tag{1.3.1}$$

holds for any $\varphi(x)\in C_0^{(1)}(a,b)$. Then $g(x)$ is constant.

Proof. For a constant c it is evident that $\int_a^b c\varphi'(x)\,dx=0$ for any $\varphi(x)\in C_0^{(1)}(a,b)$. So $g(x)$ can be an arbitrary constant. We show that there are no other forms for g. From (1.3.1) it follows that

$$\int_a^b [g(x)-c]\varphi'(x)\,dx=0. \tag{1.3.2}$$

Take $c = c_0 = (b-a)^{-1}\int_a^b g(x)\,dx$. The function $\varphi(x) = \int_a^x [g(s) - c_0]\,ds$ is continuously differentiable and satisfies $\varphi(a) = \varphi(b) = 0$. Hence we can put it into (1.3.2) and obtain

$$\int_a^b [g(x) - c_0]^2\,dx = 0,$$

from which $g(x) \equiv c$. □

We now use Lemma 1.3.1 to establish a necessary condition for a relative minimum.

Theorem 1.3.1 *Suppose that $y = y(x) \in C^{(1)}(a,b)$ locally minimizes (1.2.1) on the subset of functions in $C^{(1)}(a,b)$ satisfying (1.2.2). Then $y(x)$ is a solution of the equation*

$$\int_0^x f_y(s, y(s), y'(s))\,ds - f_{y'}(x, y(x), y'(x)) = c \tag{1.3.3}$$

with a constant c.

Proof. Let us return to the equality (1.2.7),

$$\int_a^b [f_y(x,y,y')\varphi + f_{y'}(x,y,y')\varphi']\,dx = 0,$$

which is valid here as well. Integration by parts gives

$$\int_a^b f_y(x,y(x),y'(x))\varphi(x)\,dx = -\int_a^b \int_a^x f_y(s,y(s),y'(s))\,ds\,\varphi'(x)\,dx.$$

The boundary terms were zero again because of (1.2.3). It follows that

$$\int_a^b \left[-\int_a^x f_y(s,y(s),y'(s))\,ds + f_{y'}(x,y(x),y'(x))\right]\varphi'(x)\,dx = 0.$$

This holds for all $\varphi(x) \in C_0^{(1)}(a,b)$. So by Lemma 1.3.1 we have (1.3.3). □

The integro-differential equation (1.3.3) has been called the "Euler equation in integrated form."

Corollary 1.3.1 *If*

$$f_{y'y'}(x, y(x), y'(x)) \neq 0$$

along a minimizer $y = y(x) \in C^{(1)}(a,b)$ of (1.2.1), then $y(x) \in C^{(2)}(a,b)$.

Proof. Rewrite (1.3.3) as

$$f_{y'}(x, y(x), y'(x)) = \int_0^x f_y(s, y(s), y'(s))\, ds - c.$$

The function on the right is continuously differentiable for any $y = y(x) \in C^{(1)}(a, b)$. Thus we can differentiate both sides of the last identity with respect to x and obtain

$$f_{y'x} + f_{y'y}y' + f_{y'y'}y'' = \text{a continuous function.}$$

Considering the term with $y''(x)$ on the left, we prove the claim. □

It follows that under the condition of the corollary equations (1.2.10) and (1.3.3) are equivalent; however, this is not the case when $f_{y'y'}(x, y(x), y'(x))$ can be equal to zero on a minimizer $y = y(x)$. Since $y''(x)$ does not appear in (1.3.3), it can be considered as defining a generalized solution of (1.2.10).

At times it becomes clear that we should change variables and consider a problem in another coordinate frame. For example, if we consider geodesic lines on a surface of revolution, then cylindrical coordinates may seem more appropriate than Cartesian coordinates. For the problem of minimum of a functional we have two objects: the functional itself, and the Euler equation for this functional. Let $y = y(x)$ satisfy the Euler equation in the original frame. Let us change variables, for example from (x, y) to (u, v):

$$x = x(u, v), \qquad y = y(u, v).$$

The forms of the functional and its Euler equation both change. Next we change variables for the extremal $y = y(x)$ and get a curve $v = v(u)$ in the new variables. Is $v = v(u)$ an extremal for the transformed functional? It is, provided the transformation does not degenerate in some neighborhood of the curve $y = y(x)$: that is, if the Jacobian

$$J = \begin{vmatrix} x_u & x_v \\ y_u & y_v \end{vmatrix} \neq 0$$

there. This property is called the *invariance* of the Euler equation. Roughly speaking, we can change all the variables of the problem at any stage of the solution and get the same solutions in the original coordinates. This invariance is frequently used in practice. We shall not stop to consider the issue of invariance for each type of functional we treat, but the results are roughly the same.

We have derived a necessary condition for a function to be a point of minimum or maximum of (1.2.1). In what follows we show how this is done for many other functionals. The solution of an Euler equation is the starting point for any variational investigation of a physical problem, and in practice this solution is often undertaken numerically. Let us consider some methods of doing this for (1.2.1).

1.4 Ritz's Method

We now consider a numerical approach to minimizing the functional (1.2.1) with boundary conditions (1.2.2). Corresponding techniques for other problems will be presented later; we shall benefit from a consideration of this simple problem, however, since the main ideas will be the same.

In § 1.1 we obtained the Euler equation for (1.2.1). The intermediate equations (1.1.12) with boundary conditions (1.1.13)–(1.1.14), which for this case must be replaced by the Dirichlet conditions

$$y(a) = y_0 = d_0, \qquad y(b) = y_n = d_1,$$

present us with a finite difference variational method for solving the problem (1.2.10), (1.2.2), belonging to a class of numerical methods based on the idea of representing the derivatives of $y(x)$ in finite-difference form and the functional as a finite sum. These methods differ in how the functions and integrals are discretized. Despite widespread application of the finite element and boundary element methods for the numerical solution of industrial problems, the finite-difference variational methods remain useful because of certain advantages they possess.

Other methods for minimizing a functional, and hence of solving certain boundary value problems, fall under the general heading of *Ritz's method.* Included here are the modifications of the finite element method. Ritz's method was popular before the advent of the computer, and remains so today, because it can yield accurate results for complex problems that are difficult to solve analytically.

The idea of Ritz's method is to reduce the problem of minimizing (1.2.1) on the space of all continuously differentiable functions satisfying (1.2.2) to the problem of minimizing the same functional on a finite dimensional subspace of functions that can approximate the solution. Formerly, the necessity of doing manual calculations forced engineers to choose such subspaces quite carefully, since it was important to get accurate results in as few calculations as possible. The choice of subspace remains an important issue today, because an inappropriate choice can lead to computational instability.

In Ritz's method we seek a solution to the problem of minimization of the functional (1.2.1), with boundary conditions (1.2.2), in the form

$$y_n(x) = \varphi_0(x) + \sum_{k=1}^{n} c_k \varphi_k(x). \tag{1.4.1}$$

Here $\varphi_0(x)$ satisfies (1.2.2); a common choice is the linear function $\varphi_0(x) = \alpha x + \beta$ with

$$\alpha = \frac{d_1 - d_0}{b - a}, \qquad \beta = \frac{bd_0 - ad_1}{b - a}.$$

The remaining functions, called *basis functions*, satisfy the homogeneous conditions

$$\varphi_k(a) = \varphi_k(b) = 0, \qquad k = 1, \ldots, n.$$

The c_k are constants. The function $y_n^*(x)$ that minimizes (1.2.1) on the set of all functions of the form (1.4.1) is called the *nth approximation of the solution by Ritz's method.* It satisfies the boundary conditions (1.2.2) automatically. The above mentioned subspace is the space of functions of the form $\sum_{k=0}^{n} c_k \varphi_k(x)$. For a numerical solution it is necessary that the functions $\varphi_1(x), \ldots, \varphi_n(x)$ be linearly independent, which means that

$$\sum_{k=1}^{n} c_k \varphi_k(x) = 0 \quad \text{only if } c_k = 0 \text{ for } k = 1, \ldots, n.$$

In the days of manual calculation this was supplemented by the requirement that a small value of n — say $n = 1$, 2, or 3 at most — would suffice. This requirement could be met since the corresponding boundary value problems described real objects, such as bent beams, whose shapes under load were understood. Now, to provide a theoretical justification of the method, we require that the system $\{\varphi_k(x)\}_{k=1}^{\infty}$ be *complete.* This means that given any $y = g(x) \in C_0^{(1)}(a, b)$ and $\varepsilon > 0$ we can find a finite sum $\sum_{k=1}^{n} c_k \varphi_k(x)$ such that

$$\left\| g(x) - \sum_{k=1}^{n} c_k \varphi_k(x) \right\| < \varepsilon.$$

(Here the norm is defined by (1.2.5).) It is sometimes required that $\{\varphi_k(x)\}_{k=1}^{\infty}$ be a basis of the corresponding space, but this is not needed for either the justification of the method or its numerical realization.

We have therefore come to the problem of minimum of the functional

$$\int_a^b f(x, y_n, y_n')\,dx$$

where $y_n(x)$ is given by (1.4.1). The unknowns are the c_k, so the functional becomes a function

$$\Phi(c_1, c_2, \ldots, c_n) = \int_a^b f(x, y_n, y_n')\,dx$$

in n real variables. To minimize this we solve the *Ritz system of equations of nth approximation*:

$$\frac{\partial \Phi(c_1, c_2, \ldots, c_n)}{\partial c_k} = 0, \qquad k = 1, \ldots, n. \tag{1.4.2}$$

Denoting $c_0 = 1$, we have

$$\begin{aligned}
\frac{\partial \Phi(c_1, c_2, \ldots, c_n)}{\partial c_k} &= \frac{\partial}{\partial c_k} \int_a^b f(x, y_n, y_n')\,dx \\
&= \frac{\partial}{\partial c_k} \int_a^b f\left(x, \sum_{i=0}^{n} c_i\varphi_i(x), \sum_{i=0}^{n} c_i\varphi_i'(x)\right) dx \\
&= \int_a^b f_y\left(x, \sum_{i=0}^{n} c_i\varphi_i(x), \sum_{i=0}^{n} c_i\varphi_i'(x)\right) \varphi_k(x)\,dx \\
&\quad + \int_a^b f_{y'}\left(x, \sum_{i=0}^{n} c_i\varphi_i(x), \sum_{i=0}^{n} c_i\varphi_i'(x)\right) \varphi_k'(x)\,dx,
\end{aligned}$$

hence (1.4.2) becomes

$$\begin{aligned}
&\int_a^b f_y\left(x, \sum_{i=0}^{n} c_i\varphi_i(x), \sum_{i=0}^{n} c_i\varphi_i'(x)\right) \varphi_k(x)\,dx \\
&\quad + \int_a^b f_{y'}\left(x, \sum_{i=0}^{n} c_i\varphi_i(x), \sum_{i=0}^{n} c_i\varphi_i'(x)\right) \varphi_k'(x)\,dx = 0
\end{aligned} \tag{1.4.3}$$

for $k = 1, \ldots, n$. This is a system of n simultaneous equations in the n variables $c_1, c_2, \ldots, c_n$. It is linear only if f is a quadratic form in c_k; i.e., only if the Euler equation is linear in $y(x)$. For methods of solving simultaneous equations, the reader is referred to specialized books on numerical analysis.

We note that (1.4.3) can be obtained in other ways. We could simply put $y = y_n$ and $\varphi = \varphi_k$ in (1.2.7), since during the derivation of (1.4.3) we used the same steps we used in deriving (1.2.7). Alternatively, we could put y_n into the left-hand side of the Euler equation,

$$f_y(x, y_n, y_n') - \frac{d}{dx} f_{y'}(x, y_n, y_n'), \tag{1.4.4}$$

and then require it to be "orthogonal" to each of the $\varphi_1, \ldots, \varphi_n$. That is, we could multiply (1.4.4) by φ_k, integrate the result over $[a, b]$, use integration by parts on the term with the total derivative d/dx, and equate the result to zero. This is opposite the way we derived (1.4.3). This method of approximating the solution of the boundary value problem (1.2.10), (1.4.1) is called *Galerkin's method.* In the Russian literature it is called the *Bubnov–Galerkin* method, because in 1915 I.G. Bubnov, who was reviewing a paper by S.P. Timoshenko on applications of Ritz's method to the solution of a problem for a bending beam, offered a brief remark on another method of obtaining the equations of Ritz's method. The journal in which Timoshenko's paper appeared happened to publish the comments of reviewers together with the papers (a nice way to hold reviewers responsible for their comments!). In this way Bubnov became an originator of the method. Galerkin was Bubnov's successor, and his real achievement was the development of various forms and applications of the method. In particular, there is a modification of this method wherein (1.4.4) is multiplied not by φ_k, the functions from the representation of y_n, but by other functions $\psi_1, \ldots, \psi_n$. This is sometimes a better way to minimize the "residual" (1.4.4).

We note that the most popular systems of basis functions $\{\varphi_k\}$ for use in Ritz's method for 1-D problems are trigonometric polynomials, or systems of the type $\{(x-a)(x-b)P_k(x)\}$ where the $P_k(x)$ polynomials. Here the factors $(x-a)$ and $(x-b)$ enforce the required homogeneous boundary conditions at $x = a, b$.

When deriving the equations of the Ritz (or Bubnov–Galerkin) method, we imposed no special conditions on $\{\varphi_k\}$ other than linear independence and some smoothness, that is $\varphi_k(x) \in C_0^{(1)}(a, b)$. It is seen that in general each of the equations (1.4.3) contains all of the c_k. By the integral nature of (1.4.3), we see that if we select basis functions so that each $\varphi_k(x)$ is nonzero only on some small part of $[a, b]$, we get a system in which each equation involves only a subset of $\{\varphi_i\}$. This is the background for the finite element method based on Galerkin's method: depending on the problem each equation involves just a few of the c_k (three to five, usually). Moreover, the derivation of the equations of Galerkin's method leads to the idea that it is not necessary to have basis functions with continuous derivatives — it is enough to take the functions with piecewise continuous derivatives of higher order (first order for the problem under consideration) when it is possible to calculate the terms of (1.4.3).

Ritz's method is convenient because it can use low-order approximations to obtain very good results. A disadvantage is that the calculations at a given step are almost independent from those of the previous step. The c_k do not change continuously from step to step; hence, although the next step brings a better approximation, the coefficients can change substantially. Because of accumulated errors there are some limits on the number of basis functions in practical calculations.

Example 1.4.1 Consider the problem

$$\Psi(y) = \int_0^1 \{y'^2(x) + [1 + 0.1\sin(x)]y^2(x) - 2xy(x)\}\, dx \to \min$$

subject to the boundary conditions $y(0) = 0$, $y(1) = 10$. Find the Ritz approximations for $n = 1, 3, 5$ using $\varphi_0(x) = 10x$ and each of the following sets of basis functions:

(a) $\varphi_k(x) = (1 - x)x^k$, $k \geq 1$,
(b) $\varphi_k(x) = \sin k\pi x$, $k \geq 1$.

Solution Note that $\varphi_0(x)$ was chosen to satisfy the given boundary conditions. We must now find the expansion coefficients c_k by solving the simultaneous equations

$$\frac{\partial}{\partial c_k}\Psi\left(\varphi_0(x) + \sum_{i=1}^{n} c_i\varphi_i(x)\right) = 0, \qquad i = 1, \ldots, n.$$

For brevity let us denote

$$\langle y, z\rangle = \int_0^1 \{y'(x)z'(x) + [1 + 0.1\sin(x)]y(x)z(x)\}\, dx$$

so that

$$\Psi(y) = \langle y, y\rangle - 2\int_0^1 xy(x)\,dx.$$

Using the symmetry of the form $\langle y, z\rangle$ we write out Ritz's equations:

$$\begin{aligned}
c_1\langle\varphi_1,\varphi_1\rangle + c_2\langle\varphi_2,\varphi_1\rangle + \cdots + c_n\langle\varphi_n,\varphi_1\rangle &= -\langle\varphi_0,\varphi_1\rangle + \int_0^1 x\varphi_1(x)\,dx,\\
c_1\langle\varphi_1,\varphi_2\rangle + c_2\langle\varphi_2,\varphi_2\rangle + \cdots + c_n\langle\varphi_n,\varphi_2\rangle &= -\langle\varphi_0,\varphi_2\rangle + \int_0^1 x\varphi_2(x)\,dx,\\
&\vdots\\
c_1\langle\varphi_1,\varphi_n\rangle + c_2\langle\varphi_2,\varphi_n\rangle + \cdots + c_n\langle\varphi_n,\varphi_n\rangle &= -\langle\varphi_0,\varphi_n\rangle + \int_0^1 x\varphi_n(x)\,dx.
\end{aligned} \tag{1.4.5}$$

For small n this system can be solved by hand, but for large n a computer solution becomes necessary. In the present case we find that for the first set of basis functions the Ritz approximations are

$$\begin{aligned}
y_1(x) &= 10x - 2.162x(1-x),\\
y_3(x) &= 10x + (-1.409x - 1.356x^2 - 0.246x^3)(1-x),\\
y_5(x) &= 10x + (-1.404x - 1.404x^2 - 0.140x^3 - 0.063x^4 - 0.007x^5)(1-x).
\end{aligned}$$

For the second set of basis functions we obtain the Ritz approximations

$$\begin{aligned}
z_1(x) &= 10x - 0.289\sin\pi x,\\
z_3(x) &= 10x - 0.289\sin\pi x + 0.063\sin 2\pi x - 0.017\sin 3\pi x,\\
z_5(x) &= 10x - 0.289\sin\pi x + 0.063\sin 2\pi x - 0.017\sin 3\pi x\\
&\quad + 0.008\sin 4\pi x - 0.004\sin 5\pi x,
\end{aligned}$$

as required.

In this example we introduced the bilinear form $\langle y, z\rangle$. The symmetry of this form with respect to its arguments simplified some of the required calculations. In the static problems of linear elasticity, such a form is naturally induced by the energy expression for an elastic body. Moreover, the form of the left-hand sides of (1.4.5) is the same for all such problems, whether they are 3-D problems of elasticity, or problems describing elastic beams or shells.

In Ritz's time such approximate solutions were sought for problems describing elastic beams and plates. The resulting systems of equations were fairly hard to solve by hand. The method was justified by comparison with experimental data. A full justification of Ritz's and similar methods requires the tools of functional analysis, which forms the subject of Chapter 3. However, we would like to discuss some aspects of the method on an elementary level using, in particular, Example 1.4.1 as a model.

Notes on basis functions

First let us comment on the approximations. Our working viewpoint is that normally taken in practice: we compare each pair of successive approximations and terminate our calculation process when we reach a pair whose difference is less than some predetermined tolerance ε.

For each type of approximation, if we appoint $\varepsilon = 0.01$ then we can stop at $k = 5$. Calculation out to $k = 10$ shows that the $k = 5$ approximations are both very good. However, they do differ from each other by a maximum of about 0.25. So which is "more" correct? We can answer this by substitution into the functional, which gives $\Psi(y_5) \approx 127.046$ and $\Psi(z_5) \approx 127.449$. This is evidence that polynomial approximation is preferable. It is not hard to see why: the true solution is not oscillatory, so the oscillatory behavior of the trigonometric polynomials is not helpful in this case. So our "practical" approach to terminating the numerical process may not work well for trigonometric approximation. In this particular example it can be shown that the trigonometric approximations do converge, but slowly.

We have selected the polynomial-type Ritz approximations. But our observation regarding trigonometric approximations is cause for concern since the situation with ordinary polynomials should not differ in principle from that with trigonometric polynomials. So we would like to further discuss the problem of basis functions.

In the formulation of Ritz's method we required completeness of the set of basis functions. Let us verify this notion. Weierstrass's theorem of calculus states that any function $f(x)$ continuous on $[0,1]$ can be approximated uniformly by a polynomial to within any accuracy. In other words, given $\varepsilon > 0$ we can find an nth order polynomial $P_n(x)$ such that

$$\max_{x\in[0,1]} |f(x) - P_n(x)| < \varepsilon.$$

It follows that to within any accuracy we may use a polynomial to uniformly approximate a function $f(x)$ together with its continuous derivative. Indeed, given $\varepsilon > 0$, we begin with approximation of the derivative $f'(x)$ by a polynomial $Q_n(x)$:

$$\max_{x\in[0,1]} |f'(x) - Q_n(x)| < \varepsilon/2.$$

The polynomial

$$P_n(x) = f(0) + \int_0^x Q_n(t)\,dt$$

approximates $f(x)$:

$$\begin{aligned}|f(x) - P_n(x)| &= \left|f(0) + \int_0^x f'(t)\,dt - f(0) - \int_0^x Q_n(t)\,dt\right| \\ &\le \int_0^x |f'(t) - Q_n(t)|\,dt \\ &\le \varepsilon/2 \qquad \text{for } x \in [0,1].\end{aligned}$$

In the same way it can be shown that a function n-times continuously differentiable on $[0,1]$ can be approximated to within any prescribed accuracy by a polynomial together with all n of its derivatives on $[0,1]$. Thus *the set of monomials* $\{x^k\}$ *constitutes a complete system of functions in* $C^{(n)}[0,1]$ *for any* n.

Note that Weierstrass' theorem guarantees nothing more than the existence of an approximating polynomial. When we decrease ε we get a new polynomial where the coefficient standing at each term x^k may differ significantly from the corresponding coefficient of the previous approximating polynomial. This is because the set $\{x^k\}$ does not possess the property of uniqueness required of a true basis. Moreover, in mathematical analysis it is shown that we can arbitrarily remove infinitely-many members of the family $\{x^k\}$ and still have a complete system $\{x^{k_r}\}$. For this it is necessary only to retain such members of the family that the series $\sum_{r=1}^{\infty} 1/k_r$ diverges. So the system $\{x^k\}$ contains more members than we need. Although any finite set of monomials x^k is linearly independent, as we take more and more elements the set gets closer to becoming linearly dependent; that is, given any $\varepsilon > 0$ we can find infinitely-many polynomials approximating the zero function to within ε-accuracy on $[0,1]$. This leads to instability in numerical calculation. The difficulty can be avoided by using other families of polynomials for approximation: namely, orthogonal polynomials for which numerical instability shows itself only in higher degrees of approximation.

As we know from the theory of Fourier expansion, the second system of basis functions $\{\sin k\pi x\}$ has the so-called orthonormality property. It is, moreover, a basis as we shall discuss later. This provides greater stability in calculations to within higher accuracy. However, in low-order Ritz approximations it can be worse than a polynomial approximation of the same problem, at least for many problems whose solutions do not oscillate.

There is one additional aspect of the approximation that is seen from the above results. For Ritz's approximations we compared their values. Comparing the values of their derivatives we see that much better agreement is obtained for the values of the approximating functions than for their derivatives. It is obvious that the same holds for the difference between an exact solution and the approximating functions. This property is common to all projection methods. So, for example, in solving problems of elasticity we get comparatively good results in low-order approximations for the field of displacements, whereas the fields of stresses, which are expressed through the derivatives of the displacement fields, are approximated significantly worse.

1.5 Natural Boundary Conditions

In § 1.1 we found that by using discretization on the problem of minimum of the functional (1.2.1) without boundary conditions (or "with free boundary" as we sometimes say) we obtain the Euler equation and some boundary conditions. We shall demonstrate that the same boundary conditions appear by the method of § 1.2. They are known as *natural conditions.*

We consider the minimization of (1.2.1) when there are no restrictions on the boundary for $y = y(x)$.

Theorem 1.5.1 *Let $y = y(x) \in C^{(2)}(a,b)$ be a minimizer of the functional $\int_a^b f(x,y,y')\,dx$ over the space $C^{(1)}(a,b)$. Then for $y = y(x)$ the Euler equation*

$$f_y - \frac{d}{dx} f_{y'} = 0 \qquad \text{for all } x \in (a,b) \tag{1.5.1}$$

holds along with the natural boundary conditions

$$f_{y'}\big|_{x=a} = 0, \qquad\qquad f_{y'}\big|_{x=b} = 0.$$

Proof. We can repeat the initial steps of § 1.2. Namely, consider the values of the functional on the bundle of functions $y = y(x) + t\varphi(x)$ where

$\varphi(x) \in C^{(1)}(a,b)$ is arbitrary but fixed. Here, however, there are no restrictions on $\varphi(x)$ at the endpoints of $[a,b]$.

For fixed $y(x)$ and $\varphi(x)$ the functional $\int_a^b f(x, y + t\varphi, y' + t\varphi')\, dx$ becomes a function of the real variable t, and attains its minimum at $t = 0$. Differentiating with respect to t we get

$$\int_a^b [f_y(x,y,y')\varphi + f_{y'}(x,y,y')\varphi']\, dx = 0.$$

After an integration by parts we have

$$\int_a^b \left[f_y(x,y,y') - \frac{d}{dx} f_{y'}(x,y,y') \right] \varphi\, dx + f_{y'}(x, y(x), y'(x))\varphi(x) \Big|_{x=a}^{x=b} = 0. \tag{1.5.2}$$

From this we shall derive the Euler equation for $y(x)$ and the natural boundary conditions. The procedure is as follows. Let us limit the set of all continuously differentiable functions $\varphi(x)$ to those which satisfy the condition $\varphi(a) = \varphi(b) = 0$. For these functions we have

$$\int_a^b \left[f_y(x,y,y') - \frac{d}{dx} f_{y'}(x,y,y') \right] \varphi\, dx = 0. \tag{1.5.3}$$

This equation holds for all functions $\varphi(x)$ that participate in the formulation of Lemma 1.2.1. Therefore the continuous multiplier of $\varphi(x)$ in the integrand of (1.5.3) is zero. So the Euler equation (1.5.1) holds in (a,b).

Now let us return to (1.5.2). The equality (1.5.3), because of the Euler equation, holds for all $\varphi(x)$. From (1.5.2) it follows that

$$f_{y'}(x, y(x), y'(x))\varphi(x) \Big|_{x=a}^{x=b} = 0 \tag{1.5.4}$$

for any $\varphi(x)$. Taking $\varphi(x) = x - b$ we find that $f_{y'}|_{x=a} = 0$; taking $\varphi(x) = x - a$ we find that $f_{y'}|_{x=b} = 0$. □

We would like to call attention to the way we obtained this result. First we restricted the set of all admissible functions to those for which we could get a certain intermediate result (the Euler equation); because of this result, we obtained some simplification in the first variation of the functional; finally, considering the simplified first variation on *all* the admissible functions, we obtained the rest of the results.

Natural boundary conditions are of great importance in mathematical physics. For some models of real bodies or processes it may be unclear which (and how many) boundary conditions are necessary for well-posedness of the problem. The variational approach usually clarifies the situation and provides natural boundary conditions dictated by the nature of the problem. The bending of a plate is a famous example. For her pioneering studies of this problem Sophie Germain received a prize from the French Academy of Sciences. She derived the biharmonic equation for the deflections of the midsurface of the plate, but with three boundary conditions as seemed to be in accordance with mechanical intuition; variational considerations later demonstrated that only two of these were independent.

Remark 1.5.1 In § 1.1 we discussed the question of which boundary conditions can be imposed to get a well-posed boundary value problem for minimizing the functional (1.2.1). General considerations are nice; however, consider the minimization of

$$\int_0^1 (y'^2 + 2y)\,dx \tag{1.5.5}$$

on the set of continuously differentiable functions. Its Euler equation is $y'' = 1$, thus all the extremals take the form $y = \frac{1}{2}x^2 + kx + b$. The natural boundary conditions are $y'(0) = 0$, $y'(1) = 0$. These imply $k = 0$. So the problem of minimum of (1.5.5) (with natural boundary conditions) has a family of solutions $y = \frac{1}{2}x^2 + b$ with arbitrary constant b. Thus we may impose an additional condition, say $y(0) = 2$. But in general, such a third condition for an ordinary differential equation of second order can yield a boundary value problem that has no solution.

The apparent simplicity of (1.5.5) should not cause the reader to suppose that it represents an unimportant special case, too "degenerate" to be practical: the same situation is to be found with the whole class of functionals that describe the equilibrium states of linear elastic systems in terms of displacements. If we impose no geometrical restrictions on the position of an elastic body (it is normally the case of natural boundary conditions) we can always change the coordinate frame, and all the displacements can be changed in such a way that the body appears to be shifted as a whole (i.e., to move as a "rigid body"). Depending on the model of the body there are apparently one to six free constants describing such a motion — this means that we can impose additional boundary conditions at some points and still preserve the well-posedness of the problem. In a 1-D problem (where the dimension is a spatial coordinate) the situation is exactly as it is for (1.5.5): it is possible to impose an additional boundary condition when considering the problem with "free" ends. Indeed, it is often necessary to be quite careful when applying the outcomes of very general considerations.

1.6 Some Extensions to More General Functionals

Let us consider two extensions of the above results.

The functional $\int_a^b f(x, \mathbf{y}, \mathbf{y}')\,dx$

Let us replace $y(x)$ in (1.2.1) by a vector function

$$\mathbf{y}(x) = (y_1(x), y_2(x), \ldots, y_n(x)).$$

We shall denote the integrand of the functional as

$$f(x, \mathbf{y}(x), \mathbf{y}'(x)) \quad \text{or} \quad f(x, y_1(x), y_2(x), \ldots, y_n(x), y_1'(x), y_2'(x), \ldots, y_n'(x))$$

interchangeably. So our task is to treat functionals of the form

$$F(\mathbf{y}) = \int_a^b f(x, \mathbf{y}, \mathbf{y}')\,dx. \tag{1.6.1}$$

Let us first consider the problem of minimizing (1.6.1) when $\mathbf{y}(x)$ takes boundary values

$$\mathbf{y}(a) = \mathbf{c}_0, \qquad \mathbf{y}(b) = \mathbf{c}_1, \tag{1.6.2}$$

with vector constants $\mathbf{c}_0 = (c_{01}, c_{02}, \ldots, c_{0n})$, $\mathbf{c}_1 = (c_{11}, c_{12}, \ldots, c_{1n})$.

We shall retain the scalar-type notation $C^{(k)}(a,b)$ for a vector function. Hence $\mathbf{y}(x) \in C^{(k)}(a,b)$ means that each coordinate function $y_i(x) \in C^{(k)}(a,b)$; that is, each $y_i(x)$ possesses all derivatives up to order k and these are all continuous on $[a,b]$. We impose the norm

$$\|\mathbf{y}(x)\|_{C^{(k)}(a,b)} = \sum_{i=1}^{n} \|y_i(x)\|_{C^{(k)}(a,b)}$$

on $C^{(k)}(a,b)$, and can thereby define ε-neighborhoods as needed to describe minimizers of the functional (1.6.1). So we seek a minimizer $\mathbf{y}(x)$ of (1.6.1) from among all vector functions belonging to $C^{(1)}(a,b)$ and satisfying (1.6.2).

Theorem 1.6.1 *Suppose* $\mathbf{y}(x) \in C^{(2)}(a,b)$ *locally minimizes the functional* $\int_a^b f(x, \mathbf{y}, \mathbf{y}')\,dx$ *on the subset of vector functions of* $C^{(1)}(a,b)$ *satisfying* (1.6.2). *Then* $\mathbf{y}(x)$ *is a solution of the equation*

$$\nabla_{\mathbf{y}} f - \frac{d}{dx}\nabla_{\mathbf{y}'} f = 0. \tag{1.6.3}$$

Here we use the gradient notation

$$\nabla_{\mathbf{y}} = \left(\frac{\partial}{\partial y_1}, \frac{\partial}{\partial y_2}, \ldots, \frac{\partial}{\partial y_n}\right), \qquad \nabla_{\mathbf{y}'} = \left(\frac{\partial}{\partial y_1'}, \frac{\partial}{\partial y_2'}, \ldots, \frac{\partial}{\partial y_n'}\right).$$

The vector equation (1.6.3) *can be written as* n *scalar equations*

$$f_{y_i} - \frac{d}{dx} f_{y_i'} = 0, \qquad i = 1, \ldots, n, \tag{1.6.4}$$

each having the form of the Euler equation.

Proof. We begin with the same construction of admissible functions, $\mathbf{y}(x) + t\boldsymbol{\varphi}(x)$ where $\boldsymbol{\varphi}(a) = \boldsymbol{\varphi}(b) = 0$, on which we consider (1.6.1):

$$F(\mathbf{y}(x) + t\boldsymbol{\varphi}(x)) = \int_a^b f(x, \mathbf{y} + t\boldsymbol{\varphi}, \mathbf{y}' + t\boldsymbol{\varphi}')\,dx. \tag{1.6.5}$$

For fixed $\mathbf{y}(x)$ and $\boldsymbol{\varphi}(x)$ this becomes a function of the real variable t, and takes its minimum at $t = 0$ for any $\boldsymbol{\varphi}(x)$. Take $\boldsymbol{\varphi}(x)$ of the special form $\boldsymbol{\varphi}_1(x) = (\varphi(x), 0, \ldots, 0)$ where the only nonzero component stands in the first position. Then (1.6.5) becomes

$$F(\mathbf{y}(x) + t\boldsymbol{\varphi}_1(x)) = \int_a^b f(x, y_1(x) + t\varphi(x), y_2(x), \ldots, y_n(x),$$
$$y_1'(x) + t\varphi'(x), y_2'(x), \ldots, y_n'(x))\,dx. \tag{1.6.6}$$

Now the function of t becomes a particular case of the function of § 1.2, $F(y(x) + t\varphi(x))$, with the evident notational change $y \mapsto y_1$. Thus a consequence of the minimum of (1.6.6) at $t = 0$ is the corresponding Euler equation

$$f_{y_1} - \frac{d}{dx} f_{y_1'} = 0.$$

This is the first equation of (1.6.4). Similarly, the ith equation of (1.6.4) is derived by taking $\boldsymbol{\varphi}(x)$ in the form $\boldsymbol{\varphi}_1(x) = (0, \ldots, \varphi_i(x), \ldots, 0)$, where the only nonzero component stands in the ith position. □

We would like to derive the natural boundary conditions for (1.6.1). Now we should not impose any conditions for $\mathbf{y}$ at points $x = a$ and $x = b$ in advance, and thus it is the same for $\boldsymbol{\varphi}$ at these points. For a moment consider all components of the minimizer $\mathbf{y}(x)$ other than $y_i(x)$ to be given. Then (1.6.1) can be formally considered as a particular case of (1.2.1) with respect to the ordinary function $y = y_i(x)$. Now admissible vector functions differ from $\mathbf{y}(x)$ only in the ith component:

$\boldsymbol{\varphi}(x) = \boldsymbol{\varphi}_i(x) = (0, \ldots, \varphi(x), \ldots, 0)$. We can repeat the reasoning of § 1.3. Thus considering the problem of minimum of (1.6.1) without boundary restrictions, we get n pairs of boundary conditions:

$$f_{y_i'}\Big|_{x=a} = 0, \qquad f_{y_i'}\Big|_{x=b} = 0, \qquad\qquad i = 1, \ldots, n.$$

These are natural boundary conditions for a minimizer.

The functional $\int_a^b f(x, y, y', \ldots, y^{(n)})\,dx$

Having obtained the Euler equation for (1.2.1), we now examine the problem of minimum for the functional

$$F_n(y) = \int_a^b f(x, y, y', \ldots, y^{(n)})\,dx. \tag{1.6.7}$$

We may consider this on the set of functions satisfying certain boundary conditions; alternatively we may impose no boundary conditions, and obtain natural conditions as a result.

Let us consider first the problem with given boundary equations. The corresponding Euler equation will have order $2n$, hence we take n conditions at each endpoint:

$$\begin{aligned} y(a) &= c_0^*, & y(b) &= c_0^{**}, \\ y'(a) &= c_1^*, & y'(b) &= c_1^{**}, \\ &\vdots & &\vdots \\ y^{(n-1)}(a) &= c_{n-1}^*, & y^{(n-1)}(b) &= c_{n-1}^{**}. \end{aligned} \tag{1.6.8}$$

We suppose that the integrand is sufficiently smooth for our purposes. Specifically, $f(x, y, y', \ldots, y^{(n)})$ belongs to $C^{(n)}$ on the domain of all of its variables, at least in some neighborhood of a minimizer.

Theorem 1.6.2 *Suppose $y(x) \in C^{(2n)}(a, b)$ locally minimizes $F_n(y)$ of* (1.6.7) *on the subset of vector functions of $C^{(n)}(a, b)$ satisfying* (1.6.8). *Then $y(x)$ satisfies*

$$f_y - \frac{d}{dx} f_{y'} + \frac{d^2}{dx^2} f_{y''} - \cdots + (-1)^n \frac{d^n}{dx^n} f_{y^{(n)}} = 0. \tag{1.6.9}$$

This is known as the Euler–Lagrange equation.

Proof. We begin the proof with a reminder of what it means for $y(x)$ to be a local minimizer of $F_n(y)$. We consider the bundle of functions

$y(x)+\varphi(x)$ where $\varphi(x)$ is arbitrary and belongs to $C^{(n)}(a,b)$. Because the bundle must satisfy (1.6.8) for any $\varphi(x)$, we see that $\varphi(x)$ must satisfy the homogeneous conditions

$$\begin{aligned} \varphi(a) &= 0, & \varphi(b) &= 0, \\ \varphi'(a) &= 0, & \varphi'(b) &= 0, \\ &\vdots & &\vdots \\ \varphi^{(n-1)}(a) &= 0, & \varphi^{(n-1)}(b) &= 0. \end{aligned} \tag{1.6.10}$$

Let $C_0^{(n)}(a,b)$ denote the subspace of $C^{(n)}(a,b)$ consisting of functions $\varphi(x)$ which satisfy (1.6.10). A function $y(x) \in C^{(n)}(a,b)$ satisfying (1.6.8) is called a local minimizer of $F_n(y)$ if $F_n(y+\varphi) \geq F_n(y)$ for any $\varphi(x) \in C_0^{(n)}(a,b)$ and such that $\|\varphi\|_{C^{(n)}(a,b)} < \varepsilon$ for some $\varepsilon > 0$.

As usual we introduce the parameter t and consider the values of $F_n(y)$ on the bundle $y(x)+t\varphi(x)$. Considering $F_n(y(x)+t\varphi(x))$ for a momentarily fixed $\varphi(x)$ as a function of t, we see that it takes its minimal value at $t=0$ and thus $dF_n(y(x)+t\varphi(x))/dt = 0$ at $t=0$. Let us write this out in detail:

$$\begin{aligned} &\left.\frac{dF_n(y(x)+t\varphi(x))}{dt}\right|_{t=0} \\ &= \left.\frac{d}{dt}\int_a^b f(x, y+t\varphi, y'+t\varphi', y''+t\varphi'', \dots, y^{(n)}+t\varphi^{(n)})\,dx\right|_{t=0} \\ &= \int_a^b \left(f_y\varphi + f_{y'}\varphi' + f_{y''}\varphi'' + \cdots + f_{y^{(n)}}\varphi^{(n)}\right)\,dx \end{aligned} \tag{1.6.11}$$

(in the last line of the formula the arguments are $f = f(x, y, y', \dots, y^{(n)})$). Now we must implement traditional (multiple) integration by parts in each term containing derivatives of φ in such a way that on the last step the integrand contains only φ. For the term $\int_a^b f_{y'}\varphi'\,dx$ we already have (1.5.2). For the term $\int_a^b f_{y''}\varphi''\,dx$ we produce

$$\begin{aligned} \int_a^b f_{y''}\varphi''\,dx &= -\int_a^b \varphi'\frac{d}{dx}f_{y''}\,dx + \varphi' f_{y''}\Big|_{x=a}^{x=b} \\ &= \int_a^b \varphi\frac{d^2}{dx^2}f_{y''}\,dx + \left(\varphi' f_{y''} - \varphi\frac{d}{dx}f_{y''}\right)\Big|_{x=a}^{x=b}. \end{aligned}$$

Similarly

$$\int_a^b f_{y'''}\varphi'''\,dx = -\int_a^b \varphi \frac{d^3}{dx^3} f_{y'''}\,dx + \left(\varphi'' f_{y'''} - \varphi' \frac{d}{dx} f_{y'''} + \varphi \frac{d^2}{dx^2} f_{y'''}\right)\Big|_{x=a}^{x=b}$$

and, in general,

$$\int_a^b f_{y^{(n)}}\varphi^{(n)}\,dx = (-1)^n \int_a^b \varphi \frac{d^n}{dx^n} f_{y^{(n)}}\,dx + \left(\varphi^{(n-1)} f_{y^{(n)}} - \varphi^{(n-2)} \frac{d}{dx} f_{y^{(n)}} + \cdots + (-1)^{n-1}\varphi \frac{d^{n-1}}{dx^{n-1}} f_{y^{(n)}}\right)\Big|_{x=a}^{x=b}.$$

By (1.6.9) the boundary terms all vanish, and collecting results we have

$$\int_a^b \left(f_y - \frac{d}{dx} f_{y'} + \frac{d^2}{dx^2} f_{y''} - \cdots + (-1)^n \frac{d^n}{dx^n} f_{y^{(n)}}\right) \varphi\,dx = 0. \qquad (1.6.12)$$

Since this holds for any $\varphi(x) \in C_0^{(n)}(a,b)$, we can quote the fundamental lemma to complete the proof. □

Let us investigate the natural boundary conditions for $F_n(y)$. Now $\varphi(x) \in C^{(n)}(a,b)$, and there are no boundary restrictions on it. The first steps of the previous discussion still apply; however, now there are the boundary terms in the expression for the first variation of $F_n(y)$ (this is the right-hand side of (1.6.11)), so in obtaining the equation analogous to (1.6.12) we should collect all the terms including the boundary terms. We rearrange the boundary terms, collecting coefficients of each $\varphi^{(i)}(x)$:

$$\int_a^b \left(f_y - \frac{d}{dx} f_{y'} + \frac{d^2}{dx^2} f_{y''} - \cdots + (-1)^n \frac{d^n}{dx^n} f_{y^{(n)}} \right) \varphi \, dx$$

$$+ f_{y^{(n)}} \varphi^{(n-1)} \Big|_{x=a}^{x=b}$$

$$+ \left(f_{y^{(n-1)}} - \frac{d}{dx} f_{y^{(n)}} \right) \varphi^{(n-2)} \Big|_{x=a}^{x=b}$$

$$+ \left(f_{y^{(n-2)}} - \frac{d}{dx} f_{y^{(n-1)}} + \frac{d^2}{dx^2} f_{y^{(n)}} \right) \varphi^{(n-3)} \Big|_{x=a}^{x=b}$$

$$\vdots$$

$$+ \left(f_{y'} - \frac{d}{dx} f_{y''} + \cdots + (-1)^{n-1} \frac{d^{n-1}}{dx^{n-1}} f_{y^{(n)}} \right) \varphi \Big|_{x=a}^{x=b} = 0. \qquad (1.6.13)$$

We now realize the common plan. First we consider (1.6.13) only on the subset $C_0^{(n)}(a,b)$ of all $\varphi(x) \in C^{(n)}(a,b)$. Then (1.6.13) reduces to (1.6.12), implying that (1.6.9) holds. Equation (1.6.13) then becomes

$$f_{y^{(n)}} \varphi^{(n-1)} \Big|_{x=a}^{x=b}$$

$$+ \left(f_{y^{(n-1)}} - \frac{d}{dx} f_{y^{(n)}} \right) \varphi^{(n-2)} \Big|_{x=a}^{x=b}$$

$$+ \left(f_{y^{(n-2)}} - \frac{d}{dx} f_{y^{(n-1)}} + \frac{d^2}{dx^2} f_{y^{(n)}} \right) \varphi^{(n-3)} \Big|_{x=a}^{x=b}$$

$$\vdots$$

$$+ \left(f_{y'} - \frac{d}{dx} f_{y''} + \cdots + (-1)^{n-1} \frac{d^{n-1}}{dx^{n-1}} f_{y^{(n)}} \right) \varphi \Big|_{x=a}^{x=b} = 0. \qquad (1.6.14)$$

It is easy to construct the set of polynomials $P_{ik}(x)$, $k = 0, 1$, $i = 0, \ldots, n-1$, with the following properties:

$$\left. \frac{d^j P_{i0}}{dx^j} \right|_{x=a} = \delta_i^j, \qquad \left. \frac{d^j P_{i0}}{dx^j} \right|_{x=b} = 0, \qquad j = 0, 1, \ldots, n-1,$$

$$\left. \frac{d^j P_{i1}}{dx^j} \right|_{x=a} = 0, \qquad \left. \frac{d^j P_{i1}}{dx^j} \right|_{x=b} = \delta_i^j, \qquad j = 0, 1, \ldots, n-1,$$

where δ_i^j is the *Kronecker delta* symbol defined by $\delta_i^j = 1$ for $i = j$ and

$\delta_i^j = 0$ otherwise. The reader should construct them. Substituting these polynomials into (1.6.14), we get the natural boundary conditions for a minimizer $y(x)$:

$$
\begin{aligned}
f_{y^{(n)}}\Big|_{x=a} &= 0,\\
f_{y^{(n)}}\Big|_{x=b} &= 0,\\
\left(f_{y^{(n-1)}} - \frac{d}{dx} f_{y^{(n)}}\right)\Big|_{x=a} &= 0,\\
\left(f_{y^{(n-1)}} - \frac{d}{dx} f_{y^{(n)}}\right)\Big|_{x=b} &= 0,\\
\left(f_{y^{(n-2)}} - \frac{d}{dx} f_{y^{(n-1)}} + \frac{d^2}{dx^2} f_{y^{(n)}}\right)\Big|_{x=a} &= 0,\\
\left(f_{y^{(n-2)}} - \frac{d}{dx} f_{y^{(n-1)}} + \frac{d^2}{dx^2} f_{y^{(n)}}\right)\Big|_{x=b} &= 0,\\
&\vdots\\
\left(f_{y'} - \frac{d}{dx} f_{y''} + \cdots + (-1)^{n-1} \frac{d^{n-1}}{dx^{n-1}} f_{y^{(n)}}\right)\Big|_{x=a} &= 0,\\
\left(f_{y'} - \frac{d}{dx} f_{y''} + \cdots + (-1)^{n-1} \frac{d^{n-1}}{dx^{n-1}} f_{y^{(n)}}\right)\Big|_{x=b} &= 0.
\end{aligned}
$$

Note that the last two conditions contain $y^{(2n-1)}(x)$. In general, the natural boundary conditions contain higher derivatives than do the equations (1.6.8).

What happens when we appoint some of the boundary conditions (1.6.8)? For example, let $y(a) = c_1^*$ be the only boundary restriction for an unknown minimizer. Then we need to require that $\varphi(a) = 0$, and we will get all the natural boundary conditions for $y(x)$ except the one whose expression is the multiplier of $\varphi(a)$ in the boundary sum (1.6.14). That is,

$$
\left(f_{y'} - \frac{d}{dx} f_{y''} + \cdots + (-1)^{n-1} \frac{d^{n-1}}{dx^{n-1}} f_{y^{(n)}}\right)\Big|_{x=a} = 0
$$

must be removed from the list.

The reader should consider what happens to the natural boundary conditions in case the following conditions apply (consider each case separately):

(1) $y(a) + ky'(a) = c$,
(2) $y(a) + ky(b) = c$.

Example 1.6.1 Derive the Euler–Lagrange equation and natural boundary conditions for the energy functional whose minimizer defines the equilibrium of a bent cantilever beam described by parameters E, I. Assume the beam is subjected to a distributed load $q(x)$, as well as a shear force Q^* and torque M^* applied to the end $x = l$:

$$E(y) = \frac{1}{2}\int_0^l EI(y'')^2\,dx - \int_0^l qy\,dx - Q^*y(l) - M^*y'(l),$$
$$y(0) = y'(0) = 0.$$

Note that the natural boundary conditions now have mechanical meaning: they account for the given torque and shear force at the "free" end $x = l$.

Solution In this case the energy functional involves terms outside an integral, so it makes sense to repeat the derivation of the Euler–Lagrange equation for the functional $\int_a^b f(x, y, y', \ldots, y^{(n)})\,dx$ in order to understand how M^* and Q^* come into the natural conditions. Sometimes it is useful to know not only a final formula but its derivation as well.

Supposing y to be a solution of the problem, we consider the values of the functional $E(y)$ on the bundle $y + t\varphi$ with arbitrary but fixed φ: that is, we consider $E(y + t\varphi)$ where $\varphi(0) = 0 = \varphi'(0)$. As a function of the variable t this takes a minimum value at $t = 0$, so its derivative at this point is zero. This implies the equation

$$\int_0^l EIy''\varphi''\,dx - \int_0^l q\varphi\,dx - Q^*\varphi(l) - M^*\varphi'(l) = 0.$$

Two integrations by parts in the first integral bring us to the relation

$$\int_0^l (EIy^{(4)} - q)\varphi\,dx + EIy''\varphi'\big|_0^l - EIy'''\varphi\big|_0^l - Q^*\varphi(l) - M^*\varphi'(l) = 0$$

and, because $\varphi(0) = 0 = \varphi'(0)$, we get

$$\int_0^l (EIy^{(4)} - q)\varphi\,dx + (EIy''(l) - M^*)\varphi'(l) - (EIy'''(l) + Q^*)\varphi(l) = 0.$$

Now we repeat the steps connected with the choice of φ. First we take those φ for which $\varphi(l) = 0 = \varphi'(l)$, which brings us to the equation

$$\int_0^l (EIy^{(4)} - q)\varphi \, dx = 0;$$

then, because of the arbitrariness of φ, we invoke the main lemma to arrive at the Euler–Lagrange equation

$$EIy^{(4)} - q = 0 \quad \text{on } [0, l].$$

Hence for any φ that does not vanish at $x = l$ we have

$$(EIy''(l) - M^*)\varphi'(l) - (EIy'''(l) + Q^*)\varphi(l) = 0.$$

It follows that

$$EIy''(l) = M^*, \qquad EIy'''(l) = -Q^*,$$

which are the natural boundary conditions for the cantilever beam.

From the strength of materials we know the relations between the deflection y of the beam, the torque M, and the shear force Q:

$$M = EIy'', \qquad Q = -M' = -EIy'''.$$

We see that the natural conditions really do represent the conditions on the torque and shear force given at the free end $x = l$.

Let us discuss the example further. The solution of this simple boundary value problem constitutes a considerable part of any textbook on the strength of materials. At one time people relied on graphical approaches, although it is now easy to solve the problem analytically. In practice we encounter largely piecewise continuous load functions q displaying linear and parabolic-type dependences.

But the example did force us to consider a case which was not covered by the general theory of this section: the integrand can have points of discontinuity. Essentially nothing happened though. The Euler–Lagrange equation holds everywhere except at a discontinuity of q, and at such a point a jump in q will give rise to a jump in $y^{(4)}$. However, the lower-order derivatives of y all remain continuous.

In practice it is common to introduce external point torques and shear forces on the beam. What can we say in such cases? In the strength of materials, mechanical reasoning is used to show that at such points the moments and shear forces have corresponding jumps. Can we show this using the tools of the calculus of variations?

We consider a particular problem of the bending of a beam with fixed ends. The beam carries a distributed load q and is a subjected to a point torque M^* and shear force Q^* at some point c. The total energy functional, which takes its minimum value on a solution, has the form

$$\frac{1}{2}\int_0^l EI(y'')^2\,dx - \int_0^l qy\,dx - Q^*y(c) - M^*y'(c).$$

The hypothesis for the model of a beam requires continuity of y and y' at all points including $x = c$. Let us see what actually happens at this point. As in the example above, the energy functional is be considered on the bundle $y + t\varphi$ where φ, together with its first derivative, goes to zero at the endpoints of the segment $[0, l]$. Since we are unsure of what happens at $x = c$ it makes sense to split the integral into two parts: one over the domain $[0, c]$ and the other over the domain $[c, l]$. We shall use the notation $x = c - 0$ to denote a limit taken from the left, and $x = c + 0$ to denote a limit taken from the right. The approach taken in the example brings us to the following equation:

$$\begin{aligned}\int_0^c (EIy^{(4)} - q)\varphi\,dx &+ \int_c^l (EIy^{(4)} - q)\varphi\,dx\\ &+ EIy''(c-0)\varphi(c-0) - EIy''(c+0)\varphi(c+0)\\ &- EIy'''(c-0)\varphi(c-0) + EIy'''(c+0)\varphi(c+0)\\ &- M^*\varphi'(c) - Q^*\varphi(c) = 0.\end{aligned}$$

Supposing $\varphi(c) = 0 = \varphi'(c)$, we obtain the same equation $EIy^{(4)} - q = 0$ on both segments $[0, c)$ and $(c, l]$. Returning to the above equation with φ unrestricted at $x = c$, we see that the second and third derivatives of y do indeed have jumps at $x = c$ defined by M^* and Q^*, respectively:

$$EI(y''(c-0) - y''(c+0)) = M, \quad EI(y'''(c-0) - y'''(c+0)) = -Q^*.$$

The reader may wish to consider the case in which at $x = c$ the characteristic EI of the beam changes from EI_0 to EI_1. He or she can derive the conditions for a solution to the problem of equilibrium of the beam under load at point $x = c$. The solution is a point of minimum of the above total energy functional $E(y)$.

1.7 Functionals Depending on Functions in Many Variables

Although obtaining the Euler equation has become somewhat routine for us, we will not be fully prepared to treat practical problems until we can seek unknown minimizers in many variables.

We begin with the two variable case. This case is the simplest; the extension to three or more independent variables is straightforward. Consider a functional of the form

$$F(u) = \iint_S f(x, y, u(x, y), u_x(x, y), u_y(x, y))\, dx\, dy. \tag{1.7.1}$$

Here u_x and u_y denote the partial derivatives $\partial u/\partial x$ and $\partial u/\partial y$, respectively. We confine ourselves to cases where S is simple; practical problems normally involve such domains, and we thereby avoid a great deal of complexity. We therefore assume that S is a closed domain in $\mathbb{R}^2$ with a piecewise smooth boundary ∂S. (We do not elaborate on the meaning of "smooth." Our attitude toward this issue is common among practitioners: we simply require everything we need in intermediate calculations!)

We will consider two main minimization problems for (1.7.1): the problem with the Dirichlet boundary condition

$$u(x, y)\Big|_{\partial S} = \psi(s), \tag{1.7.2}$$

and the problem "without" boundary conditions (i.e., the problem for which natural boundary conditions appear).

We first obtain the analogue to the Euler equation for (1.7.1). Our general approach is to repeat the steps of § 1.2. Specifically we (1) introduce classes of functions over which we may consider the problem of minimum, (2) formulate the fundamental lemma for the two variable case, and (3) recall how to integrate by parts in the two variable case.

We denote by $C^{(k)}(S)$ the set of functions continuous on a compact domain S together with all their derivatives up to order k. The norm with which we define a neighborhood of a function is

$$\|u\|_{C^{(k)}} = \max_{\alpha+\beta\le k}\ \max_{(x,y)\in S} \left\| \frac{\partial^{\alpha+\beta} u(x,y)}{\partial x^\alpha \partial y^\beta} \right\|.$$

$C_0^{(k)}(S)$ is the subset of $C^{(k)}(S)$ consisting of functions which, together with all their derivatives up to order $k-1$, are equal to zero on the boundary

∂S. We shall use the corresponding notations $C^{(\infty)}(S)$ and $C_0^{(\infty)}(S)$ for sets of functions infinitely differentiable on S.

Lemma 1.7.1 *Let $g(\mathbf{x})$ be continuous on S, and let*

$$\iint_S g(\mathbf{x})\varphi(\mathbf{x})\,dx\,dy = 0 \tag{1.7.3}$$

hold for any function $\varphi(x) \in C_0^{(\infty)}(S)$. Then $g(\mathbf{x}) \equiv 0$.

Proof. We imitate the proof of Lemma 1.2.1. Suppose to the contrary that at some interior point $\mathbf{x}_0$ of S we have $g(\mathbf{x}_0) \neq 0$, say $g(\mathbf{x}_0) > 0$. Then $g(\mathbf{x}) > 0$ for all $\mathbf{x}$ in some disk C_ε having radius ε and center $\mathbf{x}_0$. It is easy to construct a bell-shaped surface of revolution centered at $\mathbf{x}_0$. The corresponding function $\varphi_0(\mathbf{x} - \mathbf{x}_0) \in C_0^{(\infty)}(S)$ gives us

$$\iint_S g(\mathbf{x})\varphi_0(\mathbf{x} - \mathbf{x}_0)\,dx\,dy = \iint_{C_\varepsilon} g(\mathbf{x})\varphi_0(\mathbf{x} - \mathbf{x}_0)\,dx\,dy > 0,$$

which contradicts (1.7.3). □

To integrate by parts we use

$$\iint_S u\frac{\partial v}{\partial x_i}\,dx\,dy = -\iint_S \frac{\partial u}{\partial x_i} v\,dx\,dy + \oint_{\partial S} uv\,n_i\,ds.$$

Here n_i is the cosine of the angle between the unit outward normal $\mathbf{n}$ and the unit vector along the x_i axis ($x_i = x, y$ for $i = 1, 2$, respectively). We shall use the length variable s to parameterize the contour ∂S.

We now formulate the main result of this section. Let $f(x, y, u, p, q)$ be a continuous function having continuous first partial derivatives with respect to all of its arguments.

Theorem 1.7.1 *Let $u = u(x, y) \in C^{(2)}(S)$ be a minimizer of the functional $\iint_S f(x, y, u, u_x, u_y)\,dx\,dy$ on the subset of $C^{(1)}(S)$ consisting of those functions satisfying* (1.7.2). *Then the Euler equation*

$$f_u - \left(\frac{\partial f_{u_x}}{\partial x} + \frac{\partial f_{u_y}}{\partial y}\right) = 0 \tag{1.7.4}$$

holds in S. Here $\partial/\partial x$ and $\partial/\partial y$ are total partial derivatives, analogous to the total derivative in the one-dimensional case, when the function $u = u(x, y)$ as well as its partial derivatives u_x and u_y are considered as depending on x and y respectively.

Proof. Consider the functional on the usual bundle $u = u(x, y) + t\varphi(x, y)$ where $\varphi(x, y)$ is a function from $C_0^{(1)}(S)$; that is, it has first derivatives continuous on S and satisfies

$$\varphi(x, y)\big|_{\partial S} = 0. \tag{1.7.5}$$

The functional $F(u + t\varphi)$ for a fixed $\varphi(x, y)$ becomes a function of the real variable t and takes its minimum at $t = 0$. Thus

$$\begin{aligned} 0 &= \left.\frac{dF(u + t\varphi)}{dt}\right|_{t=0} \\ &= \frac{d}{dt}\left(\iint_S f(x, y, u + t\varphi, u_x + t\varphi_x, u_y + t\varphi_y)\, dx\, dy\right)\bigg|_{t=0} \\ &= \iint_S \left(f_u\varphi + f_{u_x}\varphi_x + f_{u_y}\varphi_y\right)\, dx\, dy. \end{aligned}$$

Integration by parts in the last two terms of the integrand gives us

$$\iint_S \left[f_u - \left(\frac{\partial f_{u_x}}{\partial x} + \frac{\partial f_{u_y}}{\partial y}\right)\right] \varphi\, dx\, dy + \oint_{\partial S} \left(f_{u_x} n_x + f_{u_y} n_y\right) \varphi\, ds = 0. \tag{1.7.6}$$

Now remembering that $\varphi(x, y)$ satisfies (1.7.5), we get

$$\iint_S \left[f_u - \left(\frac{\partial f_{u_x}}{\partial x} + \frac{\partial f_{u_y}}{\partial y}\right)\right] \varphi\, dx\, dy = 0. \tag{1.7.7}$$

Equation (1.7.4) follows from Lemma 1.7.1. □

Theorem 1.7.2 *Let $u = u(x, y) \in C^{(2)}(S)$ be a minimizer of the functional $\iint_S f(x, y, u, u_x, u_y)\, dx\, dy$ on $C^{(1)}(S)$ (without any boundary conditions). Then the Euler equation* (1.7.4) *holds in S, and $u(x, y)$ satisfies the natural boundary condition*

$$\left(f_{u_x} n_x + f_{u_y} n_y\right)\Big|_{\partial S} = 0. \tag{1.7.8}$$

Proof. We consider $F(u + t\varphi)$ on the bundle $u + t\varphi$ where $\varphi(x, y) \in C^{(1)}(S)$ is arbitrary but momentarily fixed. For all such functions we establish (1.7.6) using the same reasoning as above. Then let us restrict $\varphi(x, y)$ to the set $C_0^{(1)}(S)$. This shows that (1.7.1) holds in S. So (1.7.7) holds whether φ belongs to $C_0^{(1)}(S)$ or $C^{(1)}(S)$. Hence

$$\oint_S \left(f_{u_x} n_x + f_{u_y} n_y\right) \varphi\, ds = 0. \tag{1.7.9}$$

Now we use the fact that on S, $\varphi = \varphi(s)$ is an arbitrary differentiable function. We do not prove the corresponding fundamental lemma for such an integral, but it is clear that a proof could be patterned after that of Lemma 1.2.1. (For this we could use the function $\varphi_0(\mathbf{x} - \mathbf{x}_0)$ from the proof of Lemma 1.7.1; the point $\mathbf{x}_0$ would be a chosen point of the boundary where the corresponding multiplier $g(\mathbf{x})$ is not equal to zero, by the contrary assumption.) Hence (1.7.8) follows from (1.7.9). □

Example 1.7.1 Demonstrate that for the functional

$$\Psi(u) = \frac{1}{2}\iint_S (u_x^2 + u_y^2)\,dx\,dy - \iint_S Fu\,dx\,dy \tag{1.7.10}$$

with $F = F(x, y)$ a given continuous function, the Euler equation and the natural boundary conditions are

$$\Delta u = -F \quad \text{in } S \tag{1.7.11}$$

and

$$\left.\frac{\partial u}{\partial n}\right|_{\partial S} = 0, \tag{1.7.12}$$

respectively. Show that on a solution u^* of the latter boundary value problem, if it exists, the functional $\Psi(u)$ attains a global minimum.

Solution The derivation of (1.7.11) and (1.7.12) is straightforward. Denoting

$$f = \frac{1}{2}(u_x^2 + u_y^2) - Fu$$

we get

$$f_u - \left(\frac{\partial f_{u_x}}{\partial x} + \frac{\partial f_{u_y}}{\partial y}\right) = -F - \Delta u,$$

which leads to (1.7.11). The left-hand expression in (1.7.8) is

$$f_{u_x} n_x + f_{u_y} n_y = u_x n_x + u_y n_y,$$

which is $\partial u/\partial n$ on the boundary.

Before demonstrating the last statement in the example, we note that $\Psi(u)$ expresses the total energy of an elastic membrane. From physics we know that at points of minimum of a total energy functional for a mechanical system with conservative loads, the system is in equilibrium. In particle mechanics it is even shown that such an equilibrium state is stable at a point of strict minimum. Let us see what happens in this case of a spatially distributed object. We suppose that a solution u^* of the boundary value problem (1.7.11)–(1.7.12) exists. Consider the values of Ψ over the bundle $u^* + \varphi$, where φ is arbitrary:

$$\begin{aligned}\Psi(u^* + \varphi) &= \frac{1}{2}\iint_S \left((u_x^* + \varphi_x)^2 + (u_y^* + \varphi_y)^2\right)\,dx\,dy \\ &\quad - \iint_S F(u^* + \varphi)\,dx\,dy \\ &= \Psi(u^*) \\ &\quad + \left[\iint_S \left(u_x^*\varphi_x + u_y^*\varphi_y\right)\,dx\,dy - \iint_S F\varphi\,dx\,dy\right] \\ &\quad + \frac{1}{2}\iint_S \left(\varphi_x^2 + \varphi_y^2\right)\,dx\,dy.\end{aligned}$$

Because of (1.7.11)–(1.7.12) (which, in the above theory, were derived as a direct consequence of the following equality and thus are equivalent to it when u^* is sufficiently smooth) we see that

$$\iint_S \left(u_x^*\varphi_x + u_y^*\varphi_y\right)\,dx\,dy - \iint_S F\varphi\,dx\,dy = 0.$$

So

$$\Psi(u^* + \varphi) - \Psi(u^*) = \frac{1}{2}\iint_S \left(\varphi_x^2 + \varphi_y^2\right)\,dx\,dy \geq 0,$$

which means that $\Psi(u)$ takes its global minimum at $u = u^*$. Here we cannot say anything about the stability of the membrane. To discuss this we must consider the problem from a dynamical viewpoint.

We are in the habit of supposing that a minimizer exists for each problem we encounter. But the problem of minimizing (1.7.10), which describes the equilibrium of a membrane, demonstrates that not every problem which seems sensible at first glance has a solution. Indeed, if we take $u = c$, a constant, then the first integral in (1.7.10) is zero. If $\iint_S F\,dx\,dy \neq 0$, then by changing c we make the value of the functional any large negative number. So the problem has no solution and (at least) the condition $\iint_S F\,dx\,dy = 0$ becomes necessary for the problem to be sensible. In fact, this has a clear mechanical sense: it is the condition of self-balance of the forces. A free membrane subjected to a load F can move as a whole, but in this model we neglect its inertia, so the problem of equilibrium of the membrane without the condition of self-balance of the load is senseless as we showed formally. Later we consider this question in more detail.

1.8 A Functional with Integrand Depending on Partial Derivatives of Higher Order

Now we derive the Euler equation for a minimizer $w = w(x, y)$ of the functional of the form

$$F(w) = \iint_S f(x, y, w, w_x, w_y, w_{xx}, w_{xy}, w_{yy})\, dx\, dy \tag{1.8.1}$$

on the functions of class $C^{(2)}(S)$ satisfying the boundary conditions

$$w\big|_{\partial S} = w_0(s)$$

and

$$\left.\frac{\partial w}{\partial n}\right|_{\partial S} = w_1(s).$$

We sketch the steps that have become routine. We suppose that a minimizer $w = w(x, y) \in C^{(4)}(S)$. Let $\varphi(x, y)$ be an arbitrary but fixed function from $C_0^{(2)}(S)$, which implies in particular that

$$\varphi\big|_{\partial S} = 0, \qquad \left.\frac{\partial \varphi}{\partial n}\right|_{\partial S} = 0. \tag{1.8.2}$$

$F(w + t\varphi)$ takes its minimum at $t = 0$ and thus $dF(w + t\varphi)/dt\big|_{t=0} = 0$. This equation takes the form

$$\iint_S (f_w \varphi + f_{w_x} \varphi_x + f_{w_y} \varphi_y + f_{w_{xx}} \varphi_{xx} + f_{w_{xy}} \varphi_{xy} + f_{w_{yy}} \varphi_{yy})\, dx\, dy = 0. \tag{1.8.3}$$

Supposing f has continuous derivatives of third order, we can integrate by parts in (1.8.3) and get

$$\iint_S \left(f_w - \frac{\partial}{\partial x} f_{w_x} - \frac{\partial}{\partial y} f_{w_y} + \frac{\partial^2}{\partial x^2} f_{w_{xx}} + \frac{\partial^2}{\partial x \partial y} f_{w_{xy}} + \frac{\partial^2}{\partial y^2} f_{w_{yy}} \right) \varphi\, dx\, dy = 0. \tag{1.8.4}$$

The boundary terms vanish because of (1.8.2). Using Lemma 1.7.1 we obtain the Euler equation for the functional (1.8.1):

$$f_w - \frac{\partial}{\partial x} f_{w_x} - \frac{\partial}{\partial y} f_{w_y} + \frac{\partial^2}{\partial x^2} f_{w_{xx}} + \frac{\partial^2}{\partial x \partial y} f_{w_{xy}} + \frac{\partial^2}{\partial y^2} f_{w_{yy}} = 0,$$

valid in S. Here $\partial/\partial x$ and $\partial/\partial y$ are total partial derivatives when $w = w(x, y)$ is considered as depending on its arguments x, y.

We could derive the form of the natural boundary conditions for (1.8.1), but this is cumbersome so we prefer to treat an illustrative case. We shall consider a problem of minimizing a total energy functional, whose solution describes the equilibrium of an elastic plate with free edge.

It is time to discuss how problems of minimization arose. Some came from geometrical considerations, like the isoperimetric problem mentioned in § 1.1; some were designed specifically as exercises, written out by analogy with other, more or less easily solved, problems. But for the most part the real problems of the calculus of variations came from physics — in particular, mechanics. There it was found that minimizers or maximizers of certain functionals describe important states of physical systems. It is interesting to note how this idea progressed in importance. Early in the development of classical mechanics, variational principles were derived using the "fundamental" equations of statics and mechanics; they were regarded as consequences, although in many circumstances they were actually equivalent. It was soon found that some problems were easier solved by variational methods, and the variational approach to mechanics gained a life of its own. In the theory of elasticity, for example, a great many variational principles have been derived; moreover, the name "variational principle" is applied not only to the minimization of functionals, but to any circumstance in which an important equation can be derived from an integro-differential equation having the form of the first variation of a functional being equal to zero, even if there is no functional for which it is the first variation. For example, the Virtual Work Principle arose as a consequence of the principle of minimum of potential energy of a mechanical system. But the former continues to hold in the case of non-conservative forces where it is impossible to compose the potential energy functional.

Early in the development of linear elasticity, an energy functional was derived whose minimizer describes the equilibrium of an elastic body. The procedure was to write out the equilibrium equations, multiply by appropriate components of the vector of displacements, and integrate over the region. Using integration by parts with regard for homogeneous Dirichlet boundary conditions, from the terms with second-order partial derivatives it was possible to get a symmetrical form (in the components of the strain tensor) for potential energy. The originators of this method were comforted by the fact that the associated natural boundary conditions coincided with the boundary conditions assigned to the same problem when considered as a problem of equilibrium with applied forces given on the boundary. This led to the idea that the Principle of Minimum Potential Energy (or, correspondingly, the Virtual Work Principle) could be used to derive boundary conditions for models of elastic plates and shells. Workers investigating such models had previously run into difficulty in posing appropriate boundary conditions: upon simplification from the 3-D case, uncertainties had arisen regarding precisely what force conditions should be appointed on the boundary of an object. The variational formalism provided the needed result in a simple fashion. Why are we taking the time to discuss this now? We are going to consider the problem of equilibrium of an elastic plate from the viewpoint of the calculus of variations. The first step is to formulate the energy functional. The left-hand side of the equation describing a thin elastic plate bent under load contains a biharmonic operator. In this case there is no uniquely defined procedure to derive the energy functional. Moreover, integration by parts can yield several expressions for the energy of an elastic plate with homogeneous Dirichlet conditions (1.8.2). For each of these forms one can derive the natural boundary conditions, but only one form gives the conditions corresponding to mechanics. So to formulate the problem (i.e., the functional) properly, one should have some knowledge of mechanics — perhaps this is why so many pure mathematicians prefer to study only classical problems where everything is formulated in advance! To work purely mathematical exercises, one is seldom required to know the actual physical behavior of the object under consideration. But correct mathematical procedures often depend in large part on the details of a particular realm of application.

The energy functional of an isotropic homogeneous plate bending under load $F = F(x, y)$ is

$$
\begin{aligned}
E(w) = \frac{D}{2} \iint_S \left(w_{xx}^2 + w_{yy}^2 + 2\nu w_{xx} w_{yy} + 2(1-\nu) w_{xy}^2\right) dx\, dy \\
- \iint_S F w \, dx\, dy \qquad (1.8.5)
\end{aligned}
$$

where D is the rigidity of the plate, ν is Poisson's ratio, and $w = w(x, y)$ is the deflection at point (x, y) of S, the compact domain occupied by the mid-surface of the plate. A minimizer of $E(w)$ describes the equilibrium deflection of the mid-surface. Using the standard method, we shall derive the Euler equation for the minimizer and the corresponding natural boundary conditions.

Let $w \in C^{(4)}(S)$ minimize (1.8.5) over $C^{(2)}(S)$. Consider $E(w + t\varphi)$ at a fixed $\varphi \in C^{(2)}(S)$ as a function of the parameter t. It takes its minimum at $t = 0$, so as a consequence we have

$$D \iint_S [w_{xx}\varphi_{xx} + w_{yy}\varphi_{yy} + \nu(w_{xx}\varphi_{yy} + w_{yy}\varphi_{xx}) \\ + 2(1-\nu)w_{xy}\varphi_{xy}]\, dx\, dy - \iint_S F\varphi\, dx\, dy = 0$$

which is a particular case of (1.8.4).

Now it is necessary to integrate by parts in the first integral on the left. We get

$$D \iint_S [(w_{xx} + \nu w_{yy})\varphi_{xx} + (w_{yy} + \nu w_{xx})\varphi_{yy} + 2(1-\nu)w_{xy}\varphi_{xy}]\, dx\, dy \\ = -D \iint_S \left[\varphi_x \frac{\partial}{\partial x}(w_{xx} + \nu w_{yy}) + \varphi_y \frac{\partial}{\partial y}(w_{yy} + \nu w_{xx}) \right. \\ \left. + (1-\nu)w_{xyy}\varphi_x + (1-\nu)w_{xxy}\varphi_y \right] dx\, dy \\ + D \oint_{\partial S} [(w_{xx} + \nu w_{yy})\varphi_x n_x + (w_{yy} + \nu w_{xx})\varphi_y n_y \\ + (1-\nu)w_{xy}(\varphi_x n_y + \varphi_y n_x)]\, ds \tag{1.8.6}$$

where $\mathbf{n}$, the unit normal to the boundary ∂S, has components (n_x, n_y). Note that we have preserved the symmetry of the expressions. Integrating by parts once more in the first integral on the right, denoted by A, we get

$$A = D \iint_S [(w_{xx} + \nu w_{yy})_{xx} + (w_{yy} + \nu w_{xx})_{yy} + 2(1-\nu)w_{xxyy}]\varphi\, dx\, dy \\ - D \oint_S [(w_{xx} + \nu w_{yy})_x n_x + (w_{yy} + \nu w_{xx})_y n_y \\ + (1-\nu)(w_{xyy}n_x + w_{xxy}n_y)]\varphi\, ds.$$

The first integral in A is

$$D \iint_S (w_{xxxx} + 2w_{xxyy} + w_{yyyy})\varphi\, dx\, dy = D \iint_S \varphi \Delta^2 w\, dx\, dy.$$

Thus (1.8.6) takes the form

$$\begin{aligned}
&D \iint_S \varphi \Delta^2 w \, dx \, dy - \iint_S F \varphi \, dx \, dy \\
&+ D \oint_S [(w_{xx} + \nu w_{yy})\varphi_x n_x + (w_{yy} + \nu w_{xx})\varphi_y n_y \\
&\qquad + (1-\nu) w_{xy}(\varphi_x n_y + \varphi_y n_x)] \, ds \\
&- D \oint_S [(w_{xx} + \nu w_{yy})_x n_x + (w_{yy} + \nu w_{xx})_y n_y \\
&\qquad + (1-\nu)(w_{xyy} n_x + w_{xxy} n_y)] \varphi \, ds = 0. \qquad (1.8.7)
\end{aligned}$$

First we consider the subset of admissible functions $\varphi(x, y)$ satisfying (1.8.2). Equation (1.8.7) reduces to

$$\iint_S (D\Delta^2 w - F)\varphi \, dx \, dy = 0. \qquad (1.8.8)$$

By the fundamental lemma we obtain the Euler equation

$$D\Delta^2 w - F = 0 \quad \text{in } S. \qquad (1.8.9)$$

Because of (1.8.9) the equality (1.8.8) holds for any admissible $\varphi(x, y)$, thus the two first integrals over S disappear from (1.8.7). In equation (1.8.7) there remains the sum of two contour integrals that equals zero for any $\varphi \in C^{(2)}(S)$.

We might think that since we have three arbitrary functions φ, φ_x, φ_y on S, we could set their multipliers equal to zero and obtain three natural boundary conditions. But this is incorrect. We see this first on mechanical grounds: these "boundary conditions" would depend on x and y, hence would not be invariant under coordinate rotations. Mathematically, it appears that we cannot choose φ, φ_x, and φ_y independently on S. Indeed let us fix φ on S: then its derivative φ_τ in the tangential direction $\boldsymbol{\tau}$ is determined uniquely — only the derivative φ_n of φ in the normal direction is really independent of φ on the contour.

Thus we first need to introduce this change of coordinates, getting a local frame $(\boldsymbol{\tau}, \mathbf{n})$. The transformation formulas for derivatives are

$$\varphi_x = \varphi_n n_x - \varphi_s n_y, \qquad \varphi_y = \varphi_n n_y + \varphi_s n_x.$$

Let us put these into the integrand of the first contour integral:

$$\begin{aligned}
&(w_{xx} + \nu w_{yy})\varphi_x n_x + (w_{yy} + \nu w_{xx})\varphi_y n_y + (1-\nu)w_{xy}(\varphi_x n_y + \varphi_y n_x) \\
&= (w_{xx} + \nu w_{yy})(\varphi_n n_x - \varphi_s n_y)n_x + (w_{yy} + \nu w_{xx})(\varphi_n n_y + \varphi_s n_x)n_y \\
&\quad + (1-\nu)w_{xy}[(\varphi_n n_x - \varphi_s n_y)n_y + (\varphi_n n_y + \varphi_s n_x)n_x] \\
&= (1-\nu)\{(w_{yy} - w_{xx})n_x n_y + w_{xy}(n_x^2 - n_y^2)\}\varphi_s \\
&\quad + \{(w_{xx} + \nu w_{yy})n_x^2 + (w_{yy} + \nu w_{xx})n_y^2 + 2(1-\nu)w_{xy}n_x n_y\}\varphi_n \\
&= (1-\nu)\{(w_{yy} - w_{xx})n_x n_y + w_{xy}(n_x^2 - n_y^2)\}\varphi_s \\
&\quad + \{\nu\Delta w + (1-\nu)(w_{xx}n_x^2 + w_{yy}n_y^2 + 2w_{xy}n_x n_y)\}\varphi_n. \qquad (1.8.10)
\end{aligned}$$

Change the integrand of the first contour integral in (1.8.7) by (1.8.10) and remember that $\varphi_s = \partial\varphi/\partial s$ and $\varphi_n = \partial\varphi/\partial n$:

$$\begin{aligned}
&D\oint_{\partial S}(1-\nu)\{(w_{yy} - w_{xx})n_x n_y + w_{xy}(n_x^2 - n_y^2)\}\frac{\partial\varphi}{\partial s}\,ds \\
&+ D\oint_{\partial S}\{\nu\Delta w + (1-\nu)(w_{xx}n_x^2 + w_{yy}n_y^2 + 2w_{xy}n_x n_y)\}\frac{\partial\varphi}{\partial n}\,ds \\
&- D\oint_{\partial S}[(w_{xx} + \nu w_{yy})_x n_x + (w_{yy} + \nu w_{xx})_y n_y \\
&\qquad\qquad + (1-\nu)(w_{xyy}n_x + w_{xxy}n_y)]\varphi\,ds = 0. \qquad (1.8.11)
\end{aligned}$$

If S is smooth enough we can integrate by parts in the first integral with respect to s. This gives

$$\begin{aligned}
&D\oint_{\partial S}(1-\nu)\{(w_{yy} - w_{xx})n_x n_y + w_{xy}(n_x^2 - n_y^2)\}\frac{\partial\varphi}{\partial s}\,ds \\
&= -D(1-\nu)\oint_{\partial S}\varphi\frac{\partial}{\partial s}\{(w_{yy} - w_{xx})n_x n_y + w_{xy}(n_x^2 - n_y^2)\}\,ds.
\end{aligned}$$

It follows that

$$\begin{aligned}
&- D\oint_{\partial S}\Big[(w_{xx} + \nu w_{yy})_x n_x + (w_{yy} + \nu w_{xx})_y n_y \\
&\qquad + (1-\nu)(w_{xyy}n_x + w_{xxy}n_y)] \\
&\qquad + (1-\nu)\frac{d}{ds}[(w_{yy} - w_{xx})n_x n_y + w_{xy}(n_x^2 - n_y^2)]\Big]\varphi\,ds \\
&+ D\oint_{\partial S}\{\nu\Delta w + (1-\nu)(w_{xx}n_x^2 + w_{yy}n_y^2 + 2w_{xy}n_x n_y)\}\frac{\partial\varphi}{\partial n}\,ds = 0.
\end{aligned}$$

Since we can independently choose φ and $\partial\varphi/\partial n$, we get the following two natural boundary conditions:

$$\nu\Delta w + (1-\nu)(w_{xx}n_x^2 + w_{yy}n_y^2 + 2w_{xy}n_xn_y)\Big|_{\partial S} = 0,$$

$$[(w_{xx}+\nu w_{yy})_y n_x + (w_{yy}+\nu w_{xx})_y n_y + (1-\nu)(w_{xyy}n_x + w_{xxy}n_y)]$$
$$+ (1-\nu)\frac{d}{ds}[(w_{yy}-w_{xx})n_xn_y + w_{xy}(n_x^2 - n_y^2)] = 0. \qquad (1.8.12)$$

The first means that the shear force on the lateral surface of the plate is zero, whereas the second means that the bending moment is zero.

We have assumed that ∂S is sufficiently smooth so we could integrate by parts in (1.8.11). At corner points (1.8.12) is not valid. We leave it to the reader to derive an appropriate corner condition.

1.9 The First Variation

This book is written for those who will use the calculus of variations. Although our goal is to keep the presentation simple, continued exploitation of the same technique would prevent real progress. We need ideas applicable to more complex problems. As before, these will be extensions of elementary ideas from calculus. A principal analytical tool is the differential of a function. The first differential extracts the main part of the increment of the function when its argument changes by a small amount Δx. This main part is linear with respect to Δx. In this way, we approximate the change of a smooth function in some neighborhood of a point by an expression linear in Δx. The extension to functionals is called the first variation.

A few technical details

Definition 1.9.1 We say that $f(x) = o(g(x))$ when $x \to x_0$ if

$$\lim_{x\to x_0} \frac{f(x)}{g(x)} = 0.$$

Here x can be a real variable or an element of a more general metric or normed space; in the latter case when we write $x \to x_0$ we refer to convergence in that space. We often use the abbreviated notation $f = o(g)$ and say that f is of a *higher order of smallness* than g.

So if the o relation holds then given any $\varepsilon > 0$ we can find $\delta > 0$ such that $|f(x)/g(x)| < \varepsilon$ whenever $\|x - x_0\| < \delta$.[6] Other observations are as follows:

(1) The functions $f(x)$ and $g(x)$ are not required to possess individual limits as $x \to x_0$; only the ratio must possess a limit.
(2) In practice, $g(x)$ will usually be some power of a simple real variable x.

The statement $f(x) = o(1)$ as $x \to x_0$, for example, means nothing more than $\lim_{x\to x_0} f(x) = 0$. If $f(x) = o(x - x_0)$ as $x \to x_0$, then $f(x)$ tends to zero even faster as $x \to x_0$ since the ratio $f(x)/(x - x_0)$ tends to zero even though its denominator tends to zero as $x \to x_0$.

Definition 1.9.2 We say that $f(x) = O(g(x))$ as $x \to x_0$ if in some neighborhood of x_0 an inequality

$$\left|\frac{f(x)}{g(x)}\right| \le c$$

holds for some constant c. We often use the abbreviated notation $f = O(g)$ and say that f is of the *same order of smallness* as g.

Let us consider some examples of this notation as well. The statement $f(x) = O(1)$ as $x \to 0$ means that in some neighborhood of zero we have $|f(x)| < c$ (i.e., f is bounded in this neighborhood). If $f(x) = O(x)$ as $x \to 0$, then in some neighborhood of zero we have $|f(x)| < c|x|$. This implies that $f(x) \to 0$ as $x \to 0$, hence that $f(x) = o(1)$. But $f(x) = O(x)$ tells *how fast* $f(x)$ tends to zero.

Let $f(x)$ and its first $n+1$ derivatives be continuous in an interval about $x = x_0$. Then according to Taylor's theorem

$$f(x) = f(x_0)+f'(x_0)(x-x_0)+\cdots+\frac{f^{(n)}(x_0)}{n!}(x-x_0)^n+\frac{f^{(n+1)}(\xi)}{(n+1)!}(x-x_0)^{n+1}$$

for some ξ between x_0 and x. The last term on the right is the so-called Lagrange form of the remainder and is clearly $O(|x - x_0|^{n+1})$. Since we prefer to have this in another form, let us add and subtract the term

$$\frac{f^{(n+1)}(x_0)}{(n+1)!}(x - x_0)^{n+1}$$

[6] Here we refer to a more general vector norm. A reader unfamiliar with the subject of norms will find a more complete discussion in § 1.11. For now it is sufficient to think in terms of real numbers, where the role of norm is played by the absolute value.

to obtain

$$f(x) = f(x_0) + f'(x_0)(x - x_0) + \cdots + \frac{f^{(n+1)}(x_0)}{(n+1)!}(x - x_0)^{n+1} + \frac{[f^{(n+1)}(\xi) - f^{(n+1)}(x_0)]}{(n+1)!}(x - x_0)^{n+1}.$$

In this way we have created a Taylor expansion with one more term and a new "remainder." Now since $f^{n+1}(x)$ is continuous, the bracketed term $f^{(n+1)}(\xi) - f^{(n+1)}(x_0)$ tends to zero when $x \to x_0$ (recall that ξ is an intermediate point of (x, x_0)). This means that the ratio of the new remainder to the factor $|x - x_0|^{n+1}$ will tend to zero as $x \to x_0$ and we can write

$$f(x) = f(x_0) + f'(x_0)(x - x_0) + \cdots + \frac{f^{(n+1)}(x_0)}{(n+1)!}(x - x_0)^{n+1} + o\left(|x - x_0|^{n+1}\right).$$

Let us summarize this form of Taylor's theorem, known as *Peano's form*:

Theorem 1.9.1 *Let $f(x)$ and its first n derivatives be continuous in an interval about $x = x_0$. Then*

$$f(x) = f(x_0) + \frac{f'(x_0)}{1!}(x - x_0) + \cdots + \frac{f^{(n)}(x_0)}{n!}(x - x_0)^n + o\left(|x - x_0|^n\right).$$

With this we can say something about the behavior of the remainder term in the nth-order Taylor expansion even if we know nothing about continuity of the $(n+1)$th derivative.

Back to the first variation

In calculus we define the first differential as follows. We consider the increment $f(x + \Delta x) - f(x)$ of a function $f(x)$ of a real variable x. If it is possible to represent it in the form

$$f(x + \Delta x) - f(x) = A\Delta x + \omega(\Delta x) \tag{1.9.1}$$

where $\omega(\Delta x) = o(\Delta x)$ as $\Delta x \to 0$, then

- $A\Delta x$ is called the *first differential of f at x*, and is denoted by $df(x)$,
- A is the derivative of f at x, denoted by $f'(x)$, and
- the increment Δx of the argument x is redenoted by dx and is called the differential of the argument.

We may therefore write

$$df(x) = A\Delta x = f'(x)\,dx.$$

In the mind of a calculus student the differential dx and its corresponding $df(x)$ are extremely small quantities. Let us now banish this misconception: both dx and $df(x)$ are finite. When dx is small then so is $df(x)$ and it approximates the difference $f(x+dx) - f(x)$: the smaller the value of dx, the better the relative approximation. However, neither dx nor $df(x)$ is small in general.

Let us repeat the same steps for a functional. This is especially easy to do for a quadratic functional. These arise in physics, corresponding to natural laws that are linear in form (of course, linearity is often a condition imposed rather artificially on models of real phenomena). Let us consider, for example,

$$F(u) = \frac{1}{2}\iint_S (u_x^2 + u_y^2)\,dx\,dy - \iint_S Fu\,dx\,dy.$$

We denote the "increment" of the argument $u = u(x)$ by $\varphi(x)$. We note that $\varphi(x)$ must have certain properties; it should be admissible in the sense of § 1.5. (Later we shall soften the smoothness conditions for this problem.) In mechanics φ is usually denoted by δu; this maintains a visual similarity between the two notions of increment dx and δu, and in this notation δu is called a *virtual displacement*. Now

$$F(u+\varphi) - F(u) = \iint_S (u_x\varphi_x + u_y\varphi_y)\,dx\,dy - \iint_S F\varphi\,dx\,dy + \frac{1}{2}\iint_S (\varphi_x^2 + \varphi_y^2)\,dx\,dy. \tag{1.9.2}$$

The first two integrals on the right are linear in φ and pretend to analogy with the differential of calculus; together they are called the *first variation* of the functional $F(u)$ at u:

$$\iint_S (u_x\varphi_x + u_y\varphi_y)\,dx\,dy - \iint_S F\varphi\,dx\,dy. \tag{1.9.3}$$

The third integral in (1.9.2), quadratic in φ, is analogous to $\omega(\Delta x)$ in (1.9.1). We should introduce the smallness of the increment φ in such a way (and we did this in § 1.5!) that this quadratic term becomes infinitely small in comparison with the linear terms.

In § 1.5 we found that if $u = u(x)$ is a minimizer of $F(u)$, then the expression (1.9.3) is zero for all admissible φ:

$$\iint_S (u_x\varphi_x + u_y\varphi_y)\,dx\,dy - \iint_S F\varphi\,dx\,dy = 0. \tag{1.9.4}$$

From this we derived the Euler equation (1.7.11) for the membrane. We now derive (1.9.4) in a different way. Let us suppose that $u = u(x,y)$ is a minimizer of $F(u)$; that is, $F(u+\varphi) - F(u) \geq 0$ for any admissible φ. Assume, contrary to (1.9.4), that

$$\iint_S (u_x\varphi_x^* + u_y\varphi_y^*)\,dx\,dy - \iint_S F\varphi^*\,dx\,dy \neq 0$$

for some admissible φ^*. Then putting another admissible function $t\varphi^*$ into the inequality $F(u+\varphi) - F(u) \geq 0$, we get

$$\begin{aligned} 0 &\leq F(u+t\varphi^*) - F(u) \\ &= \iint_S (u_x t\varphi_x^* + u_y t\varphi_y^*)\,dx\,dy - \iint_S Ft\varphi^*\,dx\,dy \\ &\qquad + \frac{1}{2}\iint_S t^2({\varphi_x^*}^2 + {\varphi_y^*}^2)\,dx\,dy \\ &= t\left[\iint_S (u_x\varphi_x^* + u_y\varphi_y^*)\,dx\,dy - \iint_S F\varphi^*\,dx\,dy\right] \\ &\qquad + \frac{t^2}{2}\iint_S ({\varphi_x^*}^2 + {\varphi_y^*}^2)\,dx\,dy. \end{aligned} \tag{1.9.5}$$

Suppose the bracketed term differs from zero. If we take t such that it is sufficiently close to zero and the term $t[\cdots]$ is negative, then the term which is quadratic in t is much smaller than the term which is linear in t. Therefore $F(y+t\varphi) - F(y) < 0$, which contradicts the leftmost inequality of (1.9.5). So (1.9.4) holds for any admissible φ.

It is clear that we can repeat everything in terms of the plate problem of § 1.7. The differences are only technical.

We used the fact that at least for some (positive and negative) small t the function $t\varphi^*$ is admissible. In the membrane problem this is trivial. However, in some problems the set of admissible functions is restricted (e.g., it may be that $\varphi \geq 0$); free choice of t is thereby precluded. Such problems fall outside the scope of the classical theory, and in fact belong to the theory of variational inequalities.

We consider a general case of the simplest functional with respect to functions satisfying any of the types of boundary conditions we have discussed. Let us find its increment over the increment $\varphi(x)$ of the function $y(x)$. So we consider the increment of the functional

$$F(y) = \int_a^b f(x, y, y')\, dx$$

when the argument gets an admissible increment $\varphi = \varphi(x)$. Whether the boundary conditions are stipulated or not (free ends), we have

$$F(y + \varphi) - F(y) = \int_a^b [f(x, y + \varphi, y' + \varphi') - f(x, y, y')]\, dx.$$

Regarding the arguments of f as simple real variables, we can apply the Taylor expansion to f. If f has continuous second partial derivatives, then

$$f(x, y+\varphi, y'+\varphi') - f(x, y, y') = f_y(x, y, y')\varphi + f_{y'}(x, y, y')\varphi' + O(|\varphi|^2 + |\varphi'|^2).$$

Thus

$$F(y + \varphi) - F(y) = \int_a^b [f_y(x, y, y')\varphi + f_{y'}(x, y, y')\varphi']\, dx + O\left(\int_a^b (|\varphi|^2 + |\varphi'|^2)\, dx\right). \tag{1.9.6}$$

The last integral is of the order $O(\|\varphi\|^2_{C^{(1)}(a,b)})$ because

$$\begin{aligned}\int_a^b (|\varphi|^2 + |\varphi'|^2)\, dx &\le \int_a^b (|\varphi| + |\varphi'|)^2\, dx \\ &\le (b - a) \max_{x \in [a,b]} (|\varphi| + |\varphi'|)^2 \\ &\le (b - a) \left[\max_{x \in [a,b]} (|\varphi| + |\varphi'|)\right]^2.\end{aligned}$$

For admissible functions φ that are small in the norm of $C^{(1)}(a, b)$, the last term on the right-hand side of (1.9.6) has a higher order of smallness in φ than the integral term which is linear in φ. Thus we have a complete analogy with the first differential of a function. The expression

$$\delta F(y, \varphi) \equiv \int_a^b [f_y(x, y, y')\varphi + f_{y'}(x, y, y')\varphi']\, dx \tag{1.9.7}$$

is called the *first variation* of $F(y)$. We often denote it simply by δF.

Let $y = y(x)$ be a minimizer of $F(y)$ for some boundary conditions considered above. First we demonstrate that for any admissible function φ the equation

$$\int_a^b [f_y(x,y,y')\varphi + f_{y'}(x,y,y')\varphi']\,dx = 0 \tag{1.9.8}$$

holds. Indeed, for any admissible φ we have $F(y+\varphi) - F(y) \geq 0$. Assume that (1.9.8) fails at some admissible φ^*. We suppose that $t\varphi^*$ for small t is also admissible so that

$$\begin{aligned} 0 &\leq F(y+t\varphi^*) - F(y) \\ &= t\int_a^b [f_y(x,y,y')\varphi^* + f_{y'}(x,y,y')\varphi^{*\prime}]\,dx + O(t^2\,\|\varphi^*\|^2_{C^{(1)}(a,b)}). \end{aligned} \tag{1.9.9}$$

Now the smallness of the increment of the argument is governed by t. For small t the sign of the right-hand side of (1.9.9) is determined by the first integral term. Since we can choose t to be negative or positive and its coefficient is not zero, we can find a small t^* such that

$$t^*\int_a^b [f_y(x,y,y')\varphi^* + f_{y'}(x,y,y')\varphi^{*\prime}]\,dx + O(t^{*2}\,\|\varphi^*\|^2_{C^{(1)}(a,b)}) < 0.$$

This contradicts the leftmost inequality of (1.9.9).

Let us note that in $dF(y+t\varphi)/dt\big|_{t=0}$ we obtain the same expression (1.9.7), i.e., the first variation of the functional. The two methods of obtaining the first variation are equivalent if the integrand f is sufficiently smooth. But in the general theory of functionals our method of differentiation (i.e., the selection of the linear part of the difference $F(y+\varphi) - F(y)$) corresponds to the use of the *Fréchet derivative*, whereas the computation of $dF/dt|_{t=0}$ corresponds to the use of the *Gâteaux derivative.*

The reasoning of this section can be repeated for any of the functionals and their associated minimum problems we considered earlier. We leave this to the reader as a number of exercises.

Variational derivative

We have seen that the Euler equation is analogous to the equation $y'(x) = 0$ from elementary calculus. Let us consider another approach to deriving the Euler equation. This will provide a representation for the increment of a functional $F(y)$ under bell-shaped disturbances of $y(x)$. The resulting formula will be needed later for treatment of the isoperimetric problem.

Let us preview the approach before tackling the details. We first recall the way in which we proved the fundamental lemma of the calculus of variations. The lemma states that $f(x)$ must vanish if it is continuous and if

$$\int_a^b f(x)g(x)\,dx = 0$$

holds for an arbitrary continuous function $g(x)$ that goes to zero at the endpoints a, b. However, the proof made use of only a subset of such functions $g(x)$: those that were bell-shaped and whose supports were small enough. (The support of a function $g(x)$ is the closure of the set over which $g(x) \neq 0$.) Because of this we can reframe the problem of minimizing a functional in terms of disturbance functions taken from this subset only. Such a setup will not lend itself to proof that a solution is a minimizer, but will nonetheless provide an alternative derivation of the Euler equation.

So we take $y_0(x)$ to be a minimizer of the simplest functional, and instead of considering all disturbances $\varphi(x)$ of $y_0(x)$, consider only bell-shaped disturbances $\varphi(x)$ having small supports inside $[a, b]$. The main part of the increment of the functional will be given by the same formula as when all disturbances are considered:

$$\int_a^b \left(f_y - \frac{d}{dx} f_{y'}\right) \varphi(x)\,dx.$$

We will then restrict the set of possible disturbances $\varphi(x)$ to an even smaller subset: the bell-shaped functions having small support centered at a point $x_0 \in (a, b)$. If the support of a given $\varphi(x)$ is small enough, then the function $f_y - df_{y'}/dx$, which is supposed continuous, is almost constant on the support interval; up to infinitesimals its value can be taken to equal its value at $x = x_0$. Hence we will be able to split the above increment into two parts: one of these, the main part, is

$$\left(f_y - \frac{d}{dx} f_{y'}\right)\bigg|_{x=x_0} \int_a^b \varphi(x)\,dx.$$

The integral $\int_a^b \varphi(x)\,dx$ is the small area ΔS that lies under the bell of $\varphi(x)$. The expression in parentheses is recognized as the left-hand side of the Euler equation, and we therefore expect it to vanish along the minimizing curve and thus at $x = x_0$. As justification (and, moreover, to obtain a formula for the increment of a functional on bell-shaped small disturbances) we shall prove that the error in approximating the above main increment with this term is small in comparison with ΔS when the latter tends to zero. Unfortunately this is not valid if we take just any bell-shaped functions for the limit passage. It turns out that we must further restrict $\varphi(x)$ to those functions whose *amplitudes* tend to zero along with ΔS (note that this subset of disturbance functions would still be sufficient to prove the fundamental lemma). Taking only these functions and having the main part of the increment to zero we will obtain, aside from infinitesimals,

$$\delta F = \left(f_y - \frac{d}{dx} f_{y'}\right)\bigg|_{x=x_0} \Delta S = 0.$$

We will then divide through by ΔS and perform the limit passage. The quantity that will appear on the left-hand side, i.e., the limit of the ratio $\delta F / \Delta S$ under the conditions described above (if it exists), will become known as the *variational derivative* of $F(y)$ at the point x_0 for the curve $y_0(x)$. For any $y(x)$, not necessarily a minimizer, the main part of the increment of the functional on such $\varphi(x)$ is given as

$$\left(f_y - \frac{d}{dx} f_{y'}\right)\bigg|_{x=x_0} \Delta S,$$

and all the rest has the order of infinity smaller than ΔS. Let us now realize this plan in detail.

So we note that in the fundamental lemma, instead of having φ range over $C_0^{(\infty)}(a,b)$ we could restrict it to the class of bell-shaped functions. We could in fact restrict φ to a particular set of bell-shaped functions; the only thing needed in this is the possibility to get the support of a function of this set of any small length at any point of (a,b). Let us base this set of functions on the particular function

$$\varphi_\varepsilon(x) = \begin{cases} \exp\left(\frac{\varepsilon^2}{x^2-\varepsilon^2}\right), & |x| < \varepsilon, \\ 0, & |x| \geq \varepsilon. \end{cases}$$

The set of functions of the form $A\varphi_\varepsilon(x - x_0)$ when $a < x_0 - \varepsilon < x_0 + \varepsilon < b$ is called B. It can be shown that B is contained in $C_0^{(\infty)}(a,b)$. We repeat that we can use the class B instead of $C_0^{(\infty)}(a,b)$ in the formulation of the fundamental lemma.

Let us introduce

$$\sigma_{A,\varepsilon} = \int_a^b A\varphi_\varepsilon(x - x_0)\,dx = A\int_{x_0-\varepsilon}^{x_0+\varepsilon} \varphi_\varepsilon(x - x_0)\,dx$$

and consider the first variation of $F(y)$ at some y corresponding to an increment $A\varphi_\varepsilon(x - x_0) = \psi \in B$:

$$\begin{aligned}\delta F(y,\psi) &= \int_a^b [f_y(x,y,y')\psi + f_{y'}(x,y,y')\psi']\,dx \\ &= \int_{x_0-\varepsilon}^{x_0+\varepsilon} [f_y(x,y,y')\psi + f_{y'}(x,y,y')\psi']\,dx.\end{aligned}$$

Routine integration by parts gives

$$\delta F(y,\psi) = \int_{x_0-\varepsilon}^{x_0+\varepsilon} \left[f_y(x,y,y') - \frac{d}{dx}f_{y'}(x,y,y')\right]\psi\,dx. \qquad (1.9.10)$$

We now recall a useful result from calculus, the *second mean value theorem for integrals*:

Theorem 1.9.2 *Let $f(x)$ be continuous on $[a,b]$. If $g(x)$ is integrable and does not change sign in $[a,b]$, then*

$$\int_a^b f(x)g(x)\,dx = f(\xi)\int_a^b g(x)\,dx$$

for some $\xi \in [a,b]$.

A proof can be found in any good calculus text. Returning to (1.9.10), we assume the term in brackets is continuous and write

$$\begin{aligned}\delta F(y,\psi) &= \left[f_y(x,y,y') - \frac{d}{dx}f_{y'}(x,y,y')\right]\Bigg|_{x=\zeta} \int_{x_0-\varepsilon}^{x_0+\varepsilon} \psi\,dx \\ &= \left[f_y(x,y,y') - \frac{d}{dx}f_{y'}(x,y,y')\right]\Bigg|_{x=\zeta} \sigma_{A,\varepsilon}\end{aligned}$$

where $\zeta \in (x_0 - \varepsilon, x_0 + \varepsilon)$. If ε and $\sigma_{A,\varepsilon}$ tend to zero simultaneously, then

$$\delta F(y,\psi) = \left[f_y(x,y,y') - \frac{d}{dx}f_{y'}(x,y,y')\right]\Bigg|_{x=x_0} \sigma_{A,\varepsilon} + o(|\sigma_{A,\varepsilon}|). \quad (1.9.11)$$

We shall use this formula later. Now let us remark that from this relation we get

$$\lim_{\substack{\sigma_{A,\varepsilon}\to 0 \\ \varepsilon\to 0}} \frac{\delta F(y,\psi)}{\sigma_{A,\varepsilon}} = \left[f_y(x,y,y') - \frac{d}{dx}f_{y'}(x,y,y')\right]\Bigg|_{x=x_0}.$$

On the right we get the left-hand side of the Euler equation, so this limit deserves a special name. We call it the *variational derivative* and denote it by

$$\left.\frac{\delta F}{\delta y}\right|_{x=x_0} = \lim_{\substack{\sigma_{A,\varepsilon}\to 0\\ \varepsilon\to 0}} \frac{\delta F(y,\psi)}{\sigma_{A,\varepsilon}}.$$

Thus formula (1.9.11) can be rewritten as

$$\delta F(y,\psi) = \left(\left.\frac{\delta F}{\delta y}\right|_{x=x_0} + \alpha\right)\sigma_{A,\varepsilon} \qquad (1.9.12)$$

where $\alpha \to 0$ as $\sigma_{A,\varepsilon} \to 0$. The reader can see the convenient analogy with the notation for the calculus of functions.

The first variation is the main part of the increment of a functional. We are interested in extending (1.9.12) to the increment of $F(y)$. This is possible if we suppose that the limit passage $\sigma_{A,\varepsilon} \to 0$, $\varepsilon \to 0$ can be done in such a way that

$$\int_{x_0-\varepsilon}^{x_0+\varepsilon} (\psi^2 + \psi'^2)\,dx = o(\sigma_{A,\varepsilon}) \qquad (1.9.13)$$

when $\sigma_{A,\varepsilon} \to 0$ and $\varepsilon \to 0$. This happens, say, if ε relates to A by the formula $|A| = \varepsilon^3$. It is enough to consider the case $x_0 = 0$; after the change of variables $x = \varepsilon u$ we have

$$\sigma_{A,\varepsilon} = A\varepsilon K_1 \quad \text{where} \quad K_1 \equiv \int_{-1}^{1} \exp\left(\frac{1}{u^2-1}\right) du.$$

Observe that K_1 is a positive constant. Also

$$(\psi')^2 = \frac{4A^2\varepsilon^4 x^2}{(x^2-\varepsilon^2)^4} \exp\left(\frac{2\varepsilon^2}{x^2-\varepsilon^2}\right)$$

and we obtain

$$\int_{x_0-\varepsilon}^{x_0+\varepsilon} (\psi^2 + \psi'^2)\,dx = K_2A^2\varepsilon + K_3\frac{A^2}{\varepsilon}$$

where K_2 and K_3 are the positive constants

$$K_2 = \int_{-1}^{1} \exp\left(\frac{2}{u^2-1}\right) du, \quad K_3 = \int_{-1}^{1} \frac{4u^2}{(u^2-1)^4} \exp\left(\frac{2}{u^2-1}\right) du.$$

Hence

$$\frac{1}{\sigma_{A,\varepsilon}} \int_{x_0-\varepsilon}^{x_0+\varepsilon} (\psi^2 + \psi'^2)\,dx = \frac{K_2}{K_1}A + \frac{K_3}{K_1}\frac{A}{\varepsilon^2}.$$

The first term on the right tends to zero with ε, as does the second term if we take $|A| = \varepsilon^3$. It is therefore possible to choose A in such a way that the left-hand side tends to zero as ε tends to zero, as required by (1.9.13).

Let us denote by B_0 the subset of B consisting of those functions ψ for which $|A| = \varepsilon^3$. Note that the fundamental lemma continues to hold if φ ranges over the set B_0 rather than over the whole class $C_0^{(\infty)}(a,b)$. When $|A| = \varepsilon^3$, we denote $\sigma_{A,\varepsilon}$ by σ_ε. Note that σ_ε does not determine $\sigma_{A,\varepsilon}$ uniquely; however, the sign ambiguity is easily resolved for a given A by noting that σ_ε has the same sign as A.

From the three formulas (1.9.6), (1.9.12), and (1.9.13), we get

$$F(y+\psi) - F(y) = \left(\frac{\delta F}{\delta y}\bigg|_{x=x_0} + \beta \right) \sigma_\varepsilon \tag{1.9.14}$$

where $\beta \to 0$ when $\varepsilon \to 0$, or

$$\Delta F(y,\psi) = F(y+\psi) - F(y) = \left(f_y - \frac{d}{dx} f_{y'} + \beta \right)\bigg|_{x=x_0} \sigma_\varepsilon. \tag{1.9.15}$$

For the above limit passage we get the Euler equation as

$$\lim_{\substack{\sigma_\varepsilon \to 0 \\ \varepsilon \to 0}} \frac{\delta F(y,\psi)}{\sigma_\varepsilon} = \frac{\delta F}{\delta y}\bigg|_{x=x_0} = 0 \text{ for all } x_0 \in (a,b). \tag{1.9.16}$$

Brief review of important ideas

The *increment* $F(y+\varphi) - F(y)$ of the functional $F(y)$ can be written as

$$F(y+\varphi) - F(y) = \delta F(y,\varphi) + O(\|\varphi\|^2_{C^{(1)}(a,b)})$$

where the *first variation*

$$\delta F(y,\varphi) = \int_a^b [f_y(x,y,y')\varphi + f_{y'}(x,y,y')\varphi']\,dx$$

is the principal part (i.e., the portion of the increment that is linear in φ).

We have

$$\delta F(y,\varphi) = 0$$

when $y = y(x)$ is a minimizer of $F(y)$ for some given boundary conditions; this holds for any admissible increment φ of the function y. A functional is said to be *stationary* at y if its first variation vanishes.

The idea of the variational derivative is analogous to the idea of a partial derivative of a function of many variables. We define the variational derivative of a functional $F(y)$, at a point x_0, for a curve $y = y(x)$, as follows. We give $y(x)$ an increment $\varphi(x)$ which is nonzero only in a small neighborhood of x_0; we choose $\varphi(x)$ as a small bell-shaped bump, and denote the area between it and the x-axis by ΔS. We then get the main linear part δF of the increment ΔF under this special type of localized disturbance. By continuity of the Euler expression $f_y - \frac{d}{dx} f_{y'}$ we can approximate δF as the Euler expression times ΔS, hence can study the ratio $\delta F/\Delta S$ as $\Delta S \to 0$ in such a way that the bump disturbance shrinks in both width and height (the former requirement assures that we get localized information relevant to the point x_0, and the latter requirement assures that the remainder terms tend to zero faster than the main part). So we seek the ratio of the small change in the functional value to a small change in area under $y(x)$, when that change occurs near x_0. The variational derivative is given by

$$\left.\frac{\delta F}{\delta y}\right|_{x=x_0} = \left.\left(f_y - \frac{d}{dx} f'_y\right)\right|_{x=x_0}.$$

1.10 Isoperimetric Problems

We have found a way (1.9.16) of obtaining the Euler equation by setting the variational derivative to zero. We now apply this to the solution of an *isoperimetric problem.*

It is said that the first problem of this type was solved practically by Dido, legendary queen of ancient Carthage, who was offered as much land as she could surround with the skin of a bull. Using a fuzzy formulation of this "mathematical" problem, she cut the skin into thin bands, tied them end to end, and surrounded the town with this long "rope." Note that Dido's problem was quite hard; several issues had to be addressed, including (1) how to get the longest rope from the skin, (2) how to find the closed curve of a given length that would enclose the greatest planar area, and (3) how to choose the most desirable piece of land. We shall only be able to treat the second of these issues here! Let us begin by formulating the

Simplest Isoperimetric Problem. Find the minimum of the functional

$$F(y) = \int_a^b f(x, y, y')\, dx$$

from among the functions $y \in C^{(1)}(a,b)$ that satisfy

$$y(a) = c_0, \qquad\qquad y(b) = c_1, \tag{1.10.1}$$

and

$$G(y) = \int_a^b g(x, y, y')\,dx = l \tag{1.10.2}$$

where l is a given number.

Condition (1.10.2) is analogous to the condition that the length of a curve is given. We know a similar problem from calculus: given a restriction $g(x) = c$, find a minimum of $f(x)$. This is solved using Lagrange multipliers: there is a constant λ such that a minimizer of the problem is a stationary point of the function $f(x) + \lambda g(x)$ — that is, a solution of the equation $f'(x) + \lambda g'(x) = 0$. We correctly surmise that something similar should exist for our isoperimetric problem.

Note that our previous technique cannot be used because the restriction (1.10.2) has complicated the notion of the neighborhood of a function. Indeed, if $g(x, y, y')$ is not linear in y and y' then we cannot expect that a sum of two admissible small increments of a minimizer is also admissible: condition (1.10.2) can fail for the sum. The same comment applies to increments of the form $t\varphi$ if φ is an admissible increment. However, the technique of § 1.8 does not depend on such transformations in the set of admissible increments, so we will try to use it.

Theorem 1.10.1 *Let $y = y(x)$ be a local solution of the Simplest Isoperimetric Problem, and suppose y is not an extremal of the functional $G(z)$. Then there is a real number λ such that $y = y(x)$ is an extremal of $F(z) + \lambda G(z)$ on the set of functions from $C^{(1)}(a,b)$ satisfying* (1.10.1).

Before giving the proof let us note that the problem of finding this extremal is well defined in principle. Indeed, a solution of the Euler equation for $F(z) + \lambda G(z)$ has, in principle, three independent constants: λ, and the two independent constants expected in the general solution of the (second-order) Euler equation. These can be determined from (1.10.2) and (1.10.1).

Proof. We will try the results of § 1.8. We need to consider the set of small increments of the minimizer such that the incremented functions satisfy both (1.10.1) and (1.10.2). So we construct the set of increments by combining two bell-shaped functions of the class B_0 with centers of symmetry at x_1 and x_2, $x_1 < x_2$: that is, $A_i\varphi_{\varepsilon_i}(x - x_i)$, $|A_i| = \varepsilon_i^3$, $i = 1, 2$.

Denote this increment by $\eta(x) = \sum_i A_i \varphi_{\varepsilon_i}(x - x_i)$. We can assume that $\varepsilon_i < (x_2 - x_1)/2$, so the two nonzero domains of such an increment do not intersect (or we could argue that we introduced two bell-shaped increments of y at different points successively). Since the supports of the two bell-shaped functions do not intersect we can extend (1.9.14) to this case:

$$\Delta F(y,\eta) = \left[\left(f_y - \frac{d}{dx}f_{y'}\right)\bigg|_{x=x_1} + \alpha_1\right]\sigma_{\varepsilon_1} + \left[\left(f_y - \frac{d}{dx}f_{y'}\right)\bigg|_{x=x_2} + \alpha_2\right]\sigma_{\varepsilon_2} \tag{1.10.3}$$

where for $i = 1, 2$ we have

$$\sigma_{\varepsilon_i} = A_i \int_{x_i-\varepsilon}^{x_i+\varepsilon} \varphi_{\varepsilon_i}(x - x_i)\, dx, \qquad |A_i| = \varepsilon_i^3,$$

and $\alpha_i \to 0$ when $\sigma_{\varepsilon_i} \to 0$.

We must choose the increment η so that $y + \eta$ satisfies (1.10.2). Thus we have $G(y + \eta) - G(y) = 0$. This and the analogue of (1.9.15) for $G(y + \eta) - G(y)$ imply

$$\left[\left(g_y - \frac{d}{dx}g_{y'}\right)\bigg|_{x=x_1} + \beta_1\right]\sigma_{\varepsilon_1} + \left[\left(g_y - \frac{d}{dx}g_{y'}\right)\bigg|_{x=x_2} + \beta_2\right]\sigma_{\varepsilon_2} = 0$$

with the same σ_{ε_i} as in (1.10.3) and $\beta_i \to 0$ when $\sigma_{\varepsilon_i} \to 0$.

Since $y = y(x)$ is not an extremal of $G(z)$, there is a point $x_2 \in (a, b)$ where $g_y - \frac{d}{dx}g_{y'} \neq 0$. For sufficiently small ε_2 we get β_2 as small as we wish, thus the second square bracket is nonzero in this case and so

$$\sigma_{\varepsilon_2} = -\frac{\left(g_y - \frac{d}{dx}g_{y'}\right)\big|_{x=x_1} + \beta_1}{\left(g_y - \frac{d}{dx}g_{y'}\right)\big|_{x=x_2} + \beta_2}\sigma_{\varepsilon_1}.$$

Then

$$\Delta F(y,\eta) = \left[\left(f_y - \frac{d}{dx}f_{y'}\right)\bigg|_{x=x_1} + \alpha_1\right]\sigma_{\varepsilon_1} - \left[\left(f_y - \frac{d}{dx}f_{y'}\right)\bigg|_{x=x_2} + \alpha_2\right]\frac{\left(g_y - \frac{d}{dx}g_{y'}\right)\big|_{x=x_1} + \beta_1}{\left(g_y - \frac{d}{dx}g_{y'}\right)\big|_{x=x_2} + \beta_2}\sigma_{\varepsilon_1}. \tag{1.10.4}$$

Denoting

$$\lambda = -\frac{\left(f_y - \frac{d}{dx} f_{y'}\right)\big|_{x=x_2}}{\left(g_y - \frac{d}{dx} g_{y'}\right)\big|_{x=x_2}}$$

we get from (1.10.4)

$$\Delta F(y,\eta) = \left[\left(f_y - \frac{d}{dx} f_{y'}\right)\bigg|_{x=x_1} + \lambda \left(g_y - \frac{d}{dx} g_{y'}\right)\bigg|_{x=x_1}\right] \sigma_{\varepsilon_1} + o(|\sigma_{\varepsilon_1}|)$$

The first variation of the functional that must be zero on the solution is

$$\delta F(y,\eta) = \left[\left(f_y - \frac{d}{dx} f_{y'}\right)\bigg|_{x=x_1} + \lambda \left(g_y - \frac{d}{dx} g_{y'}\right)\bigg|_{x=x_1}\right] \sigma_{\varepsilon_1} = 0.$$

Since we can choose σ_{ε_1} arbitrarily, it follows that for any $x_1 \in (a,b)$ we have

$$\left(f_y - \frac{d}{dx} f_{y'}\right)\bigg|_{x=x_1} + \lambda \left(g_y - \frac{d}{dx} g_{y'}\right)\bigg|_{x=x_1} = 0.$$

This means $y = y(x)$ is an extremal of $F + \lambda G$. □

For an isoperimetric problem where the functional F depends on a vector function $\mathbf{y} = (y_1, \ldots, y_n)$ and there are m restrictions of integral type $G_i = \int_a^b g_i(x, \mathbf{y}, \mathbf{y}')\,dx$, $i = 1, \ldots, k$, there is a corresponding statement. For this problem a minimizer $\mathbf{y}$ is an extremal of the functional $F + \sum_{i=1}^k \lambda_k G_i$. The reader can derive the corresponding Euler equations. It is clearly impossible to satisfy k integral restrictions for $\mathbf{y}$ considering only the two-belled increments, so here it is necessary to introduce increments composed of $k+1$ bell-shaped functions. This necessitates additional technical work.

Two problems

Let us consider two special problems. The first was mentioned in § 1.1: find the plane curve enclosing the maximum possible area for a given perimeter. One approach is to examine all curves $y(x)$ that, except for their endpoints, lie in the upper half of the xy-plane, and that have endpoints $(\pm a, 0)$ and a given length l. (Note that a is not specified in advance.) In the notation of Theorem 1.10.1 we have

$$F(y) = \int_{-a}^{a} y\,dx, \qquad G(y) = \int_{-a}^{a} \sqrt{1 + (y')^2}\,dx;$$

hence

$$f(x, y, y') = y, \qquad g(x, y, y') = \sqrt{1 + (y')^2},$$

and $f + \lambda g$ does not depend on x explicitly. So we can write

$$(f + \lambda g) - (f + \lambda g)_{y'} y' = y + \lambda\sqrt{1 + (y')^2} - \frac{\lambda (y')^2}{\sqrt{1 + (y')^2}} = c_1,$$

which simplifies to

$$y - c_1 = \frac{-\lambda}{\sqrt{1 + (y')^2}}.$$

Put

$$y' = \frac{dy}{dx} = \tan t \tag{1.10.5}$$

where t is a parameter; then

$$y - c_1 = \frac{-\lambda}{\sqrt{1 + \tan^2 t}} = \frac{-\lambda}{\sec t} = -\lambda \cos t. \tag{1.10.6}$$

Now from (1.10.5) and (1.10.6)

$$dx = \frac{1}{\tan t}\, dy = \frac{1}{\tan t}\frac{dy}{dt}\, dt = \frac{1}{\tan t}\lambda \sin t\, dt = \lambda \cos t\, dt$$

so that upon integration we have $x = \lambda \sin t + c_2$. From the equations

$$x - c_2 = \lambda \sin t, \qquad y - c_1 = -\lambda \cos t,$$

we may eliminate t to produce

$$(x - c_2)^2 + (y - c_1)^2 = \lambda^2.$$

Thus all extremals of $F(y) + \lambda G(y)$ are portions of a circle. The conditions

$$(-a - c_2)^2 + (0 - c_1)^2 = \lambda^2, \qquad (a - c_2)^2 + (0 - c_1)^2 = \lambda^2,$$

may be subtracted to show that $c_2 = 0$. The vertical shift c_1 of the center and the radius λ clearly depend on the given l. Note, however, that we do not verify directly whether we have actually obtained the needed maximum. We leave this to the reader instead.

Another approach is to use polar coordinates. Calling these (r, ϕ) and placing the coordinate origin inside the desired closed curve $r = r(\phi)$, we have

$$f + \lambda g = \frac{1}{2}r^2 + \lambda\sqrt{r^2 + (r')^2}$$

and the corresponding Euler equation

$$r + \frac{\lambda r}{\sqrt{r^2 + (r')^2}} - \frac{d}{d\phi}\frac{\lambda r'}{\sqrt{r^2 + (r')^2}} = 0.$$

Performing the differentiation and simplifying we obtain

$$\frac{1}{\lambda} = \frac{rr'' - 2(r')^2 - r^2}{[r^2 + (r')^2]^{3/2}},$$

which shows that the curvature of $r(\phi)$ is a constant $1/\lambda$ and gives us a circle again.

It is worth noting that we formulated the problems for a minimum but solved for a maximum. This is analogous to the standard calculus trick of maximizing a function f by minimizing $-f$. Of even more interest is the idea of obtaining a *dual problem* by reversing the roles of the functionals F and G. For example, the maximum area that can be enclosed by a curve having length l is $l^2/4\pi$. The dual problem is to find a closed curve of minimum length that borders a flat domain with area $l^2/4\pi$. Of course, the solution is a circle having circumference l.

We now turn to another classical isoperimetric problem. Early in the development of mathematics people became curious about the precise form assumed by a chain hanging from both ends (such chains were used, for instance, as "fences" along the sides of bridges). This is a hard problem if one wishes to consider it in full detail (including friction, nonuniformities in the individual links, and so on); it is possible to show that many peculiarities arise, and even the full setup of the problem is quite cumbersome. A successful approach depended on the construction of a tractable model for the chain. First an ideal chain was introduced, consisting of extremely small elements that were all identical; this permitted the tools of calculus to be applied. An even simpler model was a uniform filamentary rope — heavy, flexible, and absolutely unstretchable. Unlike a chain, such an idealized rope could lie in a plane.

Let us therefore suppose that a uniform, flexible rope of a given fixed length hangs in equilibrium with its ends attached to two fixed points: what is the shape assumed by the rope? Denote by l the length of the rope, assume it has a unit mass density, and let the endpoints be (a, h_a) and (b, h_b). (Clearly we need $b - a \leq l$.) The y coordinate of the center of gravity is proportional to the integral $\int_a^b y(s)\,ds$ where s is arc length along the rope; since the center of gravity will find the lowest possible position, we are led to minimize the functional ($ds = \sqrt{1+(y')^2}\,dx$)

$$F(y) = \int_a^b y\sqrt{1+(y')^2}\,dx$$

subject to the side condition

$$G(y) = \int_a^b \sqrt{1+(y')^2}\,dx = l.$$

Accordingly we minimize

$$F(y) + \lambda G(y) = \int_a^b (y+\lambda)\sqrt{1+(y')^2}\,dx.$$

Since the integrand does not depend on x explicitly, we write out the first integral of the differential equation,

$$(y+\lambda)\sqrt{1+(y')^2} - \frac{(y+\lambda)(y')^2}{\sqrt{1+(y')^2}} = c_1,$$

and then simplify to obtain

$$y + \lambda = c_1\sqrt{1+(y')^2}.$$

We find a parametric representation of the solution, introducing a parameter t by the substitution $y' = \sinh t$. Then

$$y + \lambda = c_1 \cosh t$$

and we have, for the dependence of x on t,

$$dx = \frac{1}{\sinh t}\,dy = \frac{1}{\sinh t}\frac{dy}{dt}\,dt = \frac{1}{\sinh t}(c_1 \sinh t)\,dt = c_1 dt$$

so that

$$x - c_2 = c_1 t.$$

Finally, eliminating t we find

$$y + \lambda = c_1 \cosh\left(\frac{x - c_2}{c_1}\right),$$

the equation of a catenary. The given conditions can be used to determine c_1, c_2, and λ. (Of course $c_2 = 0$ if $b = -a$.)

Once again we do not provide formal verification that a minimum has actually been obtained. Indeed, with many problems that arise from geometry or physics it is intuitively clear whether we have the desired solution. For the hanging chain problem, we can assert on physical grounds that a solution exists; since the solution we obtained is unique, we can rest assured that it is the desired one.

It is possible to state other types of minimum problems with restrictions which, for their solution, require a technique similar to that of Lagrange multipliers. For example, it is possible to pose a problem of minimizing the functional $\int_{x_0}^{x_1} f(x, y, z, y', z')\, dx$ under some boundary conditions when there is a restriction $g(x, y, z) = 0$ (in more advanced books this is called minimizing a functional on a manifold). Here a minimizer is an extremal of a functional $\int_a^b [f - \lambda(x)g]\, dx$ without integral restrictions imposed by g, and $\lambda(x)$ is a new unknown function that is treated as given when we compose the Euler equations. Of course to define it one must use the equation $g(x, y, z) = 0$. Some problems in mechanics involve restrictions of even more general type; e.g., $g(x, y, z, y', z') = 0$.

Quick summary

We have concentrated on an isoperimetric problem of the following general form: find the minimizer of the simplest integral functional from among those functions y that satisfy

$$y(a) = c_0, \qquad y(b) = c_1, \qquad G(y) = \int_a^b g(x, y, y')\, dx = l$$

where $G(y)$ and l are given. A solution method is to introduce a real number λ (analogous to a Lagrange multiplier) and seek to minimize the functional $F + \lambda G$ subject to the given endpoint conditions on y.

1.11 General Form of the First Variation

We would like to consider the minimization problem for functionals of the form (1.2.1) when the endpoints of integration can change.

We have seen for various functionals that at a point of minimum the first variation is zero. Let us demonstrate this in general. First let us introduce some notions. In subsequent chapters we shall use the notion of a normed space; now we introduce only the definition. A normed space is a linear space of elements x such that for each x a function called the *norm* $\|x\|$ is defined. The norm must possess the following three properties:

(i) for any x, $\|x\| \geq 0$; $\|x\| = 0$ if and only if $x = 0$;
(ii) $\|\lambda x\| = |\lambda| \, \|x\|$ for any real number λ;
(iii) $\|x + y\| \leq \|x\| + \|y\|$.

The third property is called the triangle inequality. For example, the norm (1.2.5) for functions in $C^{(1)}(a, b)$ satisfies the above properties.

We can define a functional on a general normed space. A functional on a normed space X is a function that takes values in $\mathbb{R}$; i.e., to any $x \in X$ there corresponds no more than one real number. We call a functional $\Phi(x)$ *linear* if for any x, y belonging to its domain and any real λ, μ,

$$\Phi(\lambda x + \mu y) = \lambda \Phi(x) + \mu \Phi(y).$$

Finally, a linear functional $\Phi(x)$ is continuous in X if there is a constant c such that for any $x \in X$,

$$|\Phi(x)| \leq c \, \|x\| \,.$$

The infimum of all such c is called the *norm* of Φ and is denoted $\|\Phi\|$ (it is actually a norm according to the norm properties listed above).

Let $F(x)$ be a functional on X, and assume that in some ball about a point $x \in X$ (a ball is a set of elements $x + \delta x \in X$, where $\delta x \in X$, such that $\|\delta x\| \leq \varepsilon$ for some $\varepsilon > 0$) there is a representation

$$F(x + \delta x) - F(x) = \delta F(x, \delta x) + o(\|\delta x\|) \qquad (1.11.1)$$

where $\delta F(x, \delta x)$ is a linear functional continuous in δx. We have called it the first variation of $F(x)$, but it also has another name: the *Frechét differential* of $F(x)$ at x. Hence we have extended the definition of the first variation to abstract functionals.

Let x be a local minimizer of F: that is, $F(x + \delta x) - F(x) \geq 0$ for any $\|\delta x\| \leq \varepsilon$ with some $\varepsilon > 0$.

Theorem 1.11.1 *Let x be a minimizer of F on the set of elements $\{x + \delta x \mid \|\delta x\| \leq \varepsilon\}$, and suppose F has the first variation at x such that* (1.11.1) *holds on this set. Then $\delta F(x, \delta x) = 0$.*

Proof. Suppose to the contrary there exists an $x^* \in X$ such that $\delta F(x, x^*) \neq 0$. Then for small enough t we have

$$0 \leq F(x + tx^*) - F(x) = \delta F(x, tx^*) + o(t \|x^*\|) = t\delta F(x, x^*) + o(t).$$

For small $|t|$ the difference on the left is determined by the first term on the right. Choosing an appropriate t we get $t\delta F(x, x^*) < 0$, which contradicts the leftmost inequality. □

Thus for a problem of minimum of a functional, as a first step, we have to derive its first variation, equate it to zero, and then find solutions of this equation for any admissible disturbances (or virtual variations) δx.

We return to the beginning of this section and claim again that we would like to consider a minimization problem for a more general functional than (1.2.1), i.e., the functional

$$\int_{x_0}^{x_1} f(x, y, y')\, dx \qquad (1.11.2)$$

where the endpoints x_0 and x_1 can move. Thus we need the expression for the first variation in this case. To realize the above idea we must suppose that all changes are of the same order of smallness. Here we have not only a change φ in y to consider, but also changes δx_0 and δx_1 of the ends x_0 and x_1 respectively. Since δx_0 and δx_1 are arbitrary and we could have $\delta x_0 < 0$ or $\delta x_1 > 0$, we must agree on a way of extending a given function to points outside the segment $[x_0, x_1]$. We do this by linear extrapolation, using the tangent lines to $y = y(x)$ at x_0 and x_1 to define the values of the extension. The ends of the extended curve have coordinates $(x_0 + \delta x_0, y_0 + \delta y_0)$ and $(x_1 + \delta x_1, y_1 + \delta y_1)$.

Our problem is to derive the linear part of the increment for (1.11.2) when φ, φ', δx_0, δy_0, δx_1, and δy_1 have the same order of smallness; that is, to extract the part of the increment that is linear in each of these quantities. Denote

$$\varepsilon = \|\varphi\|_{C^{(1)}(x_0, x_1)} + |\delta x_0| + |\delta y_0| + |\delta x_1| + |\delta y_1|.$$

The increment is

$$\Delta F(y) = \int_{x_0+\delta x_0}^{x_1+\delta x_1} f(x, y + \varphi, y' + \varphi')\, dx - \int_{x_0}^{x_1} f(x, y, y')\, dx.$$

The first integral can be decomposed as

$$\int_{x_0+\delta x_0}^{x_1+\delta x_1} (\cdots)\,dx = \int_{x_0}^{x_1} (\cdots)\,dx + \int_{x_1}^{x_1+\delta x_1} (\cdots)\,dx - \int_{x_0}^{x_0+\delta x_0} (\cdots)\,dx.$$

We recall that all the functions $y = y(x)$, $\varphi = \varphi(x)$, are linearly extrapolated outside $[x_0, x_1]$, preserving continuity of the functions and their first derivatives. Thus

$$\begin{aligned}\Delta F(y) = \int_{x_0}^{x_1} [f(x, y+\varphi, y'+\varphi') - f(x,y,y')]\,dx \\ + \int_{x_1}^{x_1+\delta x_1} f(x, y+\varphi, y'+\varphi')\,dx \\ - \int_{x_0}^{x_0+\delta x_0} f(x, y+\varphi, y'+\varphi')\,dx. \qquad (1.11.3)\end{aligned}$$

The integral over $[x_0, x_1]$ can be transformed in the usual manner so we use the formula we obtained before:

$$\begin{aligned}\int_{x_0}^{x_1} [f(x, y+\varphi, y'+\varphi') - f(x,y,y')]\,dx \\ = \int_{x_0}^{x_1} \left[f_y(x,y,y') - \frac{d}{dx} f_{y'}(x,y,y') \right] \varphi\,dx \\ + f_{y'}(x, y(x), y'(x))\varphi(x) \Big|_{x=x_0}^{x=x_1} + o(\varepsilon).\end{aligned}$$

Let us represent φ at the endpoints using δy_0 and δy_1. From Fig. 1.2 we

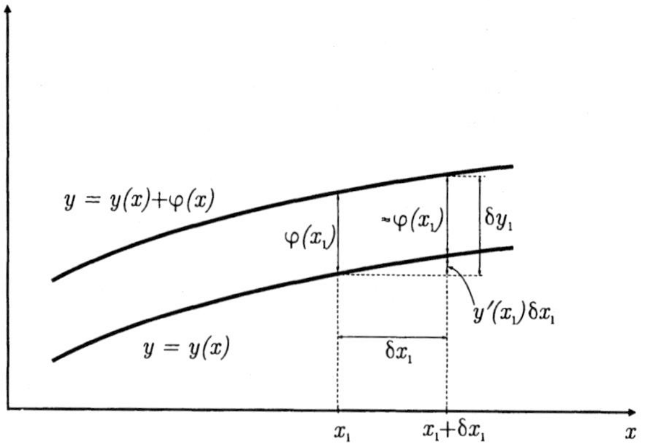

Fig. 1.2 Quantities appearing in equations (1.11.4) and (1.11.5).

see that

$$\varphi(x_1) = \delta y_1 - y'(x_1)\delta x_1 + o(\varepsilon). \tag{1.11.4}$$

We have a similar relation at x_0:

$$\varphi(x_0) = \delta y_0 - y'(x_0)\delta x_0 + o(\varepsilon). \tag{1.11.5}$$

Thus

$$\begin{aligned}
&\int_{x_0}^{x_1} [f(x, y + \varphi, y' + \varphi') - f(x, y, y')]\, dx \\
&\quad = \int_{x_0}^{x_1} \left[f_y(x, y, y') - \frac{d}{dx} f_{y'}(x, y, y') \right] \varphi\, dx \\
&\qquad + f_{y'}(x_1, y(x_1), y'(x_1))\delta y_1 - f_{y'}(x_0, y(x_0), y'(x_0))\delta y_0 \\
&\qquad - [f_{y'}(x_1, y(x_1), y'(x_1))y'(x_1)\delta x_1 \\
&\qquad - f_{y'}(x_0, y(x_0), y'(x_0))y'(x_0)\delta x_0] \\
&\qquad + o(\varepsilon).
\end{aligned}$$

Now let us consider the two other terms for ΔF in (1.11.3). Extracting the terms of the first order of smallness in ε we have

$$\begin{aligned}
\int_{x_1}^{x_1+\delta x_1} f(x, y + \varphi, y' + \varphi')\, dx &= \int_{x_1}^{x_1+\delta x_1} f(x, y, y')\, dx + o(\varepsilon) \\
&= f(x_1, y(x_1), y'(x_1))\delta x_1 + o(\varepsilon)
\end{aligned}$$

and similarly

$$\int_{x_0}^{x_0+\delta x_0} f(x, y + \varphi, y' + \varphi')\, dx = f(x_0, y(x_0), y'(x_0))\delta x_0 + o(\varepsilon).$$

Collecting terms we have

$$\begin{aligned}
\Delta F = &\int_{x_0}^{x_1} \left[f_y(x, y, y') - \frac{d}{dx} f_{y'}(x, y, y') \right] \varphi\, dx \\
&+ f_{y'}(x_1, y(x_1), y'(x_1))\delta y_1 - f_{y'}(x_0, y(x_0), y'(x_0))\delta y_0 \\
&+ [f(x_1, y(x_1), y'(x_1)) - f_{y'}(x_1, y(x_1), y'(x_1))y'(x_1)]\delta x_1 \\
&- [f(x_0, y(x_0), y'(x_0)) - f_{y'}(x_0, y(x_0), y'(x_0))y'(x_0)]\delta x_0 \\
&+ o(\varepsilon).
\end{aligned}$$

Thus we have derived the general form of the first variation of the functional when the ends of the curve can move:

$$\delta F = \int_{x_0}^{x_1} \left(f_y - \frac{d}{dx} f_{y'} \right) \varphi \, dx + f_{y'} \delta y \Big|_{x_0}^{x_1} + (f - y' f_{y'}) \, \delta x \Big|_{x_0}^{x_1}. \tag{1.11.6}$$

The reader can demonstrate that for a functional $F(\mathbf{y}) = \int_{x_0}^{x_1} f(x, \mathbf{y}, \mathbf{y}') \, dx$ with movable boundaries the general form the first variation is

$$\delta F = \sum_{i=1}^{n} \int_{x_0}^{x_1} \left(f_{y_i} - \frac{d}{dx} f_{y_i'} \right) \varphi_i \, dx + \sum_{i=1}^{n} f_{y_i'} \delta y_i \Big|_{x_0}^{x_1} + \left(f - \sum_{i=1}^{n} y_i' f_{y_i'} \right) \delta x \Big|_{x_0}^{x_1}.$$

1.12 Movable Ends of Extremals

In the previous section we found the general form (1.11.6) of the first variation of a functional when the boundaries of integration can move. Note that when the boundaries are fixed then $\delta x_i = 0$ and (1.11.6) reduces to the left-hand side of (1.5.2). Thus in this case the equation $\delta F = 0$ for a minimizer gives us the Euler equation and natural boundary conditions. The problem with natural boundary conditions can be reformulated as follows: given two vertical lines $x = a$ and $x = b$, find a minimizer of the functional (1.2.1) that starts on the line $x = a$ and ends on the line $x = b$ (or that connects these lines).

This formulation suggests that by using (1.11.6) it is possible to find equations to solve the following problem. Given two curves $y = \psi_0(x)$ and $y = \psi_1(x)$, find a minimizer of (1.2.1) that starts on $\psi_0(x)$ and ends on $\psi_1(x)$. Let us call this the "problem with movable boundaries."

We assume any other functions of interest are defined (and twice continuously differentiable) wherever the boundary functions $\psi_i(x)$ are given. (If these latter functions are not defined on the same interval, we construct an interval that encompasses all points of interest and assume that everything is defined on this larger interval.) Moreover we assume the endpoints of the minimizer are not endpoints of the graph for the $\psi_i(x)$.

So we start with

$$\delta F = \int_{x_0}^{x_1} \left(f_y - \frac{d}{dx} f_{y'} \right) \varphi \, dx + f_{y'} \delta y_i \Big|_{x_0, i=0}^{x_1, i=1} + (f - y' f_{y'}) \, \delta x_i \Big|_{x_0, i=0}^{x_1, i=1}. \tag{1.12.1}$$

We know that for admissible increments φ of a minimizer $y = y(x)$ the first

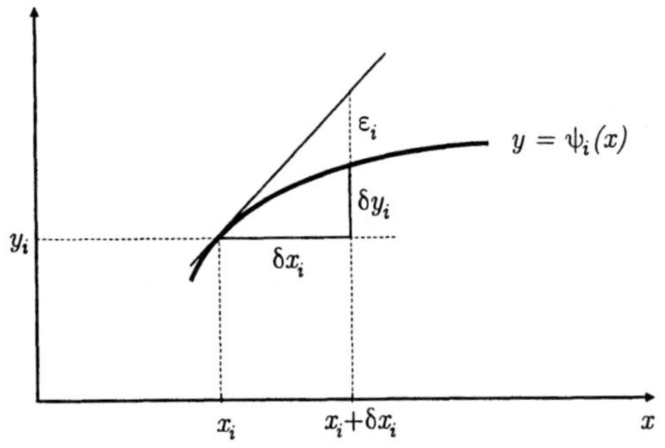

Fig. 1.3 Quantities near movable end of an extremal.

variation of the functional is equal to zero. Although the expression δF above contains all the terms of the increment of the first order of smallness, it is not the first variation in the present case. Admissible φ now are those that are continuously differentiable and such that both

$$(x_0, y(x_0)) \quad \text{and} \quad (x_0 + \delta x_0, y(x_0 + \delta x_0) + \phi(x_0 + \delta x_0))$$

belong to the curve $y = \psi_0(x)$, and both

$$(x_1, y(x_1)) \quad \text{and} \quad (x_1 + \delta x_1, y(x_1 + \delta x_1) + \phi(x_1 + \delta x_1))$$

belong to the curve $y = \psi_1(x)$.

Consider Fig. 1.3. Here each δy_i ($i = 0$ or 1) and its corresponding δx_i are no longer independent; it is clear that for small δx_i we have

$$\delta y_i = \psi_i'(x_i)\delta x_i + \varepsilon_i, \qquad i = 0, 1$$

where the ε_i are of a higher order of smallness than δx_i and δy_i. Substituting this into the right-hand side of (1.12.1), we select only the terms of the first order of smallness and get

$$\int_{x_0}^{x_1} \left(f_y - \frac{d}{dx} f_{y'} \right) \varphi \, dx + f_{y'} \psi_i' \delta x_i \bigg|_{x_0, i=0}^{x_1, i=1} + (f - y' f_{y'})\, \delta x_i \bigg|_{x_0, i=0}^{x_1, i=1}.$$

This is the first variation of the functional (note that it is equal to δF in (1.12.1) only up to terms of the first order of smallness in the norm of the increment). Thus

$$\int_{x_0}^{x_1} \left(f_y - \frac{d}{dx} f_{y'} \right) \varphi \, dx + f_{y'} \psi_i' \delta x_i \bigg|_{x_0, i=0}^{x_1, i=1} + (f - y' f_{y'}) \, \delta x_i \bigg|_{x_0, i=0}^{x_1, i=1} = 0 \tag{1.12.2}$$

for all admissible φ.

Let us derive the consequences of this equation. First, from among the admissible increments $y = \varphi(x)$ we take only those which satisfy the conditions $\varphi(x_0) = \varphi(x_1) = 0$. For any such φ we have

$$\int_{x_0}^{x_1} \left(f_y - \frac{d}{dx} f_{y'} \right) \varphi \, dx = 0$$

and thus by the fundamental lemma the Euler equation $f_y - \frac{d}{dx} f_{y'} = 0$ is fulfilled on (x_0, x_1). Hence the integral in (1.12.2) vanishes for any admissible φ, and it follows that

$$(f + (\psi_i' - y') f_{y'}) \, \delta x_i \bigg|_{x_0, i=0}^{x_1, i=1} = 0. \tag{1.12.3}$$

It is clear that we can "move" the ends of the curve independently, so (1.12.3) implies two boundary conditions for the minimizer:

$$(f + (\psi_1' - y') f_{y'}) \big|_{x_1} = 0, \qquad (f + (\psi_0' - y') f_{y'}) \big|_{x_0} = 0. \tag{1.12.4}$$

For the problem under consideration the minimizing curve $y = y(x)$ satisfies conditions (1.12.4) which are an extension of the natural boundary conditions. The way in which the minimizer intersects the boundary curves $y = \psi_i(x)$ has a special name: we say that the curve $y = y(x)$ is *transversal* to the curves $y = \psi_i(x)$, $i = 0, 1$.

Let us analyze the setting of the boundary value problem in this case. There is the Euler equation whose solution is determined up to two unknown constants (it is not always so; in nonlinear equations the situation with constants is sometimes much more complex, but when we analyze the problem qualitatively we keep in mind the terms of the linear case). The two conditions (1.12.4) could define those constants, but they contain unknown quantities x_0 and x_1 so we need to find two more equations. They are $y(x_0) = \psi_0(x_0)$ and $y(x_1) = \psi_1(x_1)$, and thus the setup of the necessary conditions for $y = y(x)$ to be a minimizer is completed.

Example 1.12.1 Show that for functionals of the form

$$\int_{x_0}^{x_1} q(x,y)\sqrt{1+(y')^2}\,dx$$

where $q(x,y) \neq 0$ at the endpoints x_0 and x_1, conditions (1.12.4) imply orthogonal intersections between $y(x)$ and the curves $\psi_0(x)$ and $\psi_1(x)$ at the points x_0 and x_1, respectively.

Solution Take, for example, the condition $(f + (\psi_1' - y')f_{y'})\big|_{x_1} = 0$. Direct substitution and a bit of simplification give

$$\left(q(x,y)\frac{1+\psi_1' y'}{\sqrt{1+(y')^2}} \right)\Bigg|_{x_1} = 0.$$

If $q(x,y)|_{x_1} \neq 0$, then $(1+\psi_1' y')|_{x_1} = 0$; i.e.,

$$y'|_{x_1} = -\frac{1}{\psi_1'|_{x_1}}.$$

Since the slopes are negative reciprocals, y is orthogonal to ψ_1 at $x = x_1$.

Quick review

The problem with movable boundaries for the simplest integral functional involves finding a minimizer that connects two given curves $y = \psi_0(x)$ and $y = \psi_1(x)$. We first solve the Euler equation, obtaining a solution in terms of two unknown constants. We then impose the transversality conditions

$$(f + (\psi_1' - y')f_{y'})\big|_{x_1} = 0, \qquad (f + (\psi_0' - y')f_{y'})\big|_{x_0} = 0;$$

here x_1 and x_0 are also unknowns. After imposing $y(x_0) = \psi(x_0)$ and $y(x_1) = \psi(x_1)$, all constants should be determined.

Special cases: (1) If one of the ψ_i is a horizontal line, say $\psi_1(x) =$ constant, then $\psi_1' \equiv 0$ and the corresponding transversality condition becomes

$$(f - y' f_{y'})\big|_{x_1} = 0.$$

If ψ_1 is a vertical line ($x =$ constant) then $f_{y'}\big|_{x_1} = 0$.

1.13 Weierstrass–Erdmann Conditions and Related Problems

We have required a minimizer $y = y(x)$ of (1.2.1) to assume given values at the endpoints of $[a, b]$. Is it possible to retain these conditions and also require that $y(x)$ assume a third given value at an interior point of $[a, b]$? That is, can we impose three conditions of the form $y(a) = c_0$, $y(b) = c_1$, and $y(\alpha) = c_2$ where $\alpha \in (a, b)$? If we require the minimizer to be in $C^{(1)}(a, b)$, then the answer is, in general, negative: a solution of the second-order Euler equation cannot be made to satisfy three conditions at once. If we omit the condition of continuity of the minimizer at $x = \alpha$, the problem can be solvable in principle. However, in this case we can consider two separate problems of minimizing two functionals, one of which is given on $[a, \alpha]$ and the other on $[\alpha, b]$. So in this case we reduce the three-point problem to the two-point problem already considered.

With some problems it is sensible to assume that a minimizing curve has a finite number of points at which continuity of its derivative fails. We cannot appoint the position of such points on (a, b) in advance. It happens that at such points the *Weierstrass–Erdmann* conditions must be fulfilled. Let us derive these, assuming the existence of one point of discontinuity of the first derivative of the minimizer. They will hold at every such point.

Suppose $x = \alpha$ is a point at which the first derivative of a minimizer is not continuous.

Theorem 1.13.1 *Let $x = \alpha \in (a, b)$ be a point at which the tangent to a minimizer $y = y(x)$ of the functional $\int_a^b f(x, y, y')\,dx$ has a break. Then y satisfies the Euler equation on the intervals (a, α) and (α, b), and at $x = \alpha$ the Weierstrass–Erdmann conditions*

$$f_{y'}\big|_{x=\alpha-0} = f_{y'}\big|_{x=\alpha+0} \tag{1.13.1}$$

and

$$(f - y' f_{y'})\big|_{x=\alpha-0} = (f - y' f_{y'})\big|_{x=\alpha+0} \tag{1.13.2}$$

hold.

Before giving the proof, let us discuss how to state the corresponding boundary value problem. On each of the intervals (a, α) and (α, b) the minimizer satisfies the Euler equation. So in general the minimizer is determined up to four unknown constants. Also unknown is α. There are five conditions to determine these constants: the two boundary conditions at a

and b, the conditions (1.13.1) and (1.13.2), and the condition of continuity $y(\alpha - 0) = y(\alpha + 0)$. Thus in principle the boundary value problem is formulated properly.

Proof. Let us consider for definiteness the boundary conditions $y(a) = c_0$, $y(b) = c_1$, for a minimizer. We require the minimizer to be continuous at $x = \alpha$. Perturbing the minimizer by an admissible φ and supposing that the point $(\alpha, y(\alpha))$ gets the increments $(\delta x, \delta y)$, we should apply the general formula for the first variation

$$\int_{x_0}^{x_1} \left(f_y - \frac{d}{dx} f_{y'} \right) \varphi \, dx + f_{y'} \delta y \Big|_{x_0}^{x_1} + (f - y' f_{y'}) \, \delta x \Big|_{x_0}^{x_1} \tag{1.13.3}$$

twice, on each of intervals (a, α) and (α, b) separately, taking into account that the increment $(\delta x, \delta y)$ at $(\alpha, y(\alpha))$ is the same on the left and the right of α. Remembering that δx and δy are zero at $x = a$ and $x = b$ for all admissible increments, we have

$$\begin{aligned} \delta F &= \delta \left(\int_a^\alpha f(x, y, y') \, dx + \int_\alpha^b f(x, y, y') \, dx \right) \\ &= \int_a^\alpha \left(f_y - \frac{d}{dx} f_{y'} \right) \varphi \, dx + f_{y'} \delta y \Big|_{x=\alpha-0} + (f - y' f_{y'}) \, \delta x \Big|_{x=\alpha-0} \\ &\quad + \int_\alpha^b \left(f_y - \frac{d}{dx} f_{y'} \right) \varphi \, dx - f_{y'} \delta y \Big|_{x=\alpha+0} - (f - y' f_{y'}) \, \delta x \Big|_{x=\alpha+0}. \end{aligned}$$

Thus for all admissible increments

$$\begin{aligned} &\int_a^\alpha \left(f_y - \frac{d}{dx} f_{y'} \right) \varphi \, dx + \int_\alpha^b \left(f_y - \frac{d}{dx} f_{y'} \right) \varphi \, dx \\ &\quad + \left[f_{y'} \Big|_{x=\alpha-0} - f_{y'} \Big|_{x=\alpha+0} \right] \delta y \\ &\quad + \left[(f - y' f_{y'}) \Big|_{x=\alpha-0} - (f - y' f_{y'}) \Big|_{x=\alpha+0} \right] \delta x = 0. \end{aligned} \tag{1.13.4}$$

Now we choose certain classes of admissible increments φ to show that each term summed in (1.13.4) is equal to zero separately. Let us take first those admissible φ that are zero on $[\alpha, b]$. Also take $\delta x = \delta y = 0$. All terms except the first integral on the left are equal to zero identically now. Thus

$$\int_a^\alpha \left(f_y - \frac{d}{dx} f_{y'} \right) \varphi \, dx = 0$$

for all differentiable functions φ that equal zero at a and α. By the fundamental lemma we see that the Euler equation $f_y - \frac{d}{dx} f_{y'} = 0$ holds on (a, α). Because of this the first integral is zero not only for those φ that satisfy $\varphi(\alpha) = 0$, but for all admissible increments. A similar choice of those φ that are zero on $[a, \alpha]$ together with the assumption $\delta x = \delta y = 0$ brings us to similar conclusions: the minimizer y satisfies the Euler equation on (α, b) and so for all admissible φ we have

$$\int_\alpha^b \left(f_y - \frac{d}{dx} f_{y'} \right) \varphi \, dx = 0.$$

It follows that

$$\left[f_{y'} \Big|_{x=\alpha-0} - f_{y'} \Big|_{x=\alpha+0} \right] \delta y + \left[(f - y' f_{y'}) \Big|_{x=\alpha-0} - (f - y' f_{y'}) \Big|_{x=\alpha+0} \right] \delta x = 0$$

for all admissible δx and δy, hence we obtain (1.13.1) and (1.13.2). $\square$

In the case of the functional $\int_a^b f(x, \mathbf{y}, \mathbf{y}') \, dx$ depending on a vector function, at a discontinuity of a component y_i we have the similar conditions

$$f_{y_i'} \big|_{x=\alpha-0} = f_{y_i'} \big|_{x=\alpha+0}, \qquad (f - y_i' f_{y_i'}) \big|_{x=\alpha-0} = (f - y_i' f_{y_i'}) \big|_{x=\alpha+0}.$$

Indeed, when deriving the corresponding equation for the first variation of the functional, we can appoint the increments of all the components except y_i to be zero, so formally the corresponding equation does not differ from (1.13.4).

The Weierstrass–Erdmann conditions are similar in form to the natural conditions for a functional. Using the idea of the proof of Theorem 1.13.1 we can find similar boundary conditions for other types of problems.

Example 1.13.1 Let us consider the problem of minimization of the functional

$$\int_a^\beta f(x, y, y') \, dx + \int_\beta^b g(x, y, y') \, dx \qquad (1.13.5)$$

where β is a fixed point of (a, b), and y is continuous on $[a, b]$, twice continuously differentiable on (a, β) and (β, b), and satisfies $y(a) = c_0$ and $y(b) = c_1$. We assume the integrand is discontinuous at $x = \beta$, hence y has no continuous derivative there.

Solution Problems of this form are frequent in physics, arising from spatial discontinuities. A specific instance of this is when a ray of light crosses the interface between two media. We are interested in how to appoint the conditions at such points, since the equation of propagation is not valid there. Variational tools can often supply us with such conditions. Let us demonstrate how this can happen.

For the functional (1.13.5) we need to derive the expression for the first variation and put it equal to zero for admissible increment-functions. For this we use (1.13.3) as above, but should take into account that β is fixed so that $\delta x = 0$ at $x = \beta$. The changes are evident:

$$\int_a^\beta \left(f_y - \frac{d}{dx}f_{y'}\right)\varphi\,dx + \int_\beta^b \left(g_y - \frac{d}{dx}g_{y'}\right)\varphi\,dx$$
$$+\left[f_{y'}\bigg|_{x=\beta-0} - g_{y'}\bigg|_{x=\beta+0}\right]\delta y = 0.$$

Thus in a similar fashion at $x = \beta$, in addition to the continuity condition $y(\beta - 0) = y(\beta + 0)$ we get

$$f_{y'}\bigg|_{x=\beta-0} = g_{y'}\bigg|_{x=\beta+0}.$$

Let us now consider a particular problem of the same nature with another type of functional. We need to determine the deflections under transverse load $q(x)$ of a system consisting of a cantilever beam with parameters E and I and whose free end connects with a string as shown in Fig. 1.4.

The models of a string and of a beam are of different natures; they are

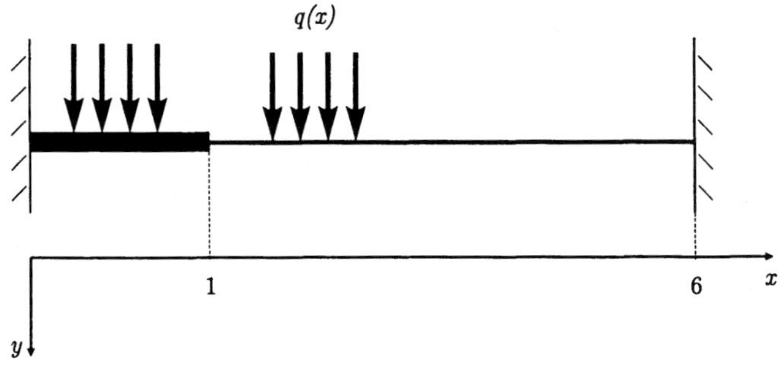

Fig. 1.4 A coupled mechanical system consisting of a beam and a string.

derived under different sets of assumptions, and the corresponding ordinary differential equations have different orders. It is clear that at the point of connection the function y describing the deflections must be continuous. However, we can imagine that the angles of inclination of the beam and the string can differ under certain loads; this means that we cannot require y' to be continuous at the point of coupling. What are the other conditions at this point? There are two ways to find them. One is to undertake a careful study of the theory of beams and strings and, understanding the mechanical meaning of each derivative at the point, to write out the conditions of equilibrium of the node (coupling unit). Another is to employ variational tools. Normally the latter is preferable, as it is less likely to yield incorrect conditions. We begin with the expression for total potential energy of the system: beam-string-load. We take the lengths of the beam and the string to be 1 m and 5 m, respectively. The stretching of the string is characterized by a parameter a:

$$E(y) = \frac{1}{2}\int_0^1 EI(y''(x))^2\,dx + \frac{a}{2}\int_1^6 (y'(x))^2\,dx - \int_0^6 q(x)y(x)\,dx.$$

We see from the figure that

$$y(0) = 0, \qquad y'(0) = 0, \qquad y(6) = 0.$$

Using tools from the first sections of the book, we obtain the first variation

$$\delta E = \int_0^1 EIy''\varphi''\,dx + a\int_1^6 y'\varphi'\,dx - \int_0^6 q(x)\varphi(x)\,dx$$

of the energy functional. For all admissible functions that necessarily satisfy $\varphi(0) = 0$, $\varphi'(0) = 0$, and $\varphi(6) = 0$, we have

$$\delta E = 0.$$

Integrating by parts we obtain

$$\int_0^1 EIy^{(4)}\varphi\,dx + EIy''\varphi'\Big|_{x=1-0} - EIy'''\varphi\Big|_{x=1-0}$$
$$- a\int_1^6 y''\varphi\,dx - ay'\varphi\Big|_{x=1+0} - \int_0^1 q\varphi\,dx - \int_1^6 q\varphi\,dx = 0.$$

We now reason as we did in the proof of Theorem 1.13.1. Putting $\varphi = 0$ on $[1, 6]$ and the "boundary" values $\varphi(1-0)$ and $\varphi'(1-0)$ equal to zero, we get

$$EIy^{(4)} - q = 0 \quad \text{on } (0,1)$$

for the beam equation; similarly, we get

$$ay'' + q = 0 \quad \text{on } (1,6)$$

for the string equation. Hence we deduce two additional boundary conditions at the point of connection:

$$EIy'''\big|_{x=1-0} = -ay'\big|_{x=1+0} \tag{1.13.6}$$

and

$$EIy''\big|_{x=1-0} = 0. \tag{1.13.7}$$

Condition (1.13.6) means that at the connection point the shear force of the beam is contracted by the force produced by the projection onto the vertical direction of the stretching force produced by the string, whereas (1.13.7) shows that the string cannot resist a torque so the moment at this point of the beam is zero.

Such constructions consisting of elements of different natures are common in practice, and now the reader knows how to set up the corresponding boundary value problems.

Quick review

In some problems it becomes necessary to extend the class of admissible functions to include those that are piecewise smooth. Let $y(x)$ be a minimizer of the simplest integral functional, and suppose $y'(x)$ is continuous on the closed intervals $[a, \alpha]$ and $[\alpha, b]$ where $\alpha \in (a, b)$ is the sole corner point. The position of α cannot be determined in advance, but is subject to the Weierstrass–Erdmann conditions

$$f_{y'}\big|_{x=\alpha-0} = f_{y'}\big|_{x=\alpha+0}, \qquad (f - y'f_{y'})\big|_{x=\alpha-0} = (f - y'f_{y'})\big|_{x=\alpha+0}.$$

In addition to the Euler equation on the intervals (a, α) and (α, b) then, y must satisfy (1) the Weierstrass–Erdmann conditions, (2) any given endpoint conditions on $y(a)$ and $y(b)$, and (3) the continuity condition $y(\alpha - 0) = y(\alpha + 0)$. A piecewise smooth extremal with a corner (or with multiple corners) is called a broken extremal.

1.14 Sufficient Conditions for Minimum

Thus far we have studied some of the techniques used to identify possible minimizers. It is also of interest to know how to solve the boundary value problems that yield corresponding extremals, although the treatment of this topic falls outside the scope of this book (and within the scope of books on ordinary and partial differential equations). But the solutions of these problems represent only the first step in a full solution of the problem of minimization; the next step is to learn whether an extremal is a minimizer. As we shall see, for many linear problems of mathematical physics an extremal satisfying boundary conditions is automatically a minimizer. Nonlinear problems, as a rule, need additional investigation. For this we need to derive sufficient conditions for an extremal to be a minimizer. First we shall derive conditions analogous to those found in the calculus of functions of many variables.

We reconsider the problem of minimum of the simplest functional $F(y) = \int_a^b f(x, y, y')\, dx$ in the class $C^{(1)}(a, b)$ under the boundary conditions $y(a) = c_0$, $y(b) = c_1$. Let y be a minimizer of the problem under consideration and let $\Delta y(x)$ be an admissible increment of y. Consider the increment of F:

$$\begin{aligned}\Delta F &= F(y + \Delta y) - F(y) \\ &= \int_a^b [f(x, y + \Delta y, y' + \Delta y') - f(x, y, y')]\, dx. \end{aligned} \qquad (1.14.1)$$

Denote $p = y(x)$, $q = y'(x)$, and $g(p, q) = f(x, p, q)$, and let Δp and Δq be the increments of p and q, respectively (in this case they are $\varphi(x)$ and $\varphi'(x)$ in our old notation). If in some small neighborhood of the point (p, q) the function g has all continuous derivatives up to the second order, then in this neighborhood we can write the Taylor expansion of g:

$$\begin{aligned} g(p + \Delta p, q + \Delta q) = g(p, q) &+ [g_p(p, q)\Delta p + g_q(p, q)\Delta q] \\ &+ \frac{1}{2!}[g_{pp}(p, q)(\Delta p)^2 + 2g_{pq}(p, q)\Delta p \Delta q \\ &+ g_{qq}(p, q)(\Delta q)^2] + \beta(p, q, \Delta p, \Delta q)[(\Delta p)^2 + (\Delta q)^2] \end{aligned}$$

where $\beta(p, q, \Delta p, \Delta q) \to 0$ when $(\Delta p)^2 + (\Delta q)^2 \to 0$. We can write out this expansion in terms of f, y, and Δy at each $x \in [a, b]$:

$$\begin{aligned} f(x, y+\Delta y, y'+\Delta y') &= f(x,y,y') + [f_y(x,y,y')\Delta y + f_{y'}(x,y,y')\Delta y'] \\ &+ \frac{1}{2!}[f_{yy}(x,y,y')(\Delta y)^2 + 2f_{yy'}(x,y,y')\Delta y \Delta y' \\ &+ f_{y'y'}(x,y,y')(\Delta y')^2] + \beta(x,y,y',\Delta y,\Delta y')[(\Delta y)^2 + (\Delta y')^2] \end{aligned} \tag{1.14.2}$$

(we keep the same notation β for the remainder function). Let us assume that for all $x \in [a, b]$ we have

$$|\beta(x,y,y',\Delta y,\Delta y')| \le \alpha(\Delta y, \Delta y')$$

where $\alpha(\Delta y, \Delta y') \to 0$ when $(\Delta y)^2 + (\Delta y')^2 \to 0$. This is an important assumption in what follows.

Let us return to our old notation $\varphi = \Delta y$ and rewrite (1.14.2) as

$$\begin{aligned} f(x, y+\varphi, y'+\varphi') &= f(x,y,y') + [f_y(x,y,y')\varphi + f_{y'}(x,y,y')\varphi'] \\ &+ \frac{1}{2!}[f_{yy}(x,y,y')\varphi^2 + 2f_{yy'}(x,y,y')\varphi\varphi' \\ &+ f_{y'y'}(x,y,y')(\varphi')^2] + o(\varphi^2 + (\varphi')^2). \end{aligned} \tag{1.14.3}$$

Here $o(\varphi^2 + (\varphi')^2)$ indicates that the term which is uniform in x is small in comparison with $\varphi^2 + (\varphi')^2$. We now apply the expansion (1.14.3) to (1.14.1):

$$\begin{aligned} \Delta F &= \int_a^b [f_y(x,y,y')\varphi + f_{y'}(x,y,y')\varphi']\,dx \\ &+ \frac{1}{2!}\int_a^b [f_{yy}(x,y,y')\varphi^2 + 2f_{yy'}(x,y,y')\varphi\varphi' + f_{y'y'}(x,y,y')(\varphi')^2]\,dx \\ &+ o\left(\int_a^b (\varphi^2 + (\varphi')^2)\,dx\right). \end{aligned}$$

Since y is a minimizer of the problem we necessarily have

$$\int_a^b [f_y(x,y,y')\varphi + f_{y'}(x,y,y')\varphi']\,dx = 0$$

(cf., § 1.1) and thus

$$\Delta F = F(y+\varphi) - F(y) = \delta^2 F + o\left(\int_a^b (\varphi^2 + (\varphi')^2)\,dx\right)$$

where $\delta^2 F$ is the *second variation* defined by

$$\delta^2 F \equiv \frac{1}{2!}\int_a^b [f_{yy}(x,y,y')\varphi^2 + 2f_{yy'}(x,y,y')\varphi\varphi' + f_{y'y'}(x,y,y')(\varphi')^2]\,dx.$$

Integration by parts gives

$$\begin{aligned}\int_a^b 2f_{yy'}(x,y,y')\varphi\varphi'\,dx &= \int_a^b f_{yy'}(x,y,y')\frac{d}{dx}\varphi^2\,dx \\ &= -\int_a^b \varphi^2 \frac{d}{dx} f_{yy'}(x,y(x),y'(x))\,dx\end{aligned}$$

since $\varphi(a) = \varphi(b) = 0$. Then

$$\delta^2 F = \frac{1}{2!}\int_a^b \left\{ \left[f_{yy}(x,y,y') - \frac{d}{dx} f_{yy'}(x,y,y')\right]\varphi^2 + f_{y'y'}(x,y,y')(\varphi')^2 \right\} dx.$$

The quantity $\delta^2 F$ is quadratic in φ and φ'. Suppose it is bounded from below as follows:

$$\delta^2 F \geq m \int_a^b (\varphi^2 + (\varphi')^2)\,dx, \tag{1.14.4}$$

where the constant $m > 0$ does not depend on the choice of admissible increment φ (note that here we do not need assumptions on the smallness of φ). It then follows that

$$F(y+\varphi) - F(y) \geq 0$$

for all admissible increments φ (i.e., $\varphi \in C_0^{(1)}(a,b)$) with sufficiently small norm $\|\varphi\|_{C^{(1)}(a,b)}$. This means that (1.14.4) is sufficient for y to be a local minimizer of the problem under consideration.

Thus we need to find conditions for (1.14.4) to hold. Let us denote

$$\begin{aligned} Q(x) &= f_{yy}(x,y(x),y'(x)) - \frac{d}{dx} f_{yy'}(x,y(x),y'(x)), \\ P(x) &= f_{y'y'}(x,y(x),y'(x)). \end{aligned}$$

The functions $Q(x)$ and $P(x)$ can be regarded as momentarily given when we study whether $y = y(x)$ is a minimizer. So we need to study the functional

$$\Phi(\varphi) = \int_a^b [P(x)\varphi'^2(x) + Q(x)\varphi^2(x)]\,dx$$

in the space $C_0^{(1)}(a,b)$.

It is easy to formulate the following restrictions:

$$P(x) \geq c \quad \text{and} \quad Q(x) \geq c > 0 \qquad \text{for all } x \in [a, b].$$

Under these the inequality (1.14.4) holds for all $\varphi \in C_0^{(1)}(a, b)$. Unfortunately these restrictions fail in many cases when $y = y(x)$ is really a minimizer, so we need to derive more useful conditions.

Note that if $y = y(x)$ is a minimizer then $\Phi(\varphi) \geq 0$ at least. For if there were an admissible increment φ such that $\Phi(\varphi) < 0$ then we could find a t_0 so small that for all $0 < t < t_0$ we would have $F(y + t\varphi) - F(y) < 0$, and y would not be a minimizer. Let us suppose $\Phi(\varphi)$ is non-negative.

Theorem 1.14.1 *Let $P(x)$ and $Q(x)$ be continuous on $[a, b]$ and $\Phi(\varphi) \geq 0$ for all $\varphi \in C_0^{(1)}(a, b)$. Then $P(x) \geq 0$ on $[a, b]$.*

Proof. Suppose to the contrary that $P(x_0) < 0$ for some x_0. Then $P(x) < \gamma < 0$ in some ε-neighborhood $[x_0 - \varepsilon, x_0 + \varepsilon]$ of x_0. Choose $\varphi(x) \in C^{(1)}(a, b)$ as the particular function

$$\varphi(x) = \begin{cases} \sin^2\left[\frac{\pi(x - x_0)}{\varepsilon}\right], & x \in [x_0 - \varepsilon, x_0 + \varepsilon], \\ 0, & \text{otherwise.} \end{cases}$$

Then for $x \in [x_0 - \varepsilon, x_0 + \varepsilon]$ we have

$$\varphi'(x) = 2 \sin\left[\frac{\pi(x - x_0)}{\varepsilon}\right] \cos\left[\frac{\pi(x - x_0)}{\varepsilon}\right] \left(\frac{\pi}{\varepsilon}\right) = \frac{\pi}{\varepsilon} \sin\left[\frac{2\pi(x - x_0)}{\varepsilon}\right]$$

and therefore

$$\begin{aligned} \Phi(\varphi) = \left(\frac{\pi}{\varepsilon}\right)^2 &\int_{x_0 - \varepsilon}^{x_0 + \varepsilon} P(x) \sin^2\left[\frac{2\pi(x - x_0)}{\varepsilon}\right] dx \\ &+ \int_{x_0 - \varepsilon}^{x_0 + \varepsilon} Q(x) \sin^4\left[\frac{\pi(x - x_0)}{\varepsilon}\right] dx. \end{aligned}$$

But

$$\int_{x_0 - \varepsilon}^{x_0 + \varepsilon} P(x) \sin^2\left[\frac{2\pi(x - x_0)}{\varepsilon}\right] dx < \gamma \int_{x_0 - \varepsilon}^{x_0 + \varepsilon} \sin^2\left[\frac{2\pi(x - x_0)}{\varepsilon}\right] dx = \gamma\varepsilon$$

and

$$\int_{x_0 - \varepsilon}^{x_0 + \varepsilon} Q(x) \sin^4\left[\frac{\pi(x - x_0)}{\varepsilon}\right] dx \leq M \int_{x_0 - \varepsilon}^{x_0 + \varepsilon} \sin^4\left[\frac{\pi(x - x_0)}{\varepsilon}\right] dx = \frac{3M\varepsilon}{4}$$

where $M = \max_{x\in[a,b]} |Q(x)|$. Hence

$$\Phi(\varphi) < \left(\frac{\pi}{\varepsilon}\right)^2 \gamma\varepsilon + \frac{3M\varepsilon}{4} = \frac{\pi^2\gamma}{\varepsilon} + \frac{3M\varepsilon}{4}.$$

Recall that $\gamma < 0$; for sufficiently small ε we can make $\Phi(\varphi) < 0$, a contradiction. □

Thus, besides the Euler equation we have established another necessary condition for y to be a minimizer of the problem under consideration: we must have

$$f_{y'y'}(x, y(x), y'(x)) \geq 0 \quad \text{for all } x \in [a,b].$$

This is *Legendre's condition.*

Legendre believed that satisfaction of the strict inequality $f_{y'y'} > 0$ for all $x \in [a,b]$ should be sufficient for y to be a minimizer, and even constructed a flawed proof. However, even the mistakes of great persons are useful — on the basis of this "proof" a useful sufficient condition was subsequently established. Jacobi proposed to study the functional $\Phi(\varphi)$ using the tools of the calculus of variations itself. The Euler equation for this functional is

$$[P(x)\varphi'(x)]' - Q(x)\varphi(x) = 0. \tag{1.14.5}$$

It is clear that this equation has the trivial solution $\varphi = 0$. Let $P(x)$ be continuously differentiable. Jacobi studied the zeros of a solution of (1.14.5) for the Cauchy problem $\varphi(0) = 0$, $\varphi'(0) = 1$. The nearest value $x_0 > a$ where $\varphi(x_0) = 0$ he called the point *conjugate* to a (with respect to the functional $\Phi(\varphi)$). This point is denoted a^* (we agree to call $a^* = \infty$ if $\varphi(x)$ has no zeros to the right of $x = a$). Jacobi established another necessary condition for y to be a minimizer: that the interval (a,b) does not contain a^*.

The following set of three conditions is sufficient for y to be a minimizer of the problem under consideration:

(1) y satisfies the Euler equation $f_y - \frac{d}{dx} f_{y'} = 0$;
(2) $f_{y'y'}(x, y(x), y'(x)) > 0$ for all $x \in [a,b]$;
(3) $[a,b]$ does not contain points conjugate to a with respect to $\Phi(\varphi)$.

We shall not offer a proof of this, but do wish to note the following. The result is beautiful, but for many years it seemed impractical: the Jacobi condition (3) was quite difficult to check before the advent of the computer.

Today, however, there are many good algorithms with which Cauchy problems for ordinary differential equations may be solved. Hence it is quite easy to check the Jacobi condition numerically.

Example 1.14.1 For which range of the constant c is an extremal of the functional

$$\int_0^1 (y'^2 - c^2y^2 - 2y)\,dx, \qquad y(0) = 0, \quad y(1) = 1,$$

a minimizer?

Solution The extremal exists, as the reader can verify. We suppose $c > 0$.

Let us check the sufficiency conditions given above. Legendre's condition holds automatically. The Jacobi equation with initial conditions is

$$y'' + c^2y = 0, \qquad y(0) = 0, \quad y'(0) = 1.$$

Its solution is $y = c^{-1}\sin cx$, hence the conjugate point occurs where $cx = \pi$. Thus, by sufficient conditions, the extremal really is a minimizer of the functional when $a^* = \pi/c > 1$, and by symmetry in c, the extremal is a minimizer when $|c| < \pi$. When $a^* < 1$, then extremal is not a minimizer and, moreover the functional has no minimizer at all (why?).

Finally we note that the Jacobi theory of conjugate points and corresponding results can be established for a functional depending on an unknown vector-function.

Some field theory

We now turn to a brief, introductory discussion of certain concepts needed to express conditions sufficient for a strong minimum. The main idea is that of a *field of extremals.*

Let D be a domain in the xy-plane. Let

$$y = y(x; \alpha)$$

be a family of curves lying in D, a separate curve being generated by each choice of the parameter α. If a unique curve from the family passes through each point of D, then we call the family a *proper field* in D. A proper field can be regarded as a sort of cover for D, associating with each point $(x, y) \in D$ a unique slope $p(x, y)$ (i.e., the slope of the particular curve passing through that point). As a simple but standard example, let D be the unit disk

$$D = \{(x, y)\colon x^2 + y^2 < 1\}$$

and let $y = y(x; \alpha) = kx + \alpha$ where k is a fixed constant. Here we have a field of parallel straight lines with slopes $p(x, y) \equiv k$.

If all curves of a family $y = y(x; \alpha)$ pass through a certain point (x_0, y_0), then the family is known as a *pencil* of curves and (x_0, y_0) is called the *center* of the pencil. For example, the family $y = \alpha x$ is a pencil having center at the origin. Of course, a pencil of curves having center $(x_0, y_0) \in D$ cannot be a proper field of curves in D. However, if a pencil of curves assigns a unique slope $p(x, y)$ to all points in D *other than* (x_0, y_0), we speak of a *central field* of curves in D.

A *field of extremals* is a family of extremal curves (for some variational problem) that generates a proper or central field in a domain D. The Euler equation for the simplest functional

$$F(y) = \int_a^b f(x, y, y')\, dx \tag{1.14.6}$$

has solutions that form a two-parameter family of curves $y = y(x; \alpha; \beta)$. (Here α and β are the integration constants in the general solution of the Euler equation.) If one of the constants, say α, is determined by imposing a given fixed endpoint condition $y(a) = c_0$ on the general solution, then all the extremals in the resulting one-parameter family will issue from the same point (a, c_0). The resulting family $y = y(x; \beta)$ may be a field (proper or central) in some specified domain D. For example, consider the functional

$$\int_a^b [y^2 - (y')^2]\, dx$$

with $a = 0$ and $y(0) = 0$. The integrand does not depend explicitly on x, so $y^2 - (y')^2 - (-2y')y' = c_1$. It follows that the extremals have the form $y = c_2 \sin(x + c_3)$, which gives us a pencil having center $(0, 0)$. Another example we mention is for the functional

$$\int_a^b (y'^2 - 1)^2\, dx.$$

The extremals are straight lines. When suitably restricted, the two-parameter family of curves $y(x) = c_1 x + c_2$ can form a field in a couple of different ways: (1) when c_1 is fixed, we obtain a family $y = y(x; c_2)$ that can form a proper field in the unit disk D; (2) when $c_2 = 0$, the resulting pencil centered at the origin can form a central field in D.

Let $y = y(x; \alpha)$ generate a field of extremals (central or proper) in some domain D. Each choice of α then gives an extremal; by setting $\alpha = \alpha_0$, we select a particular extremal $y^*(x) = y(x; \alpha_0)$ from the field. If this extremal $y^*(x)$ has no common points with the boundary of D, it is said to be *admissible in the field.* We note that a given extremal may be admissible in more than one field covering a domain D. Returning to our example in which D is the unit circle, the two fields

$$y(x; \alpha) = c_1 x + \alpha, \qquad\qquad y(x; \alpha) = \alpha x,$$

mentioned above each admit the straight line extremal $y^*(x) = c_1 x$.

Armed with an understanding of the field concept, we proceed to the next step. Let D be a domain in which there is distributed a proper field of extremals for the simplest functional $F(y)$ of equation (1.14.6). Suppose further that this field admits the particular extremal $y = y^*(x)$ satisfying given endpoint conditions $y(a) = c_0$, $y(b) = c_1$. Now let $y = y(x)$ be *any* curve that lies in D and connects the desired endpoints (a, c_0) and (b, c_1). We also assume that the integral

$$H(y) = \int_a^b [f(x, y, p) + (y' - p) f_p(x, y, p)]\, dx \tag{1.14.7}$$

exists for $y = y(x)$, where $p = p(x, y)$ is the slope function (i.e., its value at (x, y) is the slope y' of the extremal through point (x, y)) of the field in D. This integral is extremely important for the theory.

When $y(x) = y^*(x)$, the integral (1.14.7) reduces to (1.14.6) because $y' \equiv p$ in that case. It can be shown that (1.14.7) is path independent in D. For this reason it is known as *Hilbert's invariant integral.*

We use these facts as follows. Defining

$$\Delta F = F(y) - F(y^*),$$

we have $\Delta F = F(y) - H(y^*) = F(y) - H(y)$ so that

$$\begin{aligned}\Delta F &= \int_a^b f(x, y, y')\, dx - \int_a^b [f(x, y, p) + (y' - p) f_p(x, y, p)]\, dx \\ &= \int_a^b [f(x, y, y') - f(x, y, p) - (y' - p) f_p(x, y, p)]\, dx.\end{aligned}$$

Thus

$$\Delta F = \int_a^b E(x, y, y', p)\, dx$$

where the integrand

$$E(x, y, y', p) = f(x, y, y') - f(x, y, p) - (y' - p) f_p(x, y, p)$$

is known as the *Weierstrass excess function.* We may now formulate conditions sufficient for $y = y^*(x)$ to be a strong minimum of $F(y)$:

(1) The curve $y = y^*(x)$ is admissible in a field of extremals for $F(y)$, and
(2) $E(x, y, y', p) \geq 0$ for all points (x, y) lying sufficiently close to the curve $y = y^*(x)$ and for arbitrary values of y'.

Taken together, these have been called the *Weierstrass conditions.* The proof is nearly obvious. Suppose condition (1) holds, and let $y = y(x)$ be any other curve lying in the domain covered by the field of extremals and connecting the desired endpoints. Then according to condition (2),

$$\Delta F = \int_a^b E(x, y, y', p)\, dx \geq 0$$

for all curves $y = y(x)$ that connect the endpoints and lie within some neighborhood of $y^*(x)$; moreover, the slope of y need not be close to that of y^* so the minimum is strong.

Although the Weierstrass conditions are attractive because of their simplicity, we can run into trouble when attempting to apply them to certain functionals. This happens, for example, with the problem of minimizing

$$\int_0^{3/2} \frac{y}{(y')^2}\, dx, \qquad y(0) = 1, \quad y(3/2) = 1/4.$$

The difficulty is related to the fact that the family of extremals has a so-called envelope.

Our treatment of sufficient conditions for the problem of minimum has been intentionally brief. We have formulated a couple of sets of such conditions; in fact, however, these are seldom used by practitioners. Rather, necessary conditions are usually applied to obtain extremals, and then various other methods are employed in place of sufficient conditions. For example, if a functional has a unique minimum residing in a class of functions, and if a unique extremal is found for the problem, then the desired minimum must be reached on the extremal found. If several extremals qualify as candidates for the minimum, it is often possible to test each one by calculating the corresponding values taken by the functional. The true minimum may then be identified and selected. Hence sufficient conditions may be viewed as largely of theoretical interest.

1.15 Exercises

1.1 In the xy-plane, find the smooth curve between (a, y_0) and (b, y_1) which by revolution about the x-axis generates the surface of least area.

1.2 The *brachistochrone problem* is a famous classical problem in which one must find the equation of the plane curve down which a particle would slide from one given point to another in the least possible time when acted upon by gravity alone. Show that the required curve is a portion of an ordinary cycloid.

1.3 Show that if f in the simplest functional depends explicitly on y' only, then the extremals are straight lines.

1.4 During the time interval $[0, T]$ a particle having mass m is required to move along a straight line from the position $x(0) = x_0$ to the position $x(T) = x_1$. Determine the extremal for the problem of minimizing the particle's average kinetic energy. Explain your result physically.

1.5 Apply Ritz's method with basis functions of the form $\varphi_n(x) = x^2(1-x)^2x^k$ to minimize the functional

$$\int_0^1 \{(y'')^2 + [1 + 0.1\sin x](y')^2 + [1 + 0.1\cos(2x)]y^2 - 2\sin(2x)y\}\,dx.$$

The boundary conditions for the problem are $y(0) = y'(0) = y'(1) = 0$, $y(1) = 1$.

1.6 (a) Consider the problem of minimum for the simplest functional (1.1.9) with boundary condition $y(a)+y(b) = 1$. Find a supplementary natural boundary condition for this case. (b) Repeat for a condition of the more general form $\psi(y(a), y(b)) = 0$ where $\psi = \psi(\alpha, \beta)$ is a given function of two variables.

1.7 Find the equation of the plane curve down which a particle would slide from one given point (a, y_0) to cross the vertical line $x = b$ in the least possible time when acted upon by gravity alone.

1.8 Find the smooth curve of least length between two points on the surface of the cylinder of radius a.

1.9 For a functional of the form

$$F_2(y) = \int_a^b f(x, y, y', y'')\,dx,$$

find the Ritz system of equations corresponding to (1.4.3).

1.10 Consider the problem of equilibrium of a plate when there are given forces f acting on the edge of the plate. It is described as the problem of minimum of the functional

$$\frac{D}{2}\iint_S [w_{xx}^2 + w_{yy}^2 + 2\nu w_{xx}w_{yy} + 2(1-\nu)w_{xy}^2]\,dx\,dy \\ - \iint_S Fw\,dx\,dy - \oint_{\partial S} fw\,ds = 0.$$

What is the form of the Euler equation now? What are the natural boundary conditions for a minimizer?

1.11 Suppose that a plate consists of two parts with different constant rigidities D_1, D_2 that connect along the line Γ of the mid-plane. Write out the conditions on the border line assuming that w, the deflection of the plate, is a continuous function together with its first derivatives over the whole domain. Note that these conditions have the same nature as the natural conditions. They have a clear mechanical meaning.

Chapter 2

Elements of Optimal Control Theory

2.1 A Variational Problem as a Problem of Optimal Control

Let us consider a special problem of the calculus of variations:

$$\int_a^b f(x, y(x), y'(x))\, dx \to \min_{\substack{y \in C^{(1)}(a,b) \\ y(a)=y_0}} \tag{2.1.1}$$

Let $y(x)$ be fixed for a moment. We introduce an equation for a new function $z = z(x)$:

$$z'(x) = f(x, y(x), y'(x)), \qquad z(a) = 0.$$

It is clear that

$$z(b) = z(b) - z(a) = \int_a^b z'(x)\, dx = \int_a^b f(x, y(x), y'(x))\, dx.$$

Now let us introduce an additional function $u(x) = y'(x)$. In these terms the problem (2.1.1) can be formulated as follows:[1]

Problem of Terminal Control. Given two ordinary differential equations

$$y'(x) = u(x), \qquad\qquad z'(x) = f(x, y(x), u(x)),$$

and two initial conditions $y(a) = y_0$ and $z(a) = 0$, in the set of all $u \in C(a,b)$ find $u = u(x)$ at which $z(b)$ attains the minimal value.

[1]Thanks to Dr. K.V. Isaev of Rostov State University, who furnished the authors with a notebook of his lectures on control theory. The presentation of the terminal control problem follows, in large part, Dr. Isaev's lectures.

Since $z(b)$ is the value of the integral, this formulation is equivalent to the formulation of the problem of strong minimum of the functional (2.1.1). Note that the last formulation does not involve the operation of integration. It is well known that the solution of the Cauchy problem for an ordinary differential equation (ODE) is less computationally intensive than the solution of the corresponding integral equation. This transformation of a variational problem to another form is numerically advantageous; moreover, it allows us to introduce a new class of minimization problems along with new methods of solution. Note that the new formulation should still give us the Euler equation for a minimizer and the natural boundary condition at $x = b$.

The formulation (2.1.1) is equivalent to the Problem of Terminal Control if f is sufficiently smooth. But the Problem of Terminal Control has brought us to a new class of problems that fall outside the calculus of variations. These problems also fall outside classical ODE theory, since for the Cauchy problem in the latter, the number of differential equations always equals the number of unknown functions. In our formulation we have two equations and three unknowns y, z, u. But if u is given we have a Cauchy problem in which y and z are uniquely determined. We solve a special minimum problem, seeking the minimum value of z at point b, changing u in the class of continuous functions. Continuity of u was stipulated by the tools of the calculus of variations. But for many problems having the form of the Problem of Terminal Control or something similar, this condition is too restrictive. We shall consider other tools for the investigation of such problems — tools not equivalent to those of the calculus of variations.

The Problem of Terminal Control falls under the heading of *optimal control theory*. The designation "terminal control" refers to the fact that something, namely z, is to be minimized at a final time instant $x = b$. A more general formulation is presented in the next section.

We have thus examined a variational problem as a problem of optimal control. Let us take a moment to compare the setups of these two problems. Each must provide a functional to be minimized. In the variational setup this functional is an integral that incorporates some information about the system structure. In the control problem these elements are separated: the system is governed by a set of ODEs relating internal parameters y, z to an external parameter u that can be changed at will (under some restrictions of course), while the "cost functional" is formulated separately. There are advantages in choosing to disentangle the elements of the problem setup in this way; in fact, many practical problems are so posed naturally and cannot be posed as variational problems. As a familiar example, consider a child sitting on a playground swing. The amplitude of the oscillations is governed by the pendulum equation — an ODE — and the effective length $u = u(t)$ of the pendulum is under the child's control. There is no reason why this control parameter must be changed in a continuous fashion; every child knows that the best results can be obtained by sudden shifts in his or her center of gravity. Hence we should be able to accommodate discontinuities in u. Of course, it is easy to cite examples on a much larger scale of economic importance — examples ranging from space travel to the damping of a ship's oscillations in the ocean.

In short, we shall consider problems in which there is a "system" or "controlled object" having a control parameter u. In general we seek u that minimizes some cost functional G, which in turn depends on u through an initial or boundary value problem to a set of ODEs. We will not consider all aspects of standard mathematical optimal control theory, including existence theorems, etc. But we will present an introduction to practical aspects of the theory that relate closely to the numerical solution of optimal control problems. The expression for the increment of the cost functional G which we will derive is analogous to the first variation in the calculus of variations, or to the differential in calculus. Its expression provides a basis for various numerical methods approaches to optimal control problems. It also brings us the important Pontryagin's maximum principle, which allows us to determine whether a governing function u is optimal.

2.2 General Problem of Optimal Control

First we generalize the Problem of Terminal Control. A controlled system is described by $n+m$ functions, which depend on a known variable. We shall denote this latter variable by t or x and regard it as the time variable. Given are n ordinary differential equations involving the first n parameters of the system $y_1, \ldots, y_n$ and their first derivatives. These equations are written in normal form. The vector $\mathbf{y} = (y_1, \ldots, y_n)$ is often called the *state vector*, and its component functions $y_1, \ldots, y_n$ the *state variables*. The remaining m parameters $u_1, \ldots, u_m$ are considered as free parameter-functions. We call $\mathbf{u} = (u_1, \ldots, u_m)$ the *control vector*, and its component functions the *control variables.* The differential equations are

$$
\begin{aligned}
y_1'(t) &= f_1(t, y_1(t), \ldots, y_n(t), u_1(t), \ldots, u_m(t)),\\
y_2'(t) &= f_2(t, y_1(t), \ldots, y_n(t), u_1(t), \ldots, u_m(t)),\\
&\ \vdots\\
y_n'(t) &= f_n(t, y_1(t), \ldots, y_n(t), u_1(t), \ldots, u_m(t)),
\end{aligned}
$$

or

$$\mathbf{y}'(t) = \mathbf{f}(t, \mathbf{y}(t), \mathbf{u}(t)). \tag{2.2.1}$$

Equations (2.2.1) should be supplemented with some conditions at the initial time $t = t_0$:

$$\mathbf{y}(t_0) = \mathbf{y}_0, \tag{2.2.2}$$

where $\mathbf{y}_0$ is a given *initial state.*

We now consider a problem of the form

$$G(\mathbf{y}(T)) \to \min$$

over the set of admissible $\mathbf{u}$, where T is a fixed (final) time instant. The quantity $G(\mathbf{y}(T))$ is a functional dependent upon the values taken by $\mathbf{u}$ and $\mathbf{y}$ over $[t_0, T]$. The space in which these vector functions reside is an important issue to be discussed later. Whereas in variational problems we permit only smooth functions for comparison and consider non-smooth functions as exceptions, here we consider non-smooth control functions since these tend to be more useful in applications (and, often more importantly, allowed by the method of solution and investigation).

Many optimal control problems arise in classical mechanics. There a system, described by the equations of classical mechanics, can be acted upon by forces whose magnitudes and directions are subject to certain restrictions. We obtain a problem of terminal control if we attempt to minimize the value of a function, depending on the internal parameters of the system, at a certain (final) time instant. For example, we may wish to bring the system to a certain state with the best accuracy.

We can generalize the Problem of Terminal Control by supplanting the initial values (2.2.2) with n equations given at some fixed points $t_k \in [t_0, T]$:

$$B_k(\mathbf{y}(t_k)) = 0, \qquad k = 1, \ldots, n.$$

The goal function can incorporate values of $\mathbf{y}$ at other points of $[t_0, T]$:

$$G(\mathbf{y}(\tau_1), \ldots, \mathbf{y}(\tau_r)) \to \min .$$

Such a problem is solved practically by any system that has to meet some time schedule (e.g., by a flight team who must land at several airports at scheduled times during a flight).

Let us consider another type of optimal control problem:

Time-Optimal Control Problem. A system is described by (2.2.1). It is necessary to move the system from state $\mathbf{y}(t_0) = \mathbf{y}_0$ to state $\mathbf{y}(T) = \mathbf{y}_f$ in minimal time T.

Again, we leave the class of admissible $\mathbf{u}$ as an issue for the future. Note that for this problem an existence theorem is essential in many cases, since there are mechanical and other systems for which an initial-final pair of states $\mathbf{y}_0, \mathbf{y}_f$ is impossible to take on for any time.

We see that in Time-Optimal Control we have $2n$ given boundary conditions, but there is an additional unknown parameter T that must be determined as an outcome of the solution. We see a big difference in the number of boundary values for the state vector $\mathbf{y}$ in these problems. This is provided by the arbitrariness of the control vector $\mathbf{u}$, changes in which can lead to the requirement for new boundary conditions. The restriction on the number of boundary conditions r at each "boundary" (initial, final, or intermediate) point of time is that it cannot exceed n, the number of components of $\mathbf{y}$ and, in total, at any admissible fixed $\mathbf{u}$ we have to obtain a boundary value problem for our system of equations that is solvable (not necessarily uniquely). When the boundary value system has too few boundary conditions for uniqueness, then, in the same way there arise natural boundary conditions in the calculus of variations, there arise additional boundary conditions for $\mathbf{y}$ in the optimal control problems. In some versions of the numerical methods that are used for solving the corresponding problems, such natural conditions do not participate explicitly — as is also the case for natural conditions in the calculus of variations — however, an optimal solution obeys them.

These are not the only possible setups for optimal control problems. We can consider, for example, problems where the cost functional is given in an integral form which takes into account the values of $\mathbf{y}$ at all instants of time.

Above we mentioned restrictions on the control vector $\mathbf{u}$, but many problems require restrictions (frequently in the form of inequalities) on $\mathbf{y}$ as well. For example, the problem of manned spaceflight forces us to minimize expenses while restricting accelerations experienced during the flight.

Many real problems of optimal control require us to consider (nonlinear) systems of PDEs rather than ODEs. The interested reader can find this discussed elsewhere. Often, however, these problems can be reduced to the problems that appear in this chapter. Each practical problem for the same object can lead to a different mathematical setup, as well as to different theoretical and practical results. In this book we will consider only mathematical aspects of the problems of optimal control, leaving applications to many other sources. First we would like to reduce the setup of the problems under consideration a bit.

The system (2.2.1) is said to be *autonomic* if f does not depend explicitly on t. Henceforth we shall consider only autonomic systems with $t_0 = 0$. We may do this without loss of generality. First, given $t_0 \neq 0$ we may shift the time origin by putting $x = t - t_0$. Let us consider the transformation to autonomic form. In principle there is nothing to limit the number of components that $\mathbf{y}$ may have. So we can always extend it by an additional component y_{n+1}, supplementing (2.2.1) with an additional equation

$$y'_{n+1}(x) = 1$$

and initial condition

$$y_{n+1}(0) = 0.$$

Then (2.2.1) takes the form

$$\mathbf{y}'(x) = \mathbf{f}(y_{n+1}(x), \mathbf{y}(x), \mathbf{u}(x)).$$

Thus, redenoting $\mathbf{y} = (y_1, \ldots, y_{n+1})$ and the corresponding $\mathbf{f}$, we arrive again at (2.2.1) but in the form

$$\mathbf{y}'(t) = \mathbf{f}(\mathbf{y}(t), \mathbf{u}(t)).$$

This is the autonomic form we shall consider.

2.3 Simplest Problem of Optimal Control

So far we have said little about the restrictions to be placed on the behavior of $\mathbf{u}(t)$. We shall take the class of admissible controls to consist of those vector functions that are *piecewise* continuous on $[0, T]$. This is in contrast to what we saw in the calculus of variations. It is possible to relax this restriction on $\mathbf{u}(t)$, requiring it to be merely measurable in some sense, but we leave this and related questions of existence[2] for more advanced books.

What constitutes a "small" variation (increment) of a control function? In the calculus of variations we regarded a variation (increment) of a function as small if its norm in the space $C^{(1)}(0, T)$ was small. With such a small increment taken in its argument, the increment of a functional was also guaranteed to be small, and we were led to apply the tools of calculus. To obtain the Euler equation and the natural boundary conditions we linearized the functional with respect to the increment of the unknown function. Here we would like apply the same linearization idea and obtain necessary conditions for the objective functional to attain its minimal value, but at the same time introduce another notion of smallness of the increment of a control function.

When we linearize an expression we use the fact that a small increment in the independent variable brings a small increment in the value of the expression. We understand that if we change the control function in some small way then the increment of the output function will be small. But in Newtonian mechanics if a large force acts on a material point for a short time then the deviation of the point trajectory during a finite time is small — the shorter the time of action, the smaller the deviation. So "smallness" of the increment can be provided by a force of small magnitude *or* by a force of short duration. This situation is quite typical for disturbances to ODEs, and suggests a new class of "small" increments to control functions. From a more mathematical viewpoint, the norm of $C(0, T)$ is not the only norm under which we can introduce small increments while requiring that the change in a solution exhibit continuous dependence on changes in the control function. In particular, we may use the norm of $L(0, T)$.

Let us introduce a class of functions $\mathcal{U}$ in which we seek control functions $u = u(t)$. $\mathcal{U}$ is a set of functions piecewise continuous on $[0, T]$, and is restricted by some conditions: normally simultaneous linear inequalities given pointwise. An example of such a restriction is

$$0 \leq u(t) \leq 1.$$

[2]Such questions are more theoretical than we are able to treat here, but this does not mean they are unimportant. There are practical problems for which no optimal solution exists. In such cases, however, it is often possible to obtain a working approximation to an optimal solution.

The simplest problem of optimal control theory is the following problem of terminal control:

Simplest Problem of Optimal Control. Let a controlled object be described by the equation

$$y'(t) = f(y(t), u(t)) \tag{2.3.1}$$

subject to

$$y(0) = y_0. \tag{2.3.2}$$

Among all functions belonging to a class $\mathcal{U}$ described above, find a control function $u(t)$ that minimizes $g(y(t))$ at $t = T$:

$$g(y(T)) \to \min_{u(t)\in\mathcal{U}}.$$

Here $g(y)$ is a continuously differentiable function on the domain of all admissible values of $y = y(t)$.[3]

First we introduce the main elementary increment of the control function, a so-called *needle-shaped increment.* This is where optimal control theory begins to depart from the calculus of variations. We choose some $u(t) \in \mathcal{U}$ and let $t = s$ be a point at which $u(t)$ is continuous. For definiteness we consider all the functions $u(t)$ to be continuous from the left on $[0, T]$. Consider another function $u^*(t)$ that differs from $u(t)$ only on the half-open segment $(s - \varepsilon, s]$ as shown in Fig. 2.1. Analytically this function is

$$u^*(t) = \begin{cases} u(t), & t \notin (s-\varepsilon, s], \\ v, & t \in (s-\varepsilon, s], \end{cases} \tag{2.3.3}$$

[3] Rather than formulating explicit restrictions on f and g, we simply assume they are sufficiently smooth for our purposes. In particular we shall differentiate $g(y)$ and $f(y, u)$ with respect to y supposing that the corresponding derivatives are continuous, we shall assume a continuous dependence of $f(y, u)$ on u, and we shall suppose that for any fixed admissible $u(t) \in \mathcal{U}$ the Cauchy problem (2.3.1)–(2.3.2) has a unique solution that depends continuously on the initial condition y_0. All this could be formulated purely in terms of the given functions f and g and it is possible that doing so would give us even sharper results, but we choose clarity over rigor at this stage. In fact, the simple problem we have chosen to consider is not the most realistic one available. However, its investigation will open the way to general problems without obscuring the essential ideas.

where $\varepsilon > 0$ is sufficiently small. The increment

$$\delta u(t) = u^*(t) - u(t),$$

which is zero everywhere except in the interval $(s-\varepsilon, s]$ of length ε, is what we term needle-shaped. Its smallness is characterized by

$$\|\delta u\|_{L(0,T)} = \int_0^T |\delta u| \, dt,$$

which is of order ε.

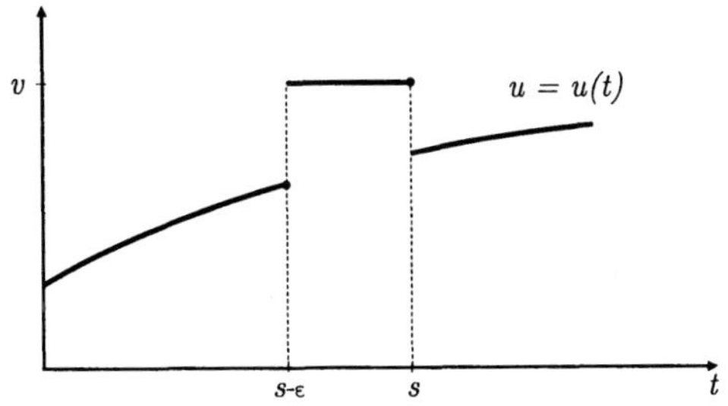

Fig. 2.1 A control function subject to a needle-shaped increment.

In what follows we suppose $u^*(t)$ belongs to $\mathcal{U}$. We also assume that together with some $u^*(t)$, defined by $\varepsilon_0 > 0$ and v_0, to $\mathcal{U}$ there belong all the $u^*(t)$ having the same final point s of the jump for which $\varepsilon < \varepsilon_0$. Since the restrictions for $\mathcal{U}$ are usually given piecewise by simultaneous linear inequalities, this assumption does not bring additional restrictions for such problems.

Many textbooks consider needle-shaped functions that are constant on the interval $(s-\varepsilon, s]$, but we consider them only for small ε so the norm in $L(0,T)$ of the difference between the above introduced and the traditional needle-shaped functions is of order higher than ε. We took our definition only for convenience. Note that we can approximate (in the uniform norm) any $u(t) \in \mathcal{U}$ with a finite linear combination of needle-shaped functions.

Since $g(y(T))$ is a number that depends on $u(t)$ through (2.3.1) and (2.3.2), we have a functional defined on $\mathcal{U}$. Our experience suggests that we apply the ideas of calculus. We need to find the increment of the functional under that of the control function, introducing something like the first differential. Now $\delta u(t)$ is an elementary needle-shaped function whose smallness is determined by ε. From the corresponding increment of $g(y(T))$

we must select the part that is proportional to ε and neglect terms of higher order in ε.

As an intermediate step we will have to obtain the increment in $y(T)$ corresponding to $\delta u(t)$. Let us denote the solution of (2.3.1)–(2.3.2) corresponding to $u^*(t)$ by $y^*(t)$:

$$y^{*\prime}(t) = f(y^*(t), u^*(t)), \qquad y^*(0) = y_0.$$

We denote

$$\Delta y(t) = y^*(t) - y(t), \qquad J(u) = g(y(T)),$$

and seek the main (in ε) part of the increment

$$\Delta J_{\varepsilon,v}(u) = J(u^*) - J(u).$$

Again, this main part must be linear in ε; we neglect terms of higher order in ε. In this, we consider $u(t)$ as given and hence $y(t)$ is known uniquely as well.

Theorem 2.3.1 *Let $t = s$ be a point of continuity of a control function $u(t)$. We have*

$$\Delta J_{s,v}(u) = \varepsilon\, \delta_{s,v} J(u) + o(\varepsilon), \qquad \varepsilon > 0, \tag{2.3.4}$$

where

$$\delta_{s,v} J(u) = \psi(s)[f(y(s), u(s)) - f(y(s), v)]$$

and where $\psi(s)$ is a solution of the following Cauchy problem (in the reverse time):

$$\psi'(s) = -\frac{\partial f(y(s), u(s))}{\partial y}\psi(s), \qquad \psi(T) = -\frac{dg(y(T))}{dy}. \tag{2.3.5}$$

The quantity $\delta_{s,v} J(u)$ in (2.3.4) is called the variational derivative of the second kind.

Proof. Take $\varepsilon > 0$ so small that all the points of $[s - \varepsilon, s]$ are points of continuity of $u(t)$. We require that $u^*(t)$, which differs from $u(t)$ by a needle-shaped increment, is admissible and has the form (2.3.3). We divide the proof into several steps.

Step 1. First let us find the main part in ε of the increment of $y(t)$, in particular at $t = T$. In Fig. 2.2 we show the behavior of $y(t)$ and $y^*(t)$. When $t < s - \varepsilon$ we have $u^*(t) = u(t)$ and thus $y^*(t) = y(t)$.

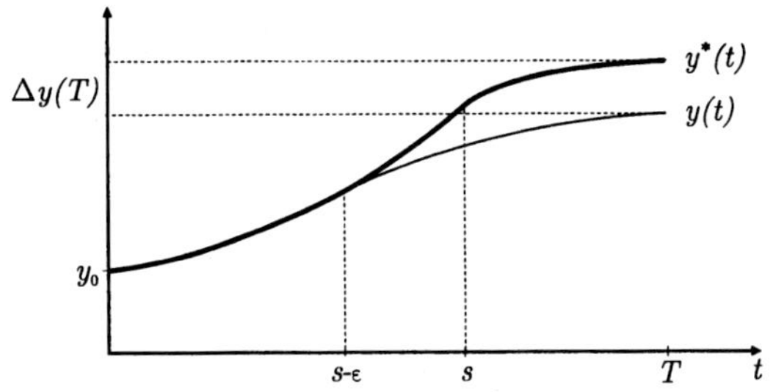

Fig. 2.2 The deviation of a trajectory $y(t)$ under a needle-shaped change of the control function on the time interval $[s-\varepsilon, s]$.

Let $t \in [s-\varepsilon, s]$. Subtracting the equations for y^* and y we get

$$y^{*\prime}(t) - y'(t) = f(y^*(t), v) - f(y(t), u(t))$$

or, since $\Delta y(t) = y^*(t) - y(t)$, the equation for the increment of the function $y(t)$ is

$$\Delta y'(t) = f(y(t) + \Delta y(t), v) - f(y(t), u(t)). \tag{2.3.6}$$

Besides, we have the "initial" condition for this interval

$$\Delta y(s-\varepsilon) = 0 \tag{2.3.7}$$

since $y^*(s-\varepsilon) = y(s-\varepsilon)$. Integration of (2.3.6) gives us an equivalent integral equation on $[s-\varepsilon, s]$:

$$\Delta y(t) - \Delta y(s-\varepsilon) = \int_{s-\varepsilon}^{t} [f(y(\tau) + \Delta y(\tau), v) - f(y(\tau), u(\tau))] \, d\tau.$$

By (2.3.7) we have

$$\Delta y(t) = \int_{s-\varepsilon}^{t} [f(y(\tau) + \Delta y(\tau), v) - f(y(\tau), u(\tau))] \, d\tau \quad \text{on } [s-\varepsilon, s].$$

We assume $f(y, u)$ is continuous and bounded on the domain where the pair (y, u) takes its value, and thus when ε is small the modulus of the integral on the right is bounded by $M\varepsilon$ for $t \in [s-\varepsilon, s]$. So this integral has the first order of smallness in ε when $t \in [s-\varepsilon, s]$, and thus the same value bounds $|\Delta y(t)|$ on the same segment. Since ε is small and $y(t), u(t)$ are continuous on $[s-\varepsilon, s]$, the integrand is continuous as well, and we introduce in the values of this integral an error of order higher than the first in ε if we replace the integrand by the constant value $f(y(s), v) - f(y(s), u(s))$. So

$$\begin{aligned}\Delta y(t) &= \int_{s-\varepsilon}^{t} [f(y(s), v) - f(y(s), u(s))]\, d\tau + o(\varepsilon) \\ &= (t - s + \varepsilon)[f(y(s), v) - f(y(s), u(s))] + o(\varepsilon),\end{aligned}$$

and thus

$$\Delta y(s) = \varepsilon[f(y(s), v) - f(y(s), u(s))] + o(\varepsilon). \tag{2.3.8}$$

This gives us the "initial" value for the solution $y^*(t)$ on $[s, T]$. Note that on $[s - \varepsilon, s]$ the change of $\Delta y(t)$ in t is almost linear, which is expected since ε is small.

On $[s, T]$, subtracting the equations for $y(t)$ and $y^*(t)$ we get

$$\Delta y'(t) = f(y(t) + \Delta y(t), u(t)) - f(y(t), u(t)). \tag{2.3.9}$$

This is supplemented by the initial condition (2.3.8), which is small when ε is small. Since y and y^* obey the same equation on $[s, T]$ but their initial values differ by a small value $\Delta y(s)$ of order ε, we can expect that there is continuous dependence of the solution on the initial data and hence that the difference between y^* and y, which is Δy, remains of order ε when T is finite. So we linearize (2.3.9) using the first-order Taylor expansion

$$f(y(t) + \Delta y(t), u(t)) - f(y(t), u(t)) = \frac{\partial f(y(t), u(t))}{\partial y} \Delta y(t) + o(|\Delta y(t)|)$$

to get

$$\Delta y'(t) = \frac{\partial f(y(t), u(t))}{\partial y} \Delta y(t) + o(\varepsilon).$$

The main part of $\Delta y(t)$, denoted by $\delta y(t)$, satisfies

$$\delta y'(t) = \frac{\partial f(y(t), u(t))}{\partial y} \delta y(t). \tag{2.3.10}$$

This can be integrated explicitly since $y(t)$ and $u(t)$ and the initial condition for $\delta y(t)$ are defined by (2.3.8) as

$$\delta y(s) = \varepsilon[f(y(s), v) - f(y(s), u(s))].$$

However, we should allow for an extension to a system of ODEs. So we shall produce a mathematical trick of "finding" the solution in other terms. At this point we must interrupt the proof and introduce some material.

2.4 Fundamental Solution of a Linear Ordinary Differential Equation

Consider a linear ODE

$$x'(t) = a(t)x(t). \tag{2.4.1}$$

This has a unique solution for any initial condition $x(s) = x_0$, $a(t)$ being a given continuous function (it can be continuous on an interval if we consider the equation on this interval or at any t). The *fundamental solution* is a function $\varphi(t, s)$ which at any fixed s satisfies

$$\frac{d\varphi(t,s)}{dt} = a(t)\varphi(t,s)$$

and the condition

$$\varphi(s,s) = 1.$$

This function in two variables has many useful properties, the first of which is trivial:

Property 2.4.1 *A solution of* (2.4.1) *satisfying the initial condition* $x(s) = x_0$ *is*

$$x(t) = x_0\varphi(t,s).$$

Property 2.4.2 *We have*

$$\varphi(t,s) = \varphi(t,\tau)\varphi(\tau,s) \tag{2.4.2}$$

for any t, s, *and* τ.

Proof. Indeed, for fixed τ, s the function $\varphi(t,\tau)\varphi(\tau,s)$ of the variable t is a solution to (2.4.1) when t is an independent variable, since $\varphi(\tau,s)$ does not depend on t. Thus we have two solutions to (2.4.1): the functions $\varphi(t,s)$ and $\varphi(t,\tau)\varphi(\tau,s)$. But for $t = \tau$ they correspondingly reduce to $\varphi(\tau,s)$ and $\varphi(\tau,\tau)\varphi(\tau,s) = 1 \cdot \varphi(\tau,s)$, and thus at $t = \tau$ they coincide. By the uniqueness of the solution of the Cauchy problem for (2.4.1) (the initial value is given at $t = \tau$) they coincide at any t. □

Since $\varphi(s,s) = 1$ we have $\varphi(s,t)\,\varphi(t,s) = 1$, hence

$$\varphi(t,s) = 1/\varphi(s,t). \tag{2.4.3}$$

In the next section we shall need $\partial\varphi(t,s)/\partial s$. By (2.4.3) we have

Property 2.4.3 *The function $\varphi(t,s)$ considered*[4] *as a function in s satisfies*

$$\frac{d\varphi(t,s)}{ds} = -a(s)\,\varphi(t,s). \tag{2.4.4}$$

Proof. Using (2.4.3) we have

$$\begin{aligned}\frac{d\varphi(t,s)}{ds} &= \frac{d(\varphi^{-1}(s,t))}{ds} = -\varphi^{-2}(s,t)\frac{d(\varphi(s,t))}{ds} = -\varphi^{-2}(s,t)\,a(s)\,\varphi(s,t)\\ &= -a(s)\varphi^{-1}(s,t) = -a(s)\varphi(t,s).\end{aligned}$$

□

Now we can continue the proof of Theorem 2.3.1.

2.5 The Simplest Problem, Continued

Setting

$$a(t) = \frac{\partial f(y(t),u(t))}{\partial y}, \tag{2.5.1}$$

we apply the notion of fundamental solution to (2.3.10). So the solution[5] of (2.3.10) on $[s,T]$ satisfying (2.4.2) is

$$\delta y(t) = \varepsilon\,[f(y(s),v) - f(y(s),u(s))]\,\varphi(t,s).$$

Hence

$$\delta y(T) = \varepsilon\,[f(y(s),v) - f(y(s),u(s))]\,\varphi(T,s)$$

and we can write

$$\Delta y(T) = \varepsilon\,[f(y(s),v) - f(y(s),u(s))]\,\varphi(T,s) + o(\varepsilon). \tag{2.5.2}$$

Step 2. The main part of the increment of $J(u) = g(y(T))$ can be found using the same idea of linearization and Taylor expansion:

$$\begin{aligned}\Delta J(u) &= J(u^*) - J(u)\\ &= g(y(T) + \Delta y(T)) - g(y(T))\\ &= \left.\frac{dg(y)}{dy}\right|_{y=y(T)} \Delta y(T) + o(|\Delta y(T)|).\end{aligned}$$

[4]Here we consider t as a fixed parameter, which is why we use the notation for an ordinary derivative rather than a partial derivative.

[5]Of course, this is really just a useful representation rather than an explicit solution.

With regard for (2.5.2) this brings us to

$$\Delta J(u) = \varepsilon \left.\frac{dg(y)}{dy}\right|_{y=y(T)} [f(y(s), v) - f(y(s), u(s))] \, \varphi(T, s) + o(\varepsilon).$$

So we have found the main part of the increment of the objective functional; however, we must still represent it in the form shown in Theorem 2.3.1.

Step 3. Let

$$\psi(s) = -\left.\frac{dg(y)}{dy}\right|_{y=y(T)} \varphi(T, s). \tag{2.5.3}$$

With this notation $\Delta J(u)$ takes the form (2.3.4). It remains only to demonstrate that $\psi(s)$ satisfies (2.3.5). The second relation of (2.3.5) holds by definition of the fundamental solution:

$$\psi(T) = -\left.\frac{dg(y)}{dy}\right|_{y=y(T)} \varphi(T, T) = -\left.\frac{dg(y)}{dy}\right|_{y=y(T)}.$$

Let us show that $\psi(s)$ satisfies the first equation of (2.3.5):

$$\begin{aligned}
\frac{d\psi(s)}{ds} &= \frac{d}{ds}\left[-\left.\frac{dg(y)}{dy}\right|_{y=y(T)} \varphi(T, s)\right] \\
&= -\left.\frac{dg(y)}{dy}\right|_{y=y(T)} \frac{d}{ds}\varphi(T, s) \\
&= -\left.\frac{dg(y)}{dy}\right|_{y=y(T)} [-a(s)\, \varphi(T, s)] \\
&= a(s) \left.\frac{dg(y)}{dy}\right|_{y=y(T)} \varphi(T, s).
\end{aligned}$$

Here we used (2.4.4) to eliminate the derivative of $\varphi(T, s)$ with respect to the second argument. Finally, remembering (2.5.3) we obtain

$$\psi'(s) = -a(s)\psi(s).$$

This is the needed equation since $a(t)$ is given by (2.5.1).

2.6 Pontryagin's Maximum Principle for the Simplest Problem

What have we established in Theorem 2.3.1? To find the increment in the goal functional under a needle-shaped increment of the control function

$u(t)$, we should do the following:

(1) Solve the Cauchy problem (2.3.1)–(2.3.2). In practice this is often done numerically (e.g., by the Runge–Kutta method).
(2) Having obtained $y(T)$, formulate equations (2.3.5) and solve this Cauchy problem with respect to $\psi(s)$ in the "reversed" time.
(3) Write out (2.3.4).

The second condition in (2.3.5) is analogous to the natural boundary condition in the calculus of variations. The first equation in (2.3.5) is called the *conjugate equation*; there is a weak analogy between this and the Euler equation. We also observe that in performing steps (1) and (2) we effectively solve a boundary value problem for the pair $y(t), \psi(s)$. A similar pair of equations arises for other types of optimal control problems, but in the terminal control problems they split.

Let us reformulate this problem, introducing a new function $H(y, \psi, u)$ in three variables:

$$H(y, \psi, u) = \psi\, f(y, u).$$

Because

$$\frac{\partial H(y, \psi, u)}{\partial \psi} = f(y, u), \qquad \frac{\partial H(y, \psi, u)}{\partial y} = \frac{\partial f(y, u)}{\partial y}\psi,$$

we can rewrite (2.3.1) and (2.3.5) as

$$y'(t) = \frac{\partial H(y(t), \psi(t), u(t))}{\partial \psi}, \qquad \psi'(t) = -\frac{\partial H(y(t), \psi(t), u(t))}{\partial y},$$

or

$$y'(t) = H_\psi(y(t), \psi(t), u(t)), \qquad \psi'(t) = -H_y(y(t), \psi(t), u(t)). \tag{2.6.1}$$

This is the so-called *Hamilton form* of a system of ODEs that is frequent in physics. L.S. Pontryagin called $H(y, \psi, u)$ the Hamilton function, but it was subsequently called the *Pontryagin function.* Again, we will obtain equations of the form (2.6.1) when we consider any sort of control problem for the system described by (2.3.1).

Let us rewrite the increment $\Delta J(u)$ under a needle-shaped increment with parameters ε, v given at $t = s$, which is presented by (2.3.4), in terms of $H(y, \psi, u)$:

$$\Delta J(u) = \varepsilon\, \left(H(y(s), \psi(s), u(s)) - H(y(s), \psi(s), v)\right) + o(\varepsilon).$$

Now we can formulate a necessary condition of minimum for $J(u)$, known as *Pontryagin's maximum principle:*

Theorem 2.6.1 *Let $u(t)$ be an optimal control function at which $J(u)$ attains its minimal value on $\mathcal{U}$, the set of all admissible control functions, and let $y(t)$ and $\psi(t)$ be solutions of the boundary value problem* (2.3.1), (2.3.2), (2.3.5). *At any point $t = s$ of continuity of $u(t)$, the function $H(y(s), \psi(s), v)$ considered as a function in the variable v takes its maximum value at $v = u(s)$.*

Proof. $J(u)$ attains its minimum at $u = u(t)$. For any admissible control function $u^*(t)$ given by (2.3.3) we have

$$J(u^*) - J(u) \geq 0.$$

For a point $t = s$ of continuity of $u = u(t)$, in terms of the Pontryagin function this is

$$\varepsilon\ (H(y(s), \psi(s), u(s)) - H(y(s), \psi(s), v)) + o(\varepsilon) \geq 0.$$

Note this is valid for any admissible v and small, nonnegative ε. It follows immediately that

$$H(y(s), \psi(s), u(s)) - H(y(s), \psi(s), v) \geq 0,$$

so for any admissible v we get $H(y(s), \psi(s), u(s)) \geq H(y(s), \psi(s), v)$. □

Let us consider the application of these results to a simple example.

Example 2.6.1 Find the form of the control function $u(t)$, $|u(t)| \leq 2$, that gives minimum deviation of $y(t)$ from 10 at $t = 1$ (described by the function $g(y(1)) = (10 - y(1))^2$) for a system governed by

$$y'(t) + y(t) = u(t), \qquad y(0) = 1.$$

Solution We stay with our previous notation. Rewrite the equation as $y' = -y + u$ and construct Pontryagin's function

$$H(y, \psi, u) = \psi(-y + u).$$

We need to learn when this function takes its maximum value with respect to u along a solution. For this we need to know some properties of ψ. Let us establish how ψ changes. The conjugate equation for ψ is

$$\psi' = -\frac{\partial H}{\partial y} = \psi.$$

Its general solution is $\psi = ce^t$. For this example we need not find (y, ψ) for any control function, so we will not formulate the final value for ψ but merely note that its sign coincides with that of the constant c. This means that along any possible solution $y = y(t)$, at any point of continuity of y, the maximum is taken when $\psi(t)u(t)$ takes its maximum. Since this expression is linear in u, the maximum is taken when u takes one of its extreme values $u = \pm 2$ and, because of the constancy of sign of ψ, it cannot change from one extreme to another.[6]

So now we must consider the governing equation in two versions, with $u = 2$ and $u = -2$. These are

$$y' = -y + 2, \qquad\qquad y' = -y - 2.$$

The initial condition leads to the respective solutions

$$y_1(t) = -e^{-t} + 2, \qquad\qquad y_2(t) = 3e^{-t} - 2.$$

Comparing the values of the cost function $g(y)$ for y_1 and y_2 at $t = 1$, we see that $u = u(t) = 2$ is the optimal control. Correspondingly $y(t) = -e^{-t} + 2$, and the minimum value of g is $g(y(1)) = (8 + e^{-1})^2$.

This example shows that not every optimal control problem has a solution. Indeed, if we pose the minimum time problem for the same equation with y beginning at $y = 1$ and ending at $y = 10$, under the restriction $|u| \le 2$, then there will be no solution; a solution starting from the point $y(0) = 1$ never takes the value 10.

Let us continue consideration of the same problem. We denote by $J(u)$ the value $g(y(1))$ so J is defined as a functional of the control function u.

Example 2.6.2 For the system of the previous example, find the main part of the increment of the goal functional under a needle-shaped disturbance of u if its value is $u(t) = 1$ for all t.

Solution The governing equation of the system for $u = 1$ is $y' = -y + 1$. The solution that satisfies the initial condition is $y = 1$. Thus the final value for ψ is

$$\psi(1) = -\frac{\partial g(y(1))}{\partial y} = -2 \cdot 9(-1) = 18,$$

[6]A reader familiar with the elements of linear programming will note that the situation is the same as in that theory. Since many optimal control problems are described by relations containing a control vector in a linear manner, the reader sees that at this stage it is necessary to solve a linear programming problem in which we must maximize a linear function over a set in a finite vector space restricted by linear inequalities.

and the corresponding solution of the conjugate equation is

$$\psi(t) = de^t, \qquad d = 18e.$$

Thus the main part of the increment of the goal functional is

$$\begin{aligned}\varepsilon\delta_{s,v}J(u) &= \psi(s)[f(y(s), u(s) - f(y(s), v)]\\ &= 18\varepsilon e^{1+t}(0 + 1 - v)\\ &= 18\varepsilon e^{1+t}(1 - v)\end{aligned}$$

for any time s. It is clear that if we wish to decrease locally at any point s the value of the functional, then we should take the maximum admissible value of v, which is $v = 2$.

This problem is important because it shows how we can improve an initial approximation to u. For sufficiently small ε, introducing a needle-shaped change of u at some s we reduce the value of $g(y(1))$. Choosing ε and s and decreasing correspondingly the value of $J(u)$ (of course, this happens only when $\varepsilon\delta_{s,v}J(u)$ has negative values on $[0, 1]$ — if there are no such values then a corresponding function u is optimal) we get a better approximation to the optimal control function. But the choice of ε, s is not uniquely defined even for this simple problem. If ε is small and fixed, it is clear that the maximal change in $J(u)$ happens (in this problem) when we take the maximum admissible value of v, that is $v = 2$. But what is the value of s? It is clear that we should introduce the needle-change into u at s where $\varepsilon\delta_{s,v}J(u)$ takes the lowest negative value. In this problem it is easy to see that it is the point $s = 1$. Changing u to 2 on $[1 - \varepsilon, 1]$ with some small ε we get a new control function u^* that is not optimal again. So we need to repeat the same steps: find $\varepsilon\delta_{s,v}J(u^*)$, choose ε and s, and introduce optimally a new needle-shaped perturbation into u to maximally decrease $J(u)$. This gives a second approximation to the optimal solution, and so on. In this simple case the approximation will be quite accurate. However, in practical problems, when we do not know the solution u in advance, it can be difficult to choose ε and s at each step.

Pontryagin's maximum principle allows us to test a given control function for optimality. In addition, we shall see later that for some relatively simple problems it can suggest an approach to finding solutions. Next we would like to note that formula (2.3.4) is the background for a class of numerical methods for finding an optimal solution. We shall discuss this for the general problem of terminal control, which should be further considered. In the next section we present some essential mathematical tools.

2.7 Some Mathematical Preliminaries

When we considered the simplest problem of control theory we used the notions of fundamental solution and linearization. To extend these to vector functions one can use the tools of matrix theory, but the resulting formulas are much more compact and clear when presented in tensor notation. We therefore pause to present a small portion of tensor analysis. In doing so we shall confine ourselves to the simplest case involving only Cartesian frames having orthonormal basis vectors $\mathbf{e}_1, \mathbf{e}_2, \ldots, \mathbf{e}_n$. In the general case the controlled functions $\mathbf{y}(t)$ take values in the n-dimensional vector space spanned by this basis, so we can represent $\mathbf{y}(t)$ as

$$\mathbf{y}(t) = \sum_{i=1}^{n} y_i(t)\,\mathbf{e}_i.$$

From now on we omit the summation symbol and write simply

$$\mathbf{y}(t) = y_i(t)\,\mathbf{e}_i.$$

This is the usual convention, due to Einstein, for dealing with Cartesian tensors: whenever we meet a repeated index (in this case i) we understand that summation is to be performed over this index from 1 to n. Now we shall demonstrate how this expansion can be used along with the dot product to produce representations of vectors, and to reproduce common operations involving vectors and matrices.

Matrices as the component representations of tensors and vectors

To perform operations with a vector $\mathbf{x}$ we must have a straightforward method of calculating its components $x_1, x_2, \ldots, x_n$ with respect to a basis $\mathbf{e}_1, \mathbf{e}_2, \ldots, \mathbf{e}_n$. This can be done through simple dot multiplication. For additional clarity let us momentarily suspend our use of the summation convention. Dotting $\mathbf{x}$ with $\mathbf{e}_1$ we have

$$\begin{aligned}\mathbf{x}\cdot\mathbf{e}_1 &= (x_1\mathbf{e}_1 + x_2\mathbf{e}_2 + \cdots + x_n\mathbf{e}_n)\cdot\mathbf{e}_1\\ &= x_1(\mathbf{e}_1\cdot\mathbf{e}_1) + x_2(\mathbf{e}_2\cdot\mathbf{e}_1) + \cdots + x_n(\mathbf{e}_n\cdot\mathbf{e}_1).\end{aligned}$$

Because $\mathbf{e}_1\cdot\mathbf{e}_1 = 1$ and $\mathbf{e}_2\cdot\mathbf{e}_1 = \mathbf{e}_3\cdot\mathbf{e}_1 = \cdots = \mathbf{e}_n\cdot\mathbf{e}_1 = 0$, we obtain

$$x_1 = \mathbf{x}\cdot\mathbf{e}_1.$$

Here the key observation is that

$$\mathbf{e}_i \cdot \mathbf{e}_j = \begin{cases} 1, & j = i, \\ 0, & j \neq i, \end{cases}$$

and this same observation can be used in similar fashion to develop the formulas

$$x_2 = \mathbf{x} \cdot \mathbf{e}_2, \qquad x_3 = \mathbf{x} \cdot \mathbf{e}_3, \qquad \ldots, \qquad x_n = \mathbf{x} \cdot \mathbf{e}_n.$$

In terms of the Kronecker delta symbol (page 38) we could have written

$$\begin{aligned} \mathbf{x} \cdot \mathbf{e}_1 &= (x_1\mathbf{e}_1 + x_2\mathbf{e}_2 + \cdots + x_n\mathbf{e}_n) \cdot \mathbf{e}_1 \\ &= x_1\delta_{11} + x_2\delta_{21} + \cdots + x_n\delta_{n1} \\ &= x_1 \end{aligned}$$

to calculate x_1. We can now return to the summation convention and repeat these calculations in tensor notation. If $\mathbf{x}$ is given by

$$\mathbf{x} = x_k\mathbf{e}_k$$

then for $i = 1, 2, \ldots, n$ we have

$$x_i = \mathbf{x} \cdot \mathbf{e}_i$$

since $\mathbf{x} \cdot \mathbf{e}_i = x_k\mathbf{e}_k \cdot \mathbf{e}_i = x_k\delta_{ki} = x_i$ for each i. Thus in a given basis $\mathbf{e}_i$ the components x_i of the vector $\mathbf{x}$ are determined uniquely, and $\mathbf{x}$ is determined by these values x_i. It is convenient to display the components of $\mathbf{x}$ in a column matrix:

$$\begin{pmatrix} x_1 \\ x_2 \\ \vdots \\ x_n \end{pmatrix}.$$

Hence a matrix can act as the component representation of a vector. It is important to understand that a vector itself is an *objective* entity: it is independent of coordinate frame. Consequently if we expand the same vector $\mathbf{x}$ relative to a different Cartesian basis $\tilde{\mathbf{e}}_1, \tilde{\mathbf{e}}_2, \ldots, \tilde{\mathbf{e}}_n$ and repeat the above steps, we will in general arrive at a matrix representation

$$\begin{pmatrix} \tilde{x}_1 \\ \tilde{x}_2 \\ \vdots \\ \tilde{x}_n \end{pmatrix}$$

whose entries $\tilde{x}_k$ differ from the x_k. We shall return to this issue later after examining tensors.

If $\mathbf{x}$ and $\mathbf{y}$ are two vectors, their dot product is a scalar:

$$c = \mathbf{x} \cdot \mathbf{y}. \tag{2.7.1}$$

When we represent each of $\mathbf{x}$ and $\mathbf{y}$ with respect to a basis $\mathbf{e}_i$ as

$$\mathbf{x} = x_i \mathbf{e}_i, \qquad \mathbf{y} = y_j \mathbf{e}_j,$$

we can easily calculate c as

$$\mathbf{x} \cdot \mathbf{y} = x_i \mathbf{e}_i \cdot y_j \mathbf{e}_j = x_i y_j (\mathbf{e}_i \cdot \mathbf{e}_j) = x_i y_j \delta_{ij} = x_i y_i.$$

Of course, this same result arises from the matrix multiplication

$$c = \begin{pmatrix} x_1 & x_2 & \cdots & x_n \end{pmatrix} \begin{pmatrix} y_1 \\ y_2 \\ \vdots \\ y_n \end{pmatrix}. \tag{2.7.2}$$

This familiar correspondence between dot multiplication of vectors and multiplication of the component matrices will be extended in what follows.

A vector is an example of a tensor of the first rank. The development of our subject will also require some simple work with tensors of the second rank. Just as a vector can be constructed as a linear combination of basis vectors $\mathbf{e}_i$, a tensor of the second rank can be constructed as a linear combination of *basis dyads.* These are in turn formed from pairs of vectors through the use of a *tensor product.* This operation, denoted $\otimes$, obeys laws analogous to those for ordinary multiplication: if $\mathbf{a}$, $\mathbf{b}$, and $\mathbf{c}$ are vectors then

$$\begin{aligned} (\lambda \mathbf{a}) \otimes \mathbf{b} &= \mathbf{a} \otimes (\lambda \mathbf{b}) = \lambda (\mathbf{a} \otimes \mathbf{b}), \\ (\mathbf{a} + \mathbf{b}) \otimes \mathbf{c} &= \mathbf{a} \otimes \mathbf{c} + \mathbf{b} \otimes \mathbf{c}, \\ \mathbf{a} \otimes (\mathbf{b} + \mathbf{c}) &= \mathbf{a} \otimes \mathbf{b} + \mathbf{a} \otimes \mathbf{c}, \end{aligned} \tag{2.7.3}$$

for any real number λ. We shall shorten the notation for the tensor product somewhat by omitting the $\otimes$ symbol: thus we write **ab** instead of $\mathbf{a} \otimes \mathbf{b}$. The quantity **ab** is an example of a *dyad* of vectors. If we expand each of the vectors **a** and **b** in terms of a basis $\mathbf{e}_i$, the dyad **ab** becomes

$$\mathbf{ab} = a_i \mathbf{e}_i b_j \mathbf{e}_j = a_i b_j \, \mathbf{e}_i \mathbf{e}_j.$$

In this way the n^2 different basis dyads $\mathbf{e}_i\mathbf{e}_j$ make their appearance. The dyads $\mathbf{e}_i\mathbf{e}_j$ form the basis for a linear space called the space of tensors of the second rank. An element **A** of this space has a representation of the form

$$\mathbf{A} = a_{ij} \mathbf{e}_i \mathbf{e}_j$$

where the a_{ij} are called the components of **A** relative to the basis $\mathbf{e}_i\mathbf{e}_j$. Here we again use Einstein's summation rule. Note that we can write out the components of **A** as a square array

$$\begin{pmatrix} a_{11} & a_{12} & \cdots & a_{1n} \\ a_{21} & a_{22} & \cdots & a_{2n} \\ \vdots & \vdots & \ddots & \vdots \\ a_{n1} & a_{n2} & \cdots & a_{nn} \end{pmatrix},$$

and thus we get a correspondence between the tensor **A** and this matrix of its components.

One goal of our discussion is to demonstrate the usefulness of the dot product. The dot product of a dyad **ab** and a vector **c** is defined by the equation

$$(\mathbf{ab}) \cdot \mathbf{c} = \mathbf{a}(\mathbf{b} \cdot \mathbf{c}). \tag{2.7.4}$$

The result is a vector directed along **a**. Analogously we can introduce the dot product from the left:

$$\mathbf{c} \cdot (\mathbf{ab}) = (\mathbf{c} \cdot \mathbf{a})\mathbf{b}. \tag{2.7.5}$$

These operations have matrix counterparts: (2.7.4) corresponds to multiplication of a matrix by a column vector and (2.7.5) corresponds to multiplication of a row vector by a matrix. For example let us write

$$\mathbf{v} = (\mathbf{ab}) \cdot \mathbf{c}, \tag{2.7.6}$$

expand $\mathbf{c}$ as $\mathbf{c} = c_k \mathbf{e}_k$, expand $\mathbf{ab}$ according to (2.7.3), and use (2.7.4) to write

$$\mathbf{v} = a_i b_j \mathbf{e}_i \mathbf{e}_j \cdot c_k \mathbf{e}_k = a_i b_j \delta_{jk} c_k \mathbf{e}_i = a_i b_j c_j \mathbf{e}_i.$$

Hence

$$v_i = a_i b_j c_j \tag{2.7.7}$$

for $i = 1, 2, \ldots, n$. Pausing to unpack the succinct tensor index notation, we see that (2.7.7) actually means the system of equalities

$$\begin{aligned} v_1 &= a_1 b_1 c_1 + a_1 b_2 c_2 + \cdots + a_1 b_n c_n, \\ v_2 &= a_2 b_1 c_1 + a_2 b_2 c_2 + \cdots + a_2 b_n c_n, \\ &\vdots \\ v_n &= a_n b_1 c_1 + a_n b_2 c_2 + \cdots + a_n b_n c_n, \end{aligned}$$

or, in matrix form,

$$\begin{pmatrix} v_1 \\ v_2 \\ \vdots \\ v_n \end{pmatrix} = \begin{pmatrix} a_1 b_1 & a_1 b_2 & \cdots & a_1 b_n \\ a_2 b_1 & a_2 b_2 & \cdots & a_2 b_n \\ \vdots & \vdots & \ddots & \vdots \\ a_n b_1 & a_n b_2 & \cdots & a_n b_n \end{pmatrix} \begin{pmatrix} c_1 \\ c_2 \\ \vdots \\ c_n \end{pmatrix}. \tag{2.7.8}$$

We now recall the analogy between (2.7.1) and (2.7.2), and examine (2.7.6) and (2.7.8) with similar thoughts in mind. Dot multiplication once again stands in correspondence with matrix multiplication; moreover, it is clear that the dyad $\mathbf{ab}$ is represented by the square matrix

$$\begin{pmatrix} a_1 b_1 & a_1 b_2 & \cdots & a_1 b_n \\ a_2 b_1 & a_2 b_2 & \cdots & a_2 b_n \\ \vdots & \vdots & \ddots & \vdots \\ a_n b_1 & a_n b_2 & \cdots & a_n b_n \end{pmatrix}.$$

We have seen that a dyad $\mathbf{ab}$ can map a vector $\mathbf{c}$ into another vector $\mathbf{v}$ through the dot product operation given in (2.7.6). This idea carries through to general tensors of the second rank, of which dyads are examples. If $\mathbf{A}$ is a tensor of second rank and $\mathbf{x}$ is a vector, then $\mathbf{A}$ can map $\mathbf{x}$ into an image vector $\mathbf{y}$ according to

$$\mathbf{y} = \mathbf{A} \cdot \mathbf{x}. \tag{2.7.9}$$

It is easy to check that the individual components of $\mathbf{A} = a_{ij}\mathbf{e}_i\mathbf{e}_j$ are given by

$$a_{ij} = \mathbf{e}_i \cdot \mathbf{A} \cdot \mathbf{e}_j,$$

and that (2.7.9) corresponds to

$$\begin{pmatrix} y_1 \\ y_2 \\ \vdots \\ y_n \end{pmatrix} = \begin{pmatrix} a_{11} & a_{12} & \cdots & a_{1n} \\ a_{21} & a_{22} & \cdots & a_{2n} \\ \vdots & \vdots & \ddots & \vdots \\ a_{n1} & a_{n2} & \cdots & a_{nn} \end{pmatrix} \begin{pmatrix} x_1 \\ x_2 \\ \vdots \\ x_n \end{pmatrix}.$$

A dot product operation known as pre-multiplication of a tensor by a vector is also considered: the quantity $\mathbf{y} \cdot \mathbf{A}$ is defined by the requirement that

$$(\mathbf{y} \cdot \mathbf{A}) \cdot \mathbf{x} = \mathbf{y} \cdot (\mathbf{A} \cdot \mathbf{x})$$

for all vectors $\mathbf{x}$. This can be also obtained as a consequence of the formal definition of left-dot-multiplication of a vector by a dyad:

$$\mathbf{a} \cdot \mathbf{bc} = (\mathbf{a} \cdot \mathbf{b})\mathbf{c}.$$

We see both dot product operations (pre-multiplication and post-multiplication) applied to the definition of the important *unit tensor* $\mathbf{E}$, which satisfies

$$\mathbf{E} \cdot \mathbf{x} = \mathbf{x} \cdot \mathbf{E} = \mathbf{x} \tag{2.7.10}$$

for any vector $\mathbf{x}$. It is easy to find the components of $\mathbf{E}$ from this definition. We start by writing $\mathbf{E} = e_{ij}\mathbf{e}_i\mathbf{e}_j$ and then apply (2.7.10) with $\mathbf{x} = \mathbf{e}_k$ to get

$$e_{ij}\mathbf{e}_i\mathbf{e}_j \cdot \mathbf{e}_k = \mathbf{e}_k.$$

Pre-multiplying by $\mathbf{e}_m$ we obtain

$$e_{ij}\delta_{mi}\delta_{jk} = \delta_{mk}$$

since $\mathbf{e}_m \cdot \mathbf{e}_i = \delta_{mi}$, $\mathbf{e}_j \cdot \mathbf{e}_k = \delta_{jk}$, and $\mathbf{e}_m \cdot \mathbf{e}_k = \delta_{mk}$. Hence $e_{mk} = \delta_{mk}$ and we have

$$\mathbf{E} = \delta_{ij}\mathbf{e}_i\mathbf{e}_j = \mathbf{e}_i\mathbf{e}_i.$$

Of course, the corresponding matrix representation is the $n \times n$ identity matrix

$$I = \begin{pmatrix} 1 & 0 & 0 & \cdots & 0 \\ 0 & 1 & 0 & \cdots & 0 \\ 0 & 0 & 1 & \cdots & 0 \\ \vdots & \vdots & \vdots & \ddots & \vdots \\ 0 & 0 & 0 & \cdots & 1 \end{pmatrix}.$$

Thus $\mathbf{E}$ is equivalent to the unit matrix.

The strong parallel that exists between tensors and matrices leads us to apply the notion of transposition to tensors of the second rank. Accordingly, if $\mathbf{A} = a_{ij}\mathbf{e}_i\mathbf{e}_j$ then we define

$$\mathbf{A}^T = a_{ji}\mathbf{e}_i\mathbf{e}_j = a_{ij}\mathbf{e}_j\mathbf{e}_i.$$

It is easy to see that

$$\mathbf{A} \cdot \mathbf{x} = \mathbf{x} \cdot \mathbf{A}^T$$

for any vector $\mathbf{x}$ and any tensor $\mathbf{A}$ of the second rank. It is even more obvious that $(\mathbf{A}^T)^T = \mathbf{A}$. If A is the matrix representation of $\mathbf{A}$, then A^T represents $\mathbf{A}^T$.

A dot product between two tensors is regarded as the composition of the two tensors viewed as operators. That is, $\mathbf{A} \cdot \mathbf{B}$ is defined by the equation

$$(\mathbf{A} \cdot \mathbf{B}) \cdot \mathbf{x} \equiv \mathbf{A} \cdot (\mathbf{B} \cdot \mathbf{x}).$$

A tensor $\mathbf{B}$ of the second rank is said to be the inverse of $\mathbf{A}$ if

$$\mathbf{A} \cdot \mathbf{B} = \mathbf{B} \cdot \mathbf{A} = \mathbf{E}.$$

In this case we write $\mathbf{B} = \mathbf{A}^{-1}$.

A central aspect of the study of tensors concerns how their components transform when the frame is changed. Although such frame transformations will not play a significant role in our discussion, the reader should understand that to express a tensor in another frame we would simply substitute the representation of the old basis vectors in terms of the new ones. As a simple example we may consider the case of a tensor of rank one: a vector. Let the components of $\mathbf{x}$ relative to the frame $\mathbf{e}_i$ be x_i so that $\mathbf{x} = x_i\mathbf{e}_i$. If a new frame $\tilde{\mathbf{e}}_i$ is introduced according to the set of linear relations $\mathbf{e}_i = A_{ij}\tilde{\mathbf{e}}_j$, then $\mathbf{x} = x_iA_{ij}\tilde{\mathbf{e}}_j$ and we see that $\mathbf{x} = \tilde{x}_j\tilde{\mathbf{e}}_j$ where $\tilde{x}_j = A_{ij}x_i$. The point is that we are not free to assign values to the $\tilde{x}_j$ in any way we wish: once the frame transformation is specified through the A_{ij}, the new components $\tilde{x}_i$ are completely determined by the old components x_i. The situation with tensors of higher order is the same.

Note that the correspondence between tensors and matrices is one-to-one only for a fixed basis. As soon as we change the basis, the matrix representation of a tensor changes by strictly defined rules. For example, if we take a non-Cartesian basis in space, the matrix representation of the tensor $\mathbf{E}$ is not the unit matrix, and thus $\mathbf{E}$ is not something we could call the unit tensor. Rather, it is known as the metric tensor.

Elements of calculus for vector and tensor fields

Now we consider how differentiation is performed on tensor and vector functions using tensor notation. Let us begin with a function $\mathbf{y}(t) = y_i(t)\,\mathbf{e}_i$. Since $\mathbf{e}_i$ does not depend on t, differentiation of $\mathbf{y}(t)$ with respect to t reduces to differentiation of the component scalar functions $y_i(t)$:

$$\mathbf{y}'(t) = y_i'(t)\,\mathbf{e}_i.$$

Similarly, the differential of a vector function $\mathbf{y}(t)$ is

$$d\mathbf{y}(t) = dy_i(t)\,\mathbf{e}_i.$$

Now suppose we wish to differentiate a composite function $f(\mathbf{y})(t)$ with respect to t. Writing this as $f(y_i(t)\,\mathbf{e}_i)$ or $f(y_1(t), \ldots, y_n(t))$, we have by the chain rule

$$\begin{aligned}\frac{d}{dt}f(\mathbf{y}(t)) &= \frac{d}{dt}f(y_1(t), \ldots, y_n(t)) \\ &= \sum_{i=1}^{n} \frac{\partial f(y_1(t), \ldots, y_n(t))}{\partial y_i} y_i'(t) \\ &= \frac{\partial f(\mathbf{y}(t))}{\partial y_i} y_i'(t). \qquad (2.7.11)\end{aligned}$$

Let us write out the right-hand side of (2.7.11) in vector form. For this we introduce ∇, a formal vector of differentiation (known as the gradient operator):

$$\nabla_{\mathbf{y}} = \sum_{i=1}^{n} \mathbf{e}_i \frac{\partial}{\partial y_i} = \mathbf{e}_i \frac{\partial}{\partial y_i}.$$

(We show the subscript $\mathbf{y}$ on ∇ to indicate the vector whose components participate in the differentiation. The subscript can be omitted if this is clear from the context.) When we apply $\nabla_{\mathbf{y}}$ to a function $f(\mathbf{y})$ we get a vector

$$\nabla_{\mathbf{y}} f(\mathbf{y}) = \sum_{i=1}^{n} \mathbf{e}_i \frac{\partial f(\mathbf{y})}{\partial y_i} = \frac{\partial f(\mathbf{y})}{\partial y_i} \mathbf{e}_i.$$

Let us dot multiply $\nabla_{\mathbf{y}} f(\mathbf{y}(t))$ by $\mathbf{y}'(t) = y_j'(t)\,\mathbf{e}_j$. Remembering that $\mathbf{e}_i \cdot \mathbf{e}_j = \delta_{ij}$, we get

$$\nabla_{\mathbf{y}} f(\mathbf{y}(t)) \cdot \mathbf{y}'(t) = \frac{\partial f(\mathbf{y}(t))}{\partial y_i} \mathbf{e}_i \cdot y_j'(t) \mathbf{e}_j = \frac{\partial f(\mathbf{y}(t))}{\partial y_i} y_j'(t) \delta_{ij} = \frac{\partial f(\mathbf{y}(t))}{\partial y_i} y_i'(t).$$

Since the right-hand side of this coincides with that of (2.7.11), we have

$$\frac{d}{dt} f(\mathbf{y}(t)) = \nabla_{\mathbf{y}} f(\mathbf{y}(t)) \cdot \mathbf{y}'(t).$$

The differential of $f(\mathbf{y}(t))$ is given by

$$df(\mathbf{y}(t)) = \nabla_{\mathbf{y}} f(\mathbf{y}(t)) \cdot d\mathbf{y}(t). \tag{2.7.12}$$

Using this formula or, equivalently, the first-order Taylor approximation, we get

$$f(\mathbf{y}(t) + \Delta\mathbf{y}(t)) - f(\mathbf{y}(t)) = \nabla_{\mathbf{y}} f(\mathbf{y}(t)) \cdot \Delta\mathbf{y}(t) + o(\|\Delta\mathbf{y}(t)\|)$$

where $\Delta\mathbf{y}(t)$ is a small increment of $\mathbf{y}(t)$.

Now we would like to present the first-order Taylor approximation of the increment of a vector function $\mathbf{f}$ that depends on a vector function $\mathbf{y}(t)$. We assume that $\mathbf{f}$ takes values in the same space as $\mathbf{y}(t)$ and thus can be represented as $\mathbf{f} = f_i\,\mathbf{e}_i$ where $f_i = f_i(\mathbf{y}(t))$. For this we find the differential of $\mathbf{f}(\mathbf{y}(t))$ at $\mathbf{y}(t)$:

$$\begin{aligned} d\mathbf{f}(\mathbf{y}(t)) &= d\left(f_j(\mathbf{y}(t))\,\mathbf{e}_j\right) = df_j(\mathbf{y}(t))\,\mathbf{e}_j \\ &= \nabla_{\mathbf{y}} f_j(\mathbf{y}(t)) \cdot d\mathbf{y}(t)\,\mathbf{e}_j \\ &= \frac{\partial f_j(\mathbf{y}(t))}{\partial y_i}\,\mathbf{e}_i \cdot dy_k(t)\,\mathbf{e}_k\,\mathbf{e}_j \end{aligned}$$

The right-hand side can be represented as

$$\left(\frac{\partial f_j(\mathbf{y}(t))}{\partial y_i}\mathbf{e}_j\mathbf{e}_i\right) \cdot dy_k(t)\,\mathbf{e}_k \quad \text{or} \quad dy_k(t)\,\mathbf{e}_k \cdot \left(\frac{\partial f_j(\mathbf{y}(t))}{\partial y_i}\mathbf{e}_i\mathbf{e}_j\right). \tag{2.7.13}$$

We see that in both brackets there is a sum of dyads so both of them are functions whose values are tensors of the second rank. A formal application of $\nabla_{\mathbf{y}}$ to $\mathbf{f}(\mathbf{y}(t))$ gives

$$\nabla_{\mathbf{y}}\mathbf{f}(\mathbf{y}(t)) = \mathbf{e}_i \frac{\partial}{\partial y_i} f_j(\mathbf{y}(t))\,\mathbf{e}_j = \frac{\partial f_j(\mathbf{y}(t))}{\partial y_i}\mathbf{e}_i\mathbf{e}_j.$$

Thus $\nabla_{\mathbf{y}}\mathbf{f}(\mathbf{y}(t))$ is the expression in brackets of the second equation (2.7.13) and the differential can be represented as

$$d\mathbf{f}(\mathbf{y}(t)) = d\mathbf{y}(t) \cdot \nabla_{\mathbf{y}}\mathbf{f}(\mathbf{y}(t)). \tag{2.7.14}$$

The term in the bracket of the first equation of (2.7.13) differs from the corresponding term of the second equation by a transposition of the vectors $\mathbf{e}_i$ and $\mathbf{e}_j$ so it can be written as $(\nabla_{\mathbf{y}}\mathbf{f}(\mathbf{y}(t)))^T$ and thus the differential can be presented in the other form

$$d\mathbf{f}(\mathbf{y}(t)) = (\nabla_{\mathbf{y}}\mathbf{f}(\mathbf{y}(t)))^T \cdot d\mathbf{y}(t). \tag{2.7.15}$$

The expression $\nabla_{\mathbf{y}}\mathbf{f}(\mathbf{y}(t))$ is called the *gradient* of $\mathbf{f}$. Let us see how it appears in more common matrix notation. We have said that a second rank tensor can be represented by a matrix of coefficients; in this representation the index i in the first position denotes the ith row of the matrix whereas the second index j denotes the jth column. Thus the matrix representation of the gradient of the vector function $\frac{\partial f_j(\mathbf{y}(t))}{\partial y_i}\mathbf{e}_i\mathbf{e}_j$ is

$$\begin{pmatrix} \dfrac{\partial f_1}{\partial y_1} & \dfrac{\partial f_2}{\partial y_1} & \cdots & \dfrac{\partial f_n}{\partial y_1} \\ \dfrac{\partial f_1}{\partial y_2} & \dfrac{\partial f_2}{\partial y_2} & \cdots & \dfrac{\partial f_n}{\partial y_2} \\ \vdots & \vdots & \ddots & \vdots \\ \dfrac{\partial f_1}{\partial y_n} & \dfrac{\partial f_2}{\partial y_n} & \cdots & \dfrac{\partial f_n}{\partial y_n} \end{pmatrix}.$$

Its determinant is the Jacobian of the transformation $\mathbf{z} = \mathbf{f}(\mathbf{y})$.

Now, using the formula for the differential (2.7.14) (or (2.7.15)) we are able to present an increment of a composite vector function $\mathbf{f}(\mathbf{y}(t))$ under the increment $\Delta\mathbf{y}(t)$ of the argument:

$$\mathbf{f}(\mathbf{y}(t) + \Delta\mathbf{y}(t)) - \mathbf{f}(\mathbf{y}(t)) = \Delta\mathbf{y}(t) \cdot \nabla_{\mathbf{y}}\mathbf{f}(\mathbf{y}(t)) + o(\|\Delta\mathbf{y}(t)\|).$$

Let the components of a tensor $\mathbf{A}(t) = a_{ij}(t)\mathbf{e}_i\mathbf{e}_j$ be continuously differentiable functions of t. Then by the rule for differentiating a matrix we have

$$\frac{d\mathbf{A}(t)}{dt} = \frac{da_{ij}(t)}{dt}\mathbf{e}_i\mathbf{e}_j.$$

The derivative of the dot product of two second-rank tensors obeys a formula similar to the ordinary product rule:

$$\frac{d}{dt}(\mathbf{A}(t) \cdot \mathbf{B}(t)) = \left(\frac{d}{dt}\mathbf{A}(t)\right) \cdot \mathbf{B}(t) + \mathbf{A}(t) \cdot \left(\frac{d}{dt}\mathbf{B}(t)\right).$$

A similar formula holds for the dot product of a tensor by a vector:

$$(\mathbf{A}(t) \cdot \mathbf{y}(t))' = \mathbf{A}'(t) \cdot \mathbf{y}(t) + \mathbf{A}(t) \cdot \mathbf{y}'(t).$$

If one factor does not depend on t then it can be removed from the symbol of differentiation:

$$(\mathbf{A} \cdot \mathbf{B}(t))' = \mathbf{A} \cdot \mathbf{B}'(t).$$

Fundamental solution of a linear system of ordinary differential equations

Consider a linear system of ODEs

$$\begin{aligned} y_1'(t) &= a_{11}(t)y_1(t) + a_{12}(t)y_2(t) + \cdots + a_{1n}(t)y_n(t), \\ y_2'(t) &= a_{21}(t)y_1(t) + a_{22}(t)y_2(t) + \cdots + a_{2n}(t)y_n(t), \\ &\vdots \\ y_n'(t) &= a_{n1}(t)y_1(t) + a_{n2}(t)y_2(t) + \cdots + a_{nn}(t)y_n(t). \end{aligned}$$

In terms of the tensor function $\mathbf{A}(t) = a_{ij}(t)\mathbf{e}_i\mathbf{e}_j$ and the vector $\mathbf{y}(t) = y_i(t)\mathbf{e}_i$ this system can be rewritten as

$$\mathbf{y}'(t) = \mathbf{A}(t) \cdot \mathbf{y}(t). \tag{2.7.16}$$

Definition 2.7.1 A tensor function $\mathbf{\Phi}(t, s)$ in two variables t, s is called the *fundamental solution*[7] of (2.7.16) if it satisfies two conditions:

(i) $\mathbf{\Phi}(t, s)$ is a solution of (2.7.16) in the first variable t:

$$\frac{d}{dt}\mathbf{\Phi}(t, s) = \mathbf{A}(t) \cdot \mathbf{\Phi}(t, s) \tag{2.7.17}$$

[7]The function $\mathbf{\Phi}(t, s)$ is also known as the *fundamental tensor* or *fundamental matrix*.

(here we use the symbol for the ordinary derivative, thinking of s as a fixed parameter).

(ii) For any s,

$$\mathbf{\Phi}(s, s) = \mathbf{E}. \tag{2.7.18}$$

This fundamental solution exists for any finite t, s if the tensor $\mathbf{A}(t)$ is continuous. The problem of finding it consists of n Cauchy problems for the same system of equations with n initial conditions given at $t = s$. Hence the fundamental solution is determined uniquely.

Now we would like to extend the results for the fundamental solution of a single linear ODE to the general case. We present them in a similar manner.

Property 2.7.1 *A solution of* (2.7.16) *satisfying the initial condition* $\mathbf{y}(s) = \mathbf{y}_0$ *is*

$$\mathbf{y}(t) = \mathbf{\Phi}(t, s) \cdot \mathbf{y}_0.$$

Indeed, dot-multiplying vector-equation (2.7.17) by $\mathbf{y}_0$ from the right we see that $\mathbf{\Phi}(t, s) \cdot \mathbf{y}_0$ satisfies (2.7.16). By (2.7.18) this solution satisfies the initial condition $\mathbf{y}(s) = \mathbf{y}_0$.

Property 2.7.2 *For any* t, s *and* τ *we have*

$$\mathbf{\Phi}(t, s) = \mathbf{\Phi}(t, \tau) \cdot \mathbf{\Phi}(\tau, s). \tag{2.7.19}$$

A consequence of this property and relation (2.7.18) *is the equation for the inverse*

$$\mathbf{\Phi}^{-1}(t, s) = \mathbf{\Phi}(s, t) \tag{2.7.20}$$

which follows when we write out a particular case of (2.7.19),

$$\mathbf{E} = \mathbf{\Phi}(t, t) = \mathbf{\Phi}(t, s) \cdot \mathbf{\Phi}(s, t).$$

Proof. Let us prove (2.7.19). Dot multiply (2.7.17) by $\mathbf{\Phi}(s, \tau)$ from the right. On the left we have

$$\left(\frac{d}{dt}\mathbf{\Phi}(t, s)\right) \cdot \mathbf{\Phi}(s, \tau) = \frac{d}{dt}\left(\mathbf{\Phi}(t, s) \cdot \mathbf{\Phi}(s, \tau)\right)$$

since $\mathbf{\Phi}(s, \tau)$ does not depend on t; on the right we have

$$\mathbf{A}(t) \cdot \mathbf{\Phi}(t, s) \cdot \mathbf{\Phi}(s, \tau) = \mathbf{A}(t) \cdot \left(\mathbf{\Phi}(t, s) \cdot \mathbf{\Phi}(s, \tau)\right).$$

So $\mathbf{\Phi}(t,s)\cdot\mathbf{\Phi}(s,\tau)$ satisfies $d\mathbf{\Psi}/dt = \mathbf{A}(t)\cdot\mathbf{\Psi}$ with parameters s,τ. Putting $t=s$ in this solution we get

$$\mathbf{\Phi}(t,s)\cdot\mathbf{\Phi}(s,\tau)|_{t=s} = \mathbf{\Phi}(s,s)\cdot\mathbf{\Phi}(s,\tau) = \mathbf{\Phi}(s,\tau).$$

So $\mathbf{\Phi}(t,s)\cdot\mathbf{\Phi}(s,\tau)$ coincides with $\mathbf{\Phi}(t,\tau)$ at $t=s$; by uniqueness of solution to the Cauchy problem, they coincide for all t. To complete the proof it remains to interchange s and τ. □

Property 2.7.3 *The equation*

$$\frac{\partial}{\partial s}\mathbf{\Phi}(t,s) = -\mathbf{\Phi}(t,s)\cdot\mathbf{A}(s)$$

holds.

Proof. It is easily verified that the derivative of the inverse to a differentiable tensor function $\mathbf{\Psi}(t)$ is given by

$$\left(\mathbf{\Psi}^{-1}(t)\right)' = -\mathbf{\Psi}^{-1}(t)\cdot\mathbf{\Psi}'(t)\cdot\mathbf{\Psi}^{-1}(t). \tag{2.7.21}$$

Hence by (2.7.20) we have

$$\begin{aligned}\frac{\partial\mathbf{\Phi}(t,s)}{\partial s} &= \frac{\partial(\mathbf{\Phi}^{-1}(s,t))}{\partial s}\\ &= -\mathbf{\Phi}^{-1}(s,t)\cdot\frac{\partial\mathbf{\Phi}(s,t)}{\partial s}\cdot\mathbf{\Phi}^{-1}(s,t)\\ &= -\mathbf{\Phi}(t,s)\cdot\frac{\partial\mathbf{\Phi}(s,t)}{\partial s}\cdot\mathbf{\Phi}(t,s).\end{aligned}$$

Finally, since s is the first argument in the derivative on the right we can change this derivative using (2.7.17):

$$\frac{\partial\mathbf{\Phi}(t,s)}{\partial s} = -\mathbf{\Phi}(t,s)\cdot\mathbf{A}(s)\cdot\mathbf{\Phi}(s,t)\cdot\mathbf{\Phi}(t,s) = -\mathbf{\Phi}(t,s)\cdot\mathbf{A}(s).$$

□

Property 2.7.4 *The solution of the Cauchy problem*

$$\mathbf{y}'(t) = \mathbf{A}(t)\cdot\mathbf{y}(t) + \mathbf{g}(t), \qquad \mathbf{y}(0) = 0,$$

with a given vector function $\mathbf{g}(t)$ *is*

$$\mathbf{y}(t) = \int_0^t \mathbf{\Phi}(t,s)\cdot\mathbf{g}(s)\,ds. \tag{2.7.22}$$

Proof. Let us find the derivative of $\mathbf{y}(t)$ given by (2.7.22):

$$\begin{aligned}\frac{d}{dt}\mathbf{y}(t) &= \frac{d}{dt}\int_0^t \mathbf{\Phi}(t,s)\cdot\mathbf{g}(s)\,ds \\ &= \mathbf{\Phi}(t,t)\cdot\mathbf{g}(t) + \int_0^t \frac{d}{dt}\mathbf{\Phi}(t,s)\cdot\mathbf{g}(s)\,ds \\ &= \mathbf{E}\cdot\mathbf{g}(t) + \int_0^t \mathbf{A}(t)\cdot\mathbf{\Phi}(t,s)\cdot\mathbf{g}(s)\,ds \\ &= \mathbf{A}(t)\cdot\int_0^t \mathbf{\Phi}(t,s)\cdot\mathbf{g}(s)\,ds + \mathbf{g}(t) \\ &= \mathbf{A}(t)\cdot\mathbf{y}(t) + \mathbf{g}(t).\end{aligned}$$

□

2.8 General Terminal Control Problem

We have stated the general problem of terminal control. Our understanding of the scope of the optimal control problem has changed, however, so it is appropriate to reexamine the setup of the terminal control problem.

The object of terminal optimal control is described by a vector function of time $\mathbf{y}(t)$ with values in Euclidean vector space E^n whose behavior is determined by a system of ODEs (or a vector ODE)

$$\mathbf{y}'(t) = \mathbf{f}(\mathbf{y}(t), \mathbf{u}(t)) \tag{2.8.1}$$

The vector function $\mathbf{f}(\mathbf{y}(t), \mathbf{u}(t))$ must be such that when the control function $\mathbf{u}(t)$ is given and admissible (i.e., belongs to the class $\mathcal{U}$), then the Cauchy problem for (2.8.1) supplemented with initial conditions has a unique continuous solution on a finite time interval $[0, T]$. Thus the history of the object determines uniquely its present state. Systems of this type are called *dynamical systems.*

The set $\mathcal{U}$ of admissible controls consists of vector functions $\mathbf{u}(t)$ taking values in the Euclidean space E^m that are piecewise continuous in t. In particular, $\mathcal{U}$ can consist of functions that take values in a finite set of vectorial values. The former is important when the control function describes several fixed positions that are taken by some governing device; it describes, say, the effect of some additional device that can exist only in "on–off" states.

Everything said so far in this section applies to all optimal control problems. The distinguishing feature of terminal control is the specification of

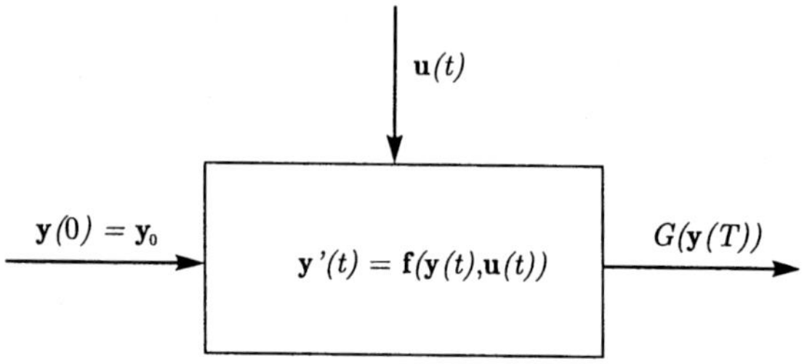

Fig. 2.3 A controlled object described by $\mathbf{y}' = f(\mathbf{y}, \mathbf{u})$: the input is $\mathbf{y}(0) = \mathbf{y}_0$, the control vector is $\mathbf{u}$, and the output is $G(\mathbf{y}(T))$.

the initial condition

$$\mathbf{y}(0) = \mathbf{y}_0 \tag{2.8.2}$$

and the form of the objective functional

$$J(\mathbf{u}) = G(\mathbf{y}(T)). \tag{2.8.3}$$

Thus we can consider terminal control as the problem of finding the minimal output value (2.8.3) when the input is determined by the initial vector $\mathbf{y}_0$ and the control function $\mathbf{u}(t)$ and the output is $G(\mathbf{y}(T))$. See Fig. 2.3. Our objective can be formulated as

$$G(\mathbf{y}(T)) \to \min_{\mathbf{u}(t)\in\mathcal{U}} . \tag{2.8.4}$$

This is known as the main setup of the problem (2.8.1)–(2.8.4).

We can reduce various other other optimal control problems to this form.

Problem. For a system described by (2.8.1) whose initial state is given by (2.8.2), among all the admissible control vectors $\mathbf{u} \in \mathcal{U}$ find such for which an objective functional

$$\int_0^T g(\mathbf{y}(t), \mathbf{u}(t))\, dt$$

takes its minimum value.

The reduction of this problem to the main form of the terminal control problem is done by introducing the additional component y_{n+1} for $\mathbf{y}$.

Namely, we introduce an additional scalar equation

$$y'_{n+1}(t) = g(\mathbf{y}(t), \mathbf{u}(t)), \qquad y'_{n+1}(0) = 0.$$

Now it is clear that

$$y_{n+1}(T) = \int_0^T g(\mathbf{y}(t), \mathbf{u}(t))\, dt \tag{2.8.5}$$

and thus the objective functional from (2.8.3) takes the form

$$J(\mathbf{u}) = y_{n+1}(T).$$

We can consider another version of the terminal control problem when it is necessary to minimize the objective functional

$$\int_0^T g(\mathbf{y}(t), \mathbf{u}(t))\, dt + G(\mathbf{y}(T))$$

for the same system described by (2.8.1)–(2.8.2). Then the same additional component for $\mathbf{y}$ given by (2.8.5) reduces the problem to the necessary form. The objective functional now is

$$J(\mathbf{u}) = y_{n+1}(T) + G(\mathbf{y}(T)).$$

Let us consider the main form of the terminal control problem (2.8.1)–(2.8.4) using an extension of the procedure for the simplest problem of optimal control. Much of the reasoning for the latter is simply reformulated to go from the scalar to the vector version. For the simplest problem, the main step involved finding the main part of the increment of $J(u)$ under a needle-shaped increment of a fixed control function. We shall do this here also. The next step involved establishing the condition under which a control function would be optimal for the problem. This led to Pontryagin's maximum principle. We shall extend this to the general problem. Finally we shall discuss how to use the formula for the increment of the functional, as well as the maximum principle, to find an optimal solution.

Let $t = s$ be a point of continuity of a control function $\mathbf{u}(t)$. Giving $\mathbf{u}(t)$ a needle-shaped increment (i.e., a vector whose components are all needle-shaped functions with perturbations in $(s - \varepsilon, s]$) we get a new control defined by

$$\mathbf{u}^*(t) = \begin{cases} \mathbf{u}(t), & t \notin (s - \varepsilon, s], \\ \mathbf{v}, & t \in (s - \varepsilon, s]. \end{cases}$$

We can continue to refer to Fig. 2.1. We can also refer to Fig. 2.3 for a representation of the function $\mathbf{y}^*(t)$ that satisfies the equation

$$(\mathbf{y}^*(t))' = \mathbf{f}(\mathbf{y}^*(t), \mathbf{u}^*(t)) \tag{2.8.6}$$

and the same initial condition $\mathbf{y}(0) = \mathbf{y}_0$. We suppose that at least for all positive ε less than some small fixed number ε_0, the incremented control function $\mathbf{u}^*(t)$ is admissible.

The main part of the increment $J(\mathbf{u}^*) - J(\mathbf{u})$, linear in small ε, is determined by

Theorem 2.8.1 *Let $t = s$ be a point of continuity of a control function $\mathbf{u}(t)$. The increment of $J(\mathbf{u})$ is*

$$J(\mathbf{u}^*) - J(\mathbf{u}) = \varepsilon\, \delta_{s,\mathbf{v}} J(\mathbf{u}) + o(\varepsilon) \tag{2.8.7}$$

where

$$\delta_{s,\mathbf{v}} J(\mathbf{u}) = \mathbf{\Psi}(s) \cdot [\mathbf{f}(\mathbf{y}(s), \mathbf{u}(s)) - \mathbf{f}(\mathbf{y}(s), \mathbf{v})] \tag{2.8.8}$$

and $\mathbf{\Psi}(s)$ is a solution of the following Cauchy problem (in the reverse time):

$$\mathbf{\Psi}'(s) = -\nabla_{\mathbf{y}} \mathbf{f}(\mathbf{y}(s), \mathbf{u}(s)) \cdot \mathbf{\Psi}(s), \qquad \mathbf{\Psi}(T) = -\nabla_{\mathbf{y}} G(\mathbf{y}(T)). \tag{2.8.9}$$

$\delta_{s,\mathbf{v}} J(\mathbf{u})$ is called the variational derivative of the second kind of the functional $J(\mathbf{u})$.

Proof. Take $\varepsilon > 0$ so small that all points of $[s - \varepsilon, s]$ are points of continuity of $\mathbf{u}(t)$ and the corresponding incremented control functions $\mathbf{u}^*(t)$ are admissible. We divide the proof into several steps.

Step 1, the main part of the increment of $\mathbf{y}(t)$. On $[0, s - \varepsilon]$ the control functions coincide. The initial conditions for $\mathbf{y}(t)$ and $\mathbf{y}^*(t)$ coincide as well, so on this segment we have $\mathbf{y}^*(t) = \mathbf{y}(t)$.

Let us find the increment of $\mathbf{y}(t)$ for $t \in [s - \varepsilon, s]$. Subtracting term by term (2.8.1) from (2.8.6) we have

$$(\mathbf{y}^*(t))' - \mathbf{y}'(t) = \mathbf{f}(\mathbf{y}^*(t), \mathbf{v}) - \mathbf{f}(\mathbf{y}(t), \mathbf{u}(t)).$$

Denoting $\Delta\mathbf{y}(t) = \mathbf{y}^*(t) - \mathbf{y}(t)$ we get

$$\Delta\mathbf{y}'(t) = \mathbf{f}(\mathbf{y}(t) + \Delta\mathbf{y}(t), \mathbf{v}) - \mathbf{f}(\mathbf{y}(t), \mathbf{u}(t)). \tag{2.8.10}$$

This equation, which holds on $(s-\varepsilon, s]$, is supplemented by the "initial" condition

$$\Delta \mathbf{y}(s-\varepsilon) = 0 \tag{2.8.11}$$

which follows from the above coincidence of $\mathbf{y}(t)$ and $\mathbf{y}^*(t)$. Let us reduce the Cauchy problem (2.8.10)–(2.8.11) for $\Delta\mathbf{y}(t)$, integrating (2.8.10) with respect to the time parameter:

$$\Delta \mathbf{y}(t) - \Delta \mathbf{y}(s-\varepsilon) = \int_{s-\varepsilon}^{t} [\mathbf{f}(\mathbf{y}(\tau) + \Delta\mathbf{y}(\tau), \mathbf{v}) - \mathbf{f}(\mathbf{y}(\tau), \mathbf{u}(\tau))]\, d\tau.$$

By (2.8.11) this reduces to

$$\Delta \mathbf{y}(t) = \int_{s-\varepsilon}^{t} [\mathbf{f}(\mathbf{y}(\tau) + \Delta\mathbf{y}(\tau), \mathbf{v}) - \mathbf{f}(\mathbf{y}(\tau), \mathbf{u}(\tau))]\, d\tau. \tag{2.8.12}$$

Since we assume $\mathbf{f}(\mathbf{y}, \mathbf{u})$ to be continuous and thus bounded, the integral on the right of (2.8.12) is of order ε and so is $\Delta\mathbf{y}(t)$. Thus replacing in the integrand the quantities $\mathbf{y}(\tau)$ and $\mathbf{u}(\tau)$ by $\mathbf{y}(s)$ and $\mathbf{u}(s)$ respectively, and placing $\Delta\mathbf{y}(\tau) = 0$, we introduce in the value of the integral an error of order $o(\varepsilon)$ for $t \in [s-\varepsilon, s]$. Hence (2.8.12) reduces to

$$\Delta \mathbf{y}(t) = \int_{s-\varepsilon}^{t} [\mathbf{f}(\mathbf{y}(s), \mathbf{v}) - \mathbf{f}(\mathbf{y}(s), \mathbf{u}(s))]\, d\tau + o(\varepsilon),$$

which can be rewritten as

$$\Delta \mathbf{y}(t) = (t - s + \varepsilon)[\mathbf{f}(\mathbf{y}(s), \mathbf{v}) - \mathbf{f}(\mathbf{y}(s), \mathbf{u}(s))] + o(\varepsilon),$$

and thus on this small segment $[s-\varepsilon, s]$ the difference $\Delta\mathbf{y}(t)$ changes almost linearly from zero, taking at $t = s$ the value

$$\Delta \mathbf{y}(s) = \varepsilon\, [\mathbf{f}(\mathbf{y}(s), \mathbf{v}) - \mathbf{f}(\mathbf{y}(s), \mathbf{u}(s))] + o(\varepsilon). \tag{2.8.13}$$

This is the initial value for the solution $\Delta\mathbf{y}(t)$ on $[s, T]$ of the equation

$$\Delta \mathbf{y}'(t) = \mathbf{f}(\mathbf{y}(t) + \Delta\mathbf{y}(t), \mathbf{u}(t)) - \mathbf{f}(\mathbf{y}(t), \mathbf{u}(t)) \tag{2.8.14}$$

(we recall that on this interval $\mathbf{u}^*(t) = \mathbf{u}(t)$ and it is considered to be known at this moment). Linearizing the right-hand side of (2.8.14) with respect to $\Delta\mathbf{y}(t)$ (taking into account (2.7.15)) we have

$$\Delta \mathbf{y}'(t) = (\nabla_{\mathbf{y}} \mathbf{f}(\mathbf{y}(t), \mathbf{u}(t)))^T \cdot \Delta\mathbf{y}(t) + o(\|\Delta\mathbf{y}(t)\|). \tag{2.8.15}$$

Because of smallness of the initial condition of $\Delta\mathbf{y}(t)$ at $t = s$ and the form of (2.8.15) we expect the solution of the corresponding Cauchy problem on the finite interval $(s, T]$ to be of order ε and, up to terms of order higher than ε, equal to the solution of the following Cauchy problem:

$$\delta\mathbf{y}'(t) = (\nabla_{\mathbf{y}}\mathbf{f}(\mathbf{y}(t), \mathbf{u}(t)))^T \cdot \delta\mathbf{y}(t), \tag{2.8.16}$$

$$\delta\mathbf{y}(s) = \varepsilon\,[\mathbf{f}(\mathbf{y}(s), \mathbf{v}) - \mathbf{f}(\mathbf{y}(s), \mathbf{u}(s))], \tag{2.8.17}$$

which is the linearization of the complete initial problem (2.8.14), (2.8.13). By the linearity of this problem its solution is proportional to ε.

To find the main part of the increment $\Delta\mathbf{y}(T)$ it remains to solve the Cauchy problem (2.8.16)–(2.8.17). This can be integrated (often numerically) but we will use the notion of the fundamental solution from the previous section.

Let us denote $\mathbf{A}(t) = (\nabla_{\mathbf{y}}\mathbf{f}(\mathbf{y}(t), \mathbf{u}(t)))^T$ and leave the notation of § 2.7 for this fundamental solution, which satisfies

$$\frac{d}{dt}\mathbf{\Phi}(t, s) = \mathbf{A}(t) \cdot \mathbf{\Phi}(t, s)$$

and the "initial" condition $\mathbf{\Phi}(s, s) = \mathbf{E}$ for all s. By Property 2.7.1 of § 2.7 the solution to (2.8.16)–(2.8.17) is

$$\delta\mathbf{y}(t) = \varepsilon\,\mathbf{\Phi}(t, s) \cdot [\mathbf{f}(\mathbf{y}(s), \mathbf{v}) - \mathbf{f}(\mathbf{y}(s), \mathbf{u}(s))]$$

and thus, assuming "good" behavior of $\Delta\mathbf{y}(t)$ we have

$$\Delta\mathbf{y}(T) = \varepsilon\,\mathbf{\Phi}(T, s) \cdot [\mathbf{f}(\mathbf{y}(s), \mathbf{v}) - \mathbf{f}(\mathbf{y}(s), \mathbf{u}(s))] + o(\varepsilon). \tag{2.8.18}$$

Step 2, the main part of the increment of $J(\mathbf{u}) = G(\mathbf{y}(T))$. We again use the formula of the differential (2.7.12) for linearization of the increment of $J(\mathbf{u})$ with respect to $\Delta\mathbf{y}(t)$:

$$\begin{aligned}\Delta J(\mathbf{u}) &= J(\mathbf{u}^*) - J(\mathbf{u}) \\ &= G(\mathbf{y}(T) + \Delta\mathbf{y}(T)) - G(\mathbf{y}(T)) \\ &= \nabla_{\mathbf{y}}G(\mathbf{y})\big|_{\mathbf{y}=\mathbf{y}(T)} \cdot \Delta\mathbf{y}(T) + o(|\Delta\mathbf{y}(T)|).\end{aligned}$$

Using (2.8.18) we get

$$\Delta J(\mathbf{u}) = \varepsilon\,\nabla_{\mathbf{y}}G(\mathbf{y})\big|_{\mathbf{y}=\mathbf{y}(T)} \cdot \mathbf{\Phi}(T, s) \cdot [\mathbf{f}(\mathbf{y}(s), \mathbf{v}) - \mathbf{f}(\mathbf{y}(s), \mathbf{u}(s))] + o(\varepsilon).$$

This is the required formula. It remains to represent it in the form asserted by the theorem.

Step 3, the final step. Let us introduce a vector function $\mathbf{\Psi}(s)$ as

$$\mathbf{\Psi}(s) = -\nabla_{\mathbf{y}} G(\mathbf{y})\big|_{y=y(T)} \cdot \mathbf{\Phi}(T,s).$$

With this notation for $\Delta J(\mathbf{u})$ we do have the representation (2.8.7)–(2.8.8), so it remains to demonstrate that $\mathbf{\Psi}(s)$ satisfies (2.8.9). The second relation of (2.8.9) is a consequence of the equality $\mathbf{\Phi}(T,T) = \mathbf{E}$; indeed,

$$\mathbf{\Psi}(T) = -\nabla_{\mathbf{y}} G(\mathbf{y})\big|_{y=y(T)} \cdot \mathbf{\Phi}(T,T) = -\nabla_{\mathbf{y}} G(\mathbf{y})\big|_{y=y(T)}.$$

Let us show that it satisfies the first equation of (2.8.9) as well. The derivative of $\mathbf{\Psi}(s)$ is

$$\frac{d\mathbf{\Psi}(s)}{ds} = \frac{d}{ds}\left[-\nabla_{\mathbf{y}} G(\mathbf{y})\big|_{y=y(T)} \cdot \mathbf{\Phi}(T,s)\right] = -\nabla_{\mathbf{y}} G(\mathbf{y})\big|_{y=y(T)} \cdot \frac{d}{ds}\mathbf{\Phi}(T,s).$$

Let us now use the equation for the derivative with respect to the second argument of the fundamental solution, which is given by Property 2.7.3:

$$\begin{aligned}\frac{d\mathbf{\Psi}(s)}{ds} &= -\nabla_{\mathbf{y}} G(\mathbf{y})\big|_{y=y(T)} \cdot (-\mathbf{\Phi}(T,s)\cdot \mathbf{A}(s)) \\ &= -\left(-\nabla_{\mathbf{y}} G(\mathbf{y})\big|_{y=y(T)} \cdot \mathbf{\Phi}(T,s)\right)\cdot \mathbf{A}(s) \\ &= -\mathbf{\Psi}(s)\cdot \mathbf{A}(s) = -(\mathbf{A}(s))^T \cdot \mathbf{\Psi}(s).\end{aligned}$$

Remembering the above notation for $\mathbf{A}(s)$ we complete the proof. □

2.9 Pontryagin's Maximum Principle for the Terminal Optimal Problem

First we would like to discuss the statement of Theorem 2.8.1. When we seek a response of an object described by the problem

$$\mathbf{y}'(t) = \mathbf{f}(\mathbf{y}(t), \mathbf{u}(t)), \qquad \mathbf{y}(0) = \mathbf{y}_0, \tag{2.9.1}$$

to a needle-shaped disturbance of the control function $\mathbf{u}(t)$ we obtain a dual problem

$$\mathbf{\Psi}'(s) = -\nabla_{\mathbf{y}} \mathbf{f}(\mathbf{y}(s), \mathbf{u}(s)) \cdot \mathbf{\Psi}(s), \tag{2.9.2}$$

$$\mathbf{\Psi}(T) = -\nabla_{\mathbf{y}} G(\mathbf{y}(T)). \tag{2.9.3}$$

The dual equation (2.9.2) plays a role like that of the Euler equation of the calculus of variations, and the condition (2.9.3) is the condition of transversality. Together (2.9.1)–(2.9.3) compose a boundary value problem having a unique solution when $\mathbf{u}(t)$ is given. This splits into two "initial value problems" for $\mathbf{y}(t)$ and $\mathbf{\Psi}(s)$. For problems other than the problem of terminal control, other types of boundary conditions are given but the equations yielding a response to a needle-shaped disturbance are the same. Let us introduce an equivalent form of the equations for this boundary value problem. We introduce a scalar function in three variables $\mathbf{y}$, $\mathbf{\Psi}$, and $\mathbf{u}(t)$, called Pontryagin's function

$$H(\mathbf{y}, \mathbf{\Psi}, \mathbf{u}) = \mathbf{f}(\mathbf{y}, \mathbf{u}) \cdot \mathbf{\Psi}.$$

Simple calculation demonstrates that

$$\nabla_{\mathbf{y}} H(\mathbf{y}, \mathbf{\Psi}, \mathbf{u}) = \nabla_{\mathbf{y}} \mathbf{f}(\mathbf{y}, \mathbf{u}) \cdot \mathbf{\Psi} \quad \text{and} \quad \nabla_{\mathbf{\Psi}} H(\mathbf{y}, \mathbf{\Psi}, \mathbf{u}) = \mathbf{f}(\mathbf{y}, \mathbf{u})$$

where the second relation is a consequence of the equality

$$\nabla_{\mathbf{x}} \mathbf{x} = \mathbf{e}_i \frac{\partial}{\partial x_i}(x_j \mathbf{e}_j) = \mathbf{e}_i \mathbf{e}_i = \mathbf{E}.$$

It follows that (2.9.1) and (2.9.3) can be written as

$$\mathbf{y}'(t) = \nabla_{\mathbf{\Psi}} H(\mathbf{y}(t), \mathbf{\Psi}(t), \mathbf{u}(t)) \quad \text{and} \quad \mathbf{\Psi}'(t) = -\nabla_{\mathbf{y}} H(\mathbf{y}(t), \mathbf{\Psi}(t), \mathbf{u}(t)).$$

This is the Hamiltonian form.

In terms of Pontryagin's function the second kind derivative of $J(\mathbf{u})$ (2.8.8) can be written as

$$\delta_{s,\mathbf{v}} J(\mathbf{u}) = H(\mathbf{y}(s), \mathbf{\Psi}(s), \mathbf{u}(s)) - H(\mathbf{y}(s), \mathbf{\Psi}(s), \mathbf{v}). \tag{2.9.4}$$

Now we can formulate the Pontryagin's principle of maximum.

Theorem 2.9.1 *Let $\mathbf{u}(t)$ be an optimal control function at which $J(\mathbf{u})$ attains its minimal value on $\mathcal{U}$, the set of all admissible control functions and $\mathbf{y}(t)$ and $\mathbf{\Psi}(t)$ be a solution of the boundary value problem (2.9.1)–(2.9.3). At any point $t = s$ of continuity of $\mathbf{u}(t)$ the Pontryagin function $H(\mathbf{y}(t), \mathbf{\Psi}(t), \mathbf{v})$ considered as a function of the third argument $\mathbf{v}$ takes its maximum value at $\mathbf{v} = \mathbf{u}(s)$.*

Proof. Since $J(\mathbf{u})$ attains its minimum at $\mathbf{u}(t)$ then for any admissible control function $\mathbf{u}^*(t)$ we have

$$J(\mathbf{u}^*) - J(\mathbf{u}) \geq 0.$$

In particular it is valid for an admissible $\mathbf{u}^*(t)$ that is a disturbance of $\mathbf{u}(t)$ by a needle-shaped vector function

$$\mathbf{u}^*(t) = \begin{cases} \mathbf{u}(t), & t \notin (s-\varepsilon, s], \\ \mathbf{v}, & t \in (s-\varepsilon, s], \end{cases}$$

and thus, for sufficiently small ε because of (2.8.7) and (2.9.4) we have

$$J(\mathbf{u}^*) - J(\mathbf{u}) = \varepsilon\,(H(\mathbf{y}(s), \mathbf{\Psi}(s), \mathbf{u}(s)) - H(\mathbf{y}(s), \mathbf{\Psi}(s), \mathbf{v})) + o(\varepsilon) \geq 0.$$

From this it follows that $H(\mathbf{y}(s), \mathbf{\Psi}(s), \mathbf{u}(s)) - H(\mathbf{y}(s), \mathbf{\Psi}(s), \mathbf{v}) \geq 0$. □

Pontryagin's principle of maximum gives us an effective tool to check whether $\mathbf{u}(t)$ is a needed control function at which $J(\mathbf{u})$ attains its minimum, but it does not show, except for quite simple problems, how to find this. However, (2.8.7) is the background of various numerical methods used to find this minimum. We shall discuss them in brief.

The formula (2.8.7) for the increment of $J(\mathbf{u})$, which can be rewritten as

$$J(\mathbf{u}) \approx J(\mathbf{u}^*) - \varepsilon\,\delta_{s,\mathbf{v}} J(\mathbf{u}), \tag{2.9.5}$$

generates an iterative procedure that begins with selection of a finite number of the time instants $(\tau_1, \ldots, \tau_r)$ at which one may introduce needle-shaped disturbances for finding a more effective control function. Next one must find an instant τ_i and a corresponding admissible value of $\mathbf{v}$, which we denote by $\mathbf{v}_i$, at which the maximum of the numerical set

$$\{\delta_{\tau_1,\mathbf{v}} J(\mathbf{u}), \ldots, \delta_{\tau_r,\mathbf{v}} J(\mathbf{u})\}$$

is attained. Denoting the control parameters of the previous step as $\mathbf{u}^{(i)}(t)$ and $\mathbf{u}^{(i)*}(t)$ where $\mathbf{u}^{(i)*}(t)$ is just determined, one must choose the value of ε, denoted by ε_i, at which (2.9.5) provides a sufficiently precise approximation. Then the next approximation of the value of $J(\mathbf{u})$ is given by the formula

$$J(\mathbf{u}^{(i+1)}) = J(\mathbf{u}^{(i)*}) - \varepsilon_i\,\delta_{\tau_i,\mathbf{v}_i} J(\mathbf{u}^{(i)}).$$

Versions of this procedure differ in their methods of determining each step, in particular the points τ_i. They are called the methods of coordinate-by-coordinate descent.

A modification is called the group descent procedure. We have found the main linear part of the increment of $J(\mathbf{u})$ under a needle-shaped disturbance of $\mathbf{u}(t)$ at $t = s$, which is characterized by the pair of parameters

$\varepsilon, \mathbf{v}$. This means that if $\mathbf{u}(t)$ is disturbed by a finite set of N such needle-shaped variations, the ith of which is lumped at a point s_i of continuity of $\mathbf{u}(t)$ and is characterized by the pair $\varepsilon_i, \mathbf{v}_i$, then denoting by $\mathbf{u}^{**}(t)$ the corresponding control function we get the main part of the increment as the sum of increments of $J(\mathbf{u})$ due to each of the needle-shaped increments of $\mathbf{u}(t)$:

$$J(\mathbf{u}^{**}) - J(\mathbf{u}) = \sum_{i=1}^{N} \varepsilon_i [H(\mathbf{y}(s), \mathbf{\Psi}(s), \mathbf{u}(s)) - H(\mathbf{y}(s), \mathbf{\Psi}(s), \mathbf{v}_i)] + o\left(\max(\varepsilon_1, \ldots, \varepsilon_N)\right). \tag{2.9.6}$$

Then we can decrease the value of $J(\mathbf{u})$ on the next step of approximation using a group of needle-shaped increments and the formula (2.9.6).

2.10 Generalization of the Terminal Control Problem

Let us consider a generalized terminal control problem whose setup coincides with that of the usual problem except for the form of the objective function (functional). This set up is

Definition 2.10.1 From among the piecewise continuous control functions $\mathbf{u}(t) \in \mathcal{U}$ on $[0, T]$, find one that minimizes the functional $\mathcal{I}(\mathbf{u})$,

$$\mathcal{I}(\mathbf{u}) \to \min_{\mathbf{u} \in \mathcal{U}},$$

when $\mathcal{I}(\mathbf{u})$ is defined as

$$\mathcal{I}(\mathbf{u}) = G(\mathbf{y}(s_1), \mathbf{y}(s_2), \ldots, \mathbf{y}(s_N)),$$

$G(\mathbf{y}(s_1), \mathbf{y}(s_2), \ldots, \mathbf{y}(s_N))$ being a function continuously differentiable in all its variables, $0 < s_1 < s_2 < \cdots < s_N = T$ some fixed points of time, and $\mathbf{y}(t)$ satisfying the equations

$$\mathbf{y}'(t) = \mathbf{f}(\mathbf{y}(t), \mathbf{u}(t)), \qquad \mathbf{y}(0) = \mathbf{y}_0.$$

Such a form of the objective function can appear, for example, if the objective functional contains an integral depending on $\mathbf{y}(t)$ which is discretized according to some simple method such as Simpson's rule or the rectangular rule. To proceed further we need some additional material. We shall obtain a nonstandard Cauchy problem and then find a way to present it in a form that resembles the usual form for such a problem. For this we digress briefly to discuss the Dirac δ-function.

The δ-function concept was originated by physicists and used for many years before being given a rigorous footing (called the *theory of distributions*) by mathematicians. Although rigor has certain advantages, the heuristic viewpoint of the early physicists will be adequate for our purposes. This viewpoint rests on the notion that $\delta(t)$ is a function of the argument t, taking the value zero for $t \neq 0$ and an infinite "value" at $t = 0$ such that

$$\int_{-\infty}^{+\infty} \delta(t)\, dt = 1.$$

Now from a mathematical viewpoint we are in trouble already because it can be shown that there is no such function. But we nonetheless proceed formally with the understanding that every step we take can be justified rigorously (with tremendous effort and with full chapters of extra explanation which, unfortunately, would not lend clarity to the topic for our purposes).

The δ-function is a generalized derivative of the step function $h(t)$ given by

$$h(t) = \begin{cases} 1, & t \geq 0, \\ 0, & t < 0, \end{cases}$$

and we shall exploit this property. The introduction of the generalized derivative uses the main lemma of the calculus of variations and the formula for integration by parts. Let $\varphi(t)$ be a function infinitely differentiable on $(-\infty, +\infty)$ and with compact support (the support of $\phi(t)$ is the closure of the set of all t for which $\varphi(t) \neq 0$). Let us denote this class by $\mathcal{D}$. For any differentiable function $f(t)$ the formula for integration by parts holds:

$$\int_{-\infty}^{+\infty} f(t)\varphi'(t)\, dt = -\int_{-\infty}^{+\infty} f'(t)\varphi(t)\, dt.$$

The main lemma of the calculus of variations states that if the equality

$$\int_{-\infty}^{+\infty} f(t)\varphi'(t)\, dt = -\int_{-\infty}^{+\infty} g(t)\varphi(t)\, dt \qquad (2.10.1)$$

holds for any $\varphi(t) \in \mathcal{D}$ then $g(t) = f'(t)$. This is valid for a differentiable function $f(t)$, but the same equation introduces the generalized derivative of an integrable function $f(t)$: a function $g(t)$ is called the generalized derivative of $f(t)$ if (2.10.1) holds for any $\varphi(t) \in \mathcal{D}$. The generalized derivative is denoted by the usual differentiation symbols. The main lemma of the calculus of variations (more precisely, its variant) provides uniqueness of definition of the generalized derivative. Let us check that $h'(t) = \delta(t)$ in the generalized sense. Indeed,

$$\int_{-\infty}^{+\infty} h(t)\varphi'(t)\,dt = \int_0^{\infty} h(t)\varphi'(t)\,dt = \int_0^{\infty} \varphi'(t)\,dt = -\varphi(0)$$

and by the definition of δ-function

$$\int_{-\infty}^{+\infty} \delta(t)\varphi(t)\,dt = \varphi(0).$$

Thus for the pair $h(t), \delta(t)$ the definition of generalized derivative is valid and so $h'(t) = \delta(t)$. Using this property we can write out the Cauchy problem

$$y'(t) = g(t, y(t)), \qquad y(0) = y_0, \tag{2.10.2}$$

in an equivalent form

$$y'(t) = f(t, y(t)) + y_0\delta(t), \qquad y(t)\big|_{t\to -0} = 0. \tag{2.10.3}$$

Indeed, integration of (2.10.3) with respect to t (the starting point is $t = -0$) implies the equation

$$y(t) = \int_0^t f(s, y(s))\,ds + y_0 h(t),$$

which is equivalent to (2.10.2).

Now let us formulate the main theorem of this section, in which we keep the notation of § 2.8 for $\mathbf{u}^*(t)$ and $\mathbf{y}^*(t)$.

Theorem 2.10.1 *Let $t = s$ be a point of continuity of a control function $\mathbf{u}(t)$ that is different from $s_1, s_2, \ldots, s_N = T$. The increment of $\mathcal{I}(\mathbf{u})$ is*

$$\mathcal{I}(\mathbf{u}^*) - \mathcal{I}(\mathbf{u}) = \varepsilon\, \delta_{s,\mathbf{v}}\mathcal{I}(\mathbf{u}) + o(\varepsilon)$$

where

$$\delta_{s,\mathbf{v}}\mathcal{I}(\mathbf{u}) = \boldsymbol{\Psi}(s) \cdot [\mathbf{f}(\mathbf{y}(s), \mathbf{u}(s)) - \mathbf{f}(\mathbf{y}(s), \mathbf{v})] \tag{2.10.4}$$

and $\mathbf{\Psi}(s)$ is a solution of the following Cauchy problem (in the reverse time)

$$\mathbf{\Psi}'(s) = -\nabla_{\mathbf{y}}\mathbf{f}(\mathbf{y}(s), \mathbf{u}(s)) \cdot \mathbf{\Psi}(s) + \sum_{i=1}^{N} \delta(s_i - s)\nabla_{\mathbf{y}(s_i)} G(\mathbf{y}(s_1), \mathbf{y}(s_2), \ldots, \mathbf{y}(s_N)),$$
$$\mathbf{y}(T+0) = 0. \tag{2.10.5}$$

Comparison with Theorem 2.8.1 shows that the current theorem differs only in the form of the problem for $\mathbf{\Psi}(s)$.

Proof. It is clear that $\mathbf{y}^*(t)$ for this problem coincides with that of § 2.8, so we can use the corresponding formulas of that section. In particular, for $t > s$ the main part of the increment $\Delta\mathbf{y}(t)$ of the corresponding solution $\mathbf{y}(t)$ on $(s, T]$, under the needle-shaped increment of the control vector $\mathbf{u}$, is

$$\delta\mathbf{y}(t) = \varepsilon\, \mathbf{\Phi}(t, s) \cdot [\mathbf{f}(\mathbf{y}(s), \mathbf{v}) - \mathbf{f}(\mathbf{y}(s), \mathbf{u}(s))]. \tag{2.10.6}$$

So we immediately go to the increment of the goal function. First we use the formula for the complete differential to get the main part of the increment of $\mathcal{I}(\mathbf{u}) = G(\mathbf{y}(s_1), \mathbf{y}(s_2), \ldots, \mathbf{y}(s_N))$, which is

$$\begin{aligned}\Delta\mathcal{I}(\mathbf{u}) &= \mathcal{I}(\mathbf{u}^*) - \mathcal{I}(\mathbf{u}) \\ &= G(\mathbf{y}(s_1) + \Delta\mathbf{y}(s_1), \mathbf{y}(s_2) + \Delta\mathbf{y}(s_2), \ldots, \mathbf{y}(s_N) + \Delta\mathbf{y}(s_N)) \\ &\quad - G(\mathbf{y}(s_1), \mathbf{y}(s_2), \ldots, \mathbf{y}(s_N)) \\ &= \sum_{i=1}^{N} \nabla_{\mathbf{y}(s_i)} G(\mathbf{y}(s_1), \mathbf{y}(s_2), \ldots, \mathbf{y}(s_N)) \cdot \Delta\mathbf{y}(s_i) \\ &\quad + o\left(\max_j \|\Delta\mathbf{y}(s_j)\|\right)\end{aligned} \tag{2.10.7}$$

To implement (2.10.6) we rewrite it in the form

$$\delta\mathbf{y}(t) = \varepsilon\, \mathbf{\Phi}(t, s) \cdot [\mathbf{f}(\mathbf{y}(s), \mathbf{v}) - \mathbf{f}(\mathbf{y}(s), \mathbf{u}(s))]\, h(t - s)$$

so it becomes valid for use in (2.10.7) for all $t \in [0, T]$ when the interval $[s - \varepsilon, s]$ does not contain any s_i (assumed). Then the increment of $\mathcal{I}(\mathbf{u})$

can be rewritten as

$$\begin{aligned}\Delta\mathcal{I}(\mathbf{u}) &= \mathcal{I}(\mathbf{u}^*) - \mathcal{I}(\mathbf{u}) \\ &= \varepsilon\left\{\sum_{i=1}^{N}\nabla_{\mathbf{y}(s_i)}G(\mathbf{y}(s_1),\mathbf{y}(s_2),\ldots,\mathbf{y}(s_N))\cdot\mathbf{\Phi}(s_i,s)h(s_i-s)\right\} \\ &\quad\cdot[\mathbf{f}(\mathbf{y}(s),\mathbf{v})-\mathbf{f}(\mathbf{y}(s),\mathbf{u}(s))] + o(\varepsilon).\end{aligned}$$

Denoting

$$\mathbf{\Psi}(s) = -\sum_{i=1}^{N}\nabla_{\mathbf{y}(s_i)}G(\mathbf{y}(s_1),\mathbf{y}(s_2),\ldots,\mathbf{y}(s_N))\cdot\mathbf{\Phi}(s_i,s)h(s_i-s) \quad (2.10.8)$$

we get, as in § 2.8,

$$\delta_{s,\mathbf{v}}\mathcal{I}(\mathbf{u}) = \mathbf{\Psi}(s)\cdot[\mathbf{f}(\mathbf{y}(s),\mathbf{u}(s))-\mathbf{f}(\mathbf{y}(s),\mathbf{v})]$$

and for the increment of objective functional

$$\mathcal{I}(\mathbf{u}^*) - \mathcal{I}(\mathbf{u}) = \varepsilon\,\delta_{s,\mathbf{v}}\mathcal{I}(\mathbf{u}) + o(\varepsilon).$$

Note that the presence of $h(s_i - s)$ in the sum of the definition (2.10.8) means that at $s = s_i$ the value of $\mathbf{\Psi}(s)$ has some step change for an additional term in the sum.

It remains only to check the validity of (2.10.5). When $s > s_N = T$ we get $\mathbf{\Psi}(s) = 0$ so the second of (2.10.5) holds. To show that the first is valid let us find the derivative of $\mathbf{\Psi}(s)$. Taking into account Property 2.7.3 which in our terms is

$$\frac{d}{ds}\mathbf{\Phi}(s_i,s) = -\mathbf{\Phi}(s_i,s)\cdot(\nabla_{\mathbf{y}}\mathbf{f}(\mathbf{y}(s),\mathbf{u}(s)))^T$$

we get

$$\begin{aligned}\frac{d\mathbf{\Psi}(s)}{ds} &= \sum_{i=1}^{N}\nabla_{\mathbf{y}(s_i)}G(\mathbf{y}(s_1),\mathbf{y}(s_2),\ldots,\mathbf{y}(s_N)) \\ &\quad\cdot h(s_i-s)\mathbf{\Phi}(s_i,s)\cdot(\nabla_{\mathbf{y}}\mathbf{f}(\mathbf{y}(s),\mathbf{u}(s)))^T \\ &\quad+\sum_{i=1}^{N}\nabla_{\mathbf{y}(s_i)}G(\mathbf{y}(s_1),\mathbf{y}(s_2),\ldots,\mathbf{y}(s_N))\cdot\mathbf{\Phi}(s_i,s)\delta(s_i-s) \\ &= -\mathbf{\Psi}(s)\cdot(\nabla_{\mathbf{y}}\mathbf{f}(\mathbf{y}(s),\mathbf{u}(s)))^T \\ &\quad+\sum_{i=1}^{N}\nabla_{\mathbf{y}(s_i)}G(\mathbf{y}(s_1),\mathbf{y}(s_2),\ldots,\mathbf{y}(s_N))\delta(s_i-s).\end{aligned}$$

In the last transformation we used $\mathbf{\Phi}(s_i, s)\delta(s_i - s) = \mathbf{E}\delta(s_i - s)$. □

The form of Pontryagin's maximum principle for the generalized terminal control problem is the same as in the previous section. We leave its formulation to the reader.

This kind of generalized terminal control problem is used in practice and, as a rule, requires numerical solution of the problems when the formula for the increment (2.10.4) of the goal functional is used.

2.11 Small Variations of Control Function for Terminal Control Problem

The form of the increment of the objective functional for the generalized terminal control problem provides a hint that the conjugate equations and similar material should enter not only for needle-shaped variations of the control function, but for any small variations. We will see that this is really so, and for this case we will find the expression for the increment of the objective functional under the increment of control vector of other type. We reconsider the terminal control problem described by the dynamical system

$$\mathbf{y}'(t) = \mathbf{f}(\mathbf{y}(t), \mathbf{u}(t)), \qquad \mathbf{y}(0) = \mathbf{y}_0.$$

We wish to find the increment of the objective functional $J(\mathbf{u}) = G(\mathbf{y}(T))$ under a small increment $\Delta\mathbf{u}(t)$ of the control function $\mathbf{u}(t)$.

We demonstrated that one of the problems of the calculus of variations was covered by the setup of a problem of optimal control, but did not use the type of variations used in the calculus of variations until now. Here we will demonstrate how it can be done.

Let us define $\mathbf{v}(t) = \mathbf{u}(t) + \Delta\mathbf{u}(t)$ and require that $\mathbf{v}(t)$ is admissible. Smallness of $\Delta\mathbf{u}(t)$ means that $\sup_{[0,T]} \|\Delta\mathbf{u}(t)\|$ is sufficiently small. We suppose that the changed value $\mathbf{y}^*(t)$ satisfying the Cauchy problem

$$(\mathbf{y}^*(t))' = \mathbf{f}(\mathbf{y}^*(t), \mathbf{v}(t)), \qquad \mathbf{y}^*(0) = \mathbf{y}_0,$$

is such that $\Delta\mathbf{y}(t) = \mathbf{y}^*(t) - \mathbf{y}(t)$ is also small enough, that is $\max_{[0,T]} \|\Delta\mathbf{y}(t)\|$ is small.

Now we would like to find the increment of $J(\mathbf{u})$ under such a small admissible increment of $\mathbf{u}(t)$. The answer is given by

Theorem 2.11.1 *Suppose that* $\sup_{[0,T]} \|\Delta \mathbf{u}(t)\| = \varepsilon$. *Then the increment of* $J(\mathbf{u})$ *is*

$$J(\mathbf{u}^*) - J(\mathbf{u}) = \delta J(\mathbf{u}) + o(\varepsilon)$$

where

$$\delta J(\mathbf{u}) = \int_0^T \boldsymbol{\Psi}(t) \cdot [\mathbf{f}(\mathbf{y}(t), \mathbf{u}(t)) - \mathbf{f}(\mathbf{y}(t), \mathbf{v}(t))]\, dt$$

and $\boldsymbol{\Psi}(s)$ *is a solution of the following Cauchy problem (in the reverse time):*

$$\boldsymbol{\Psi}'(s) = -\nabla_{\mathbf{y}} \mathbf{f}(\mathbf{y}(s), \mathbf{u}(s)) \cdot \boldsymbol{\Psi}(s), \qquad \boldsymbol{\Psi}(T) = -\nabla_{\mathbf{y}} G(\mathbf{y}(T)). \tag{2.11.1}$$

Proof. Let us note first that the conjugate equation (2.11.1) for $\boldsymbol{\Psi}(s)$ coincides with the conjugate equation we established for the terminal control problem in § 2.8. Much of the reasoning used in that section will apply here. Suppose for simplicity of notation that $\Delta \mathbf{y}(t)$ for all $t \in [0, T]$ is of order ε. The problem defining the increment $\Delta \mathbf{y}(t)$ is

$$\Delta \mathbf{y}'(t) = \mathbf{f}(\mathbf{y}(t) + \Delta \mathbf{y}(t), \mathbf{v}(t)) - \mathbf{f}(\mathbf{y}(t), \mathbf{u}(t)), \tag{2.11.2}$$

$$\Delta \mathbf{y}(0) = 0.$$

We need to find the main part of $\Delta \mathbf{y}(t)$ at $t = T$. Let us transform the right-hand side of (2.11.2):

$$\begin{aligned}
\mathbf{f}(\mathbf{y} + \Delta \mathbf{y}, \mathbf{v}) - \mathbf{f}(\mathbf{y}, \mathbf{u}) &= \mathbf{f}(\mathbf{y} + \Delta \mathbf{y}, \mathbf{v}) - \mathbf{f}(\mathbf{y}, \mathbf{v}) + [\mathbf{f}(\mathbf{y}, \mathbf{v}) - \mathbf{f}(\mathbf{y}, \mathbf{u})] \\
&= \nabla_{\mathbf{y}} \mathbf{f}(\mathbf{y}, \mathbf{v}) \cdot \Delta \mathbf{y} + [\mathbf{f}(\mathbf{y}, \mathbf{v}) - \mathbf{f}(\mathbf{y}, \mathbf{u})] + o(\|\Delta \mathbf{y}\|) \\
&= \nabla_{\mathbf{y}} \mathbf{f}(\mathbf{y}, \mathbf{u}) \cdot \Delta \mathbf{y} + [\mathbf{f}(\mathbf{y}, \mathbf{v}) - \mathbf{f}(\mathbf{y}, \mathbf{u})] + o(\|\Delta \mathbf{y}\|).
\end{aligned}$$

Thus (2.11.2) becomes

$$\begin{aligned}
(\Delta \mathbf{y}(t))' = \nabla_{\mathbf{y}} \mathbf{f}(\mathbf{y}(t), \mathbf{u}(t)) \cdot \Delta \mathbf{y}(t) &+ [\mathbf{f}(\mathbf{y}(t), \mathbf{v}(t)) - \mathbf{f}(\mathbf{y}(t), \mathbf{u}(t))] \\
&+ o(\|\Delta \mathbf{y}(t)\|).
\end{aligned}$$

The main linear part of $\Delta \mathbf{y}(t)$ is described by the following problem:

$$\begin{aligned}
&(\delta \mathbf{y}(t))' = \nabla_{\mathbf{y}} \mathbf{f}(\mathbf{y}(t), \mathbf{u}(t)) \cdot \delta \mathbf{y}(t) + [\mathbf{f}(\mathbf{y}(t), \mathbf{v}(t)) - \mathbf{f}(\mathbf{y}(t), \mathbf{u}(t))], \\
&\delta \mathbf{y}(0) = 0.
\end{aligned}$$

Now we can use Property 2.7.4 and write out the form of the solution:

$$\delta \mathbf{y}(t) = \int_0^t \mathbf{\Phi}(t,s) \cdot [\mathbf{f}(\mathbf{y}(s), \mathbf{v}(s)) - \mathbf{f}(\mathbf{y}(s), \mathbf{u}(s))]\, ds.$$

So the main linear part of $\Delta \mathbf{y}(T)$ is

$$\delta \mathbf{y}(T) = \int_0^T \mathbf{\Phi}(T,s) \cdot [\mathbf{f}(\mathbf{y}(s), \mathbf{v}(s)) - \mathbf{f}(\mathbf{y}(s), \mathbf{u}(s))]\, ds.$$

Now we can find the main linear part of the increment of the objective functional $J(\mathbf{u})$:

$$\begin{aligned}\Delta J(\mathbf{u}) &= \nabla_{\mathbf{y}} G(\mathbf{y}(T)) \cdot \Delta \mathbf{y}(T) + o(\|\Delta \mathbf{y}(T)\|) \\ &= \int_0^T \nabla_{\mathbf{y}} G(\mathbf{y}(T)) \cdot \mathbf{\Phi}(T,s) \cdot [\mathbf{f}(\mathbf{y}(s), \mathbf{v}(s)) - \mathbf{f}(\mathbf{y}(s), \mathbf{u}(s))]\, ds + o(\varepsilon).\end{aligned}$$

Denote $\mathbf{\Psi}(s) = -\nabla_{\mathbf{y}} G(\mathbf{y}(T)) \cdot \mathbf{\Phi}(T,s)$. Then the last relation takes the form

$$\Delta J(\mathbf{u}) = \int_0^T \mathbf{\Psi}(s) \cdot [\mathbf{f}(\mathbf{y}(s), \mathbf{u}(s)) - \mathbf{f}(\mathbf{y}(s), \mathbf{v}(s))]\, ds + o(\varepsilon)$$

as stated by the theorem. Since $\mathbf{\Psi}(s)$ is defined exactly as in § 2.8, we have completed the proof. □

2.12 A Discrete Version of Small Variations of Control Function for Generalized Terminal Control Problem

The formulas presented above for finding the change of the goal functional of a problem are used in practical calculations, but the problem itself should be discretized for this. Following the lecture of Dr. K.V. Isaev (Rostov State University) but in vector notation, let us consider one of the versions of possible discretization of the generalized terminal control problem. Let us recall the original problem. Given the governing equation

$$\mathbf{y}'(t) = \mathbf{f}(\mathbf{y}(t), \mathbf{u}(t)) \tag{2.12.1}$$

for $\mathbf{y} = \mathbf{y}(t)$ with the initial value $\mathbf{y}(0) = \mathbf{y}_0$, find an admissible control function $\mathbf{u} = \mathbf{u}(t)$ such that

$$\mathcal{I}(\mathbf{u}) \to \min_{\mathbf{u} \in U}$$

where

$$\mathcal{I}(\mathbf{u}) = G(\mathbf{y}(s_1), \mathbf{y}(s_2), \dots, \mathbf{y}(s_N)). \tag{2.12.2}$$

We suppose that $\mathbf{u}(t)$ changes by a small variation $\delta\mathbf{u}(t)$ and would like to find the main part of the increment $\Delta\mathcal{I}(\mathbf{u}) = \mathcal{I}(\mathbf{u}+\delta\mathbf{u}) - \mathcal{I}(\mathbf{u})$ that is linear in $\delta\mathbf{u}$. We will not find the solution for this problem but will discretize the problem in whole and formulate the result for the latter.

Let us partition the interval $[0, s_N]$ by points $t_0 = 0 < t_1 < \dots < t_R$, $t_R = s_N$, in such a way that the distance between two nearby points is small and the set $\{t_i\}$ contains all the points s_j from (2.12.2). On the segment $(t_{i-1}, t_i]$ we will approximate the control function $\mathbf{u}(t)$ by a constant value denoted $\mathbf{u}[i]$. Similarly, let us denote $\mathbf{y}[i] = \mathbf{y}(t_i)$. Considering $\mathbf{y}[i-1]$ as the initial value for equation (2.12.1) on $[t_{i-1}, t_i]$ with $\mathbf{u}(t) = \mathbf{u}[i]$, we can find the value $\mathbf{y}[i]$ that can be considered as a functional relation

$$\mathbf{y}[i] = \varphi_i(\mathbf{y}[i-1], \mathbf{u}[i]). \tag{2.12.3}$$

If all the $\mathbf{u}[i]$ are given, then starting with $\mathbf{y}[0] = \mathbf{y}_0$ we get, by (2.12.3), all the uniquely defined values $\mathbf{y}[i]$. In this way a discrete dynamical system is introduced. Note that it is not necessary to obtain (2.12.3) from (2.12.1); it can be formulated independently, and so the reasoning below is valid in a more general case that is not a consequence of the continuous dynamical system (2.12.1). The restriction for control function $\mathbf{u} \in U$ for discrete control functions is rewritten as $\mathbf{u} \in U^*$. Correspondingly the discrete generalized control problem can be reformulated as:

Problem. Given

$$\begin{aligned} \mathbf{y}[i] &= \varphi_i(\mathbf{y}[i-1], \mathbf{u}[i]), \quad \mathbf{y}[0] = \mathbf{y}_0, \\ \mathcal{I}(\mathbf{u}) &= G(\mathbf{y}[i_1], \mathbf{y}[i_2], \dots, \mathbf{y}[i_N]), \end{aligned} \tag{2.12.4}$$

find $\mathbf{u} \in U^*$ such that

$$\mathcal{I}(\mathbf{u}) \to \min_{\mathbf{u} \in U^*}.$$

The main part of the increment of $\mathcal{I}(\mathbf{u})$ that is linear in $\delta\mathbf{u}$ is given by the following

Theorem 2.12.1 *The main part of the increment of $\mathcal{I}(\mathbf{u})$ that is linear in $\delta\mathbf{u} = \delta\mathbf{u}[i]$ is*

$$\delta\mathcal{I}(\mathbf{u}) = \sum_{i=1}^{R} \nabla_{\mathbf{u}[i]}\mathcal{I}(\mathbf{u}) \cdot \delta\mathbf{u}[i] \tag{2.12.5}$$

where

$$\nabla_{\mathbf{u}[i]}\mathcal{I}(\mathbf{u}) = \left(\nabla_{\mathbf{u}[i]}\varphi_i(\mathbf{y}[i-1], \mathbf{u}[i])\right) \cdot \boldsymbol{\psi}[i] \tag{2.12.6}$$

and $\boldsymbol{\psi}[i]$ satisfy the equations

$$\begin{aligned} \boldsymbol{\psi}[i] &= \left(\nabla_{\mathbf{y}[i]}\varphi_{i+1}(\mathbf{y}[i], \mathbf{u}[i+1])\right) \cdot \boldsymbol{\psi}[i+1] \\ &\quad + \nabla_{\mathbf{y}[i]}Q(\mathbf{y}[i_1], \mathbf{y}[i_2], \ldots, \mathbf{y}[i_N]), \quad i = R-1, R-2, \ldots, 1, \\ \boldsymbol{\psi}[R] &= \nabla_{\mathbf{y}[R]}Q(\mathbf{y}[i_1], \mathbf{y}[i_2], \ldots, \mathbf{y}[i_N]). \end{aligned} \tag{2.12.7}$$

Proof. Before giving the proof we would like to point out the similarity between this and the result for the corresponding continuous control problem; in particular, there arises a system of equations for the complementary function $\boldsymbol{\psi}$ of the parameter i, whose solutions should be found in the reverse order, from $\boldsymbol{\psi}[R]$ to $\boldsymbol{\psi}[1]$. It is clear that it does not matter on which step and how we discretize the problem, the main features of solution should be the same. First let us mention that now $\mathcal{I}(\mathbf{u})$ is an ordinary function in many variables $\mathbf{u}[i]$ so all we need to find is the first differential of $\mathcal{I}(\mathbf{u})$ under constraints from (2.12.4). Thus the formula for the first differential gives us

$$\delta\mathcal{I}(\mathbf{u}) = \sum_{i=1}^{R} \nabla_{\mathbf{u}[i]}\mathcal{I}(\mathbf{u}) \cdot \delta\mathbf{u}[i]$$

which is (2.12.5). Next

$$\begin{aligned} \nabla_{\mathbf{u}[i]}\mathcal{I}(\mathbf{u}) &= \nabla_{\mathbf{u}[i]}Q(\mathbf{y}[i_1], \mathbf{y}[i_2], \ldots, \mathbf{y}[i_N]) \\ &= \sum_{j=1}^{R} \nabla_{\mathbf{u}[i]}\mathbf{y}[j] \cdot \nabla_{\mathbf{y}[j]}Q(\mathbf{y}[i_1], \mathbf{y}[i_2], \ldots, \mathbf{y}[i_N]). \end{aligned} \tag{2.12.8}$$

Here we used the chain rule for differentiation, formulated for the gradient. Let us find $\nabla_{\mathbf{u}[i]}\mathbf{y}[j]$. For this we introduce a new vector function $\mathbf{F}_{ji}$ induced by (2.12.3) that is defined for $j \geq i$:

$$\mathbf{y}[j] = \mathbf{F}_{ji}(\mathbf{y}[i]).$$

Let us formulate the properties of $\mathbf{F}_{ji}$. It is obvious that

$$\mathbf{F}_{ii}(\mathbf{y}[i]) = \mathbf{y}[i],$$
$$\mathbf{F}_{i+1\ i}(\mathbf{y}[i]) = \mathbf{y}[i+1] = \boldsymbol{\varphi}_{i+1}(\mathbf{y}[i], \mathbf{u}[i+1]).$$

Finally, it follows by the definition that

$$\mathbf{F}_{ji}(\mathbf{y}[i]) = \mathbf{F}_{j\ i+1}(\mathbf{y}[i+1]) = \mathbf{F}_{j\ i+1}\left(\boldsymbol{\varphi}_{i+1}(\mathbf{y}[i], \mathbf{u}[i+1])\right). \tag{2.12.9}$$

It is evident that the components of $\mathbf{F}_{ji}$ depend only on the components $\mathbf{u}[i+1], \mathbf{u}[i+2], \ldots, \mathbf{u}[j]$ and do not depend on the rest of the components of $\mathbf{u}$. Let us return to finding $\nabla_{\mathbf{u}[i]}\mathbf{y}[j]$ using the chain rule again:

$$\begin{aligned}\nabla_{\mathbf{u}[i]}\mathbf{y}[j] &= \nabla_{\mathbf{u}[i]}\mathbf{F}_{ji}(\mathbf{y}[i]) = \nabla_{\mathbf{u}[i]}\mathbf{y}[i] \cdot \nabla_{\mathbf{y}[i]}\mathbf{F}_{ji}(\mathbf{y}[i]) \\ &= \nabla_{\mathbf{u}[i]}\boldsymbol{\varphi}_i\left(\mathbf{y}[i-1], \mathbf{u}[i]\right) \cdot \nabla_{\mathbf{y}[i]}\mathbf{F}_{ji}(\mathbf{y}[i]).\end{aligned}$$

Returning to (2.12.8) we get

$$\begin{aligned}\nabla_{\mathbf{u}[i]}\mathcal{I}(\mathbf{u}) = \sum_{j=i}^{R} &\nabla_{\mathbf{u}[i]}\boldsymbol{\varphi}_i\left(\mathbf{y}[i-1], \mathbf{u}[i]\right) \cdot \nabla_{\mathbf{y}[i]}\mathbf{F}_{ji}(\mathbf{y}[i]) \\ &\cdot \nabla_{\mathbf{y}[j]}Q(\mathbf{y}[i_1], \mathbf{y}[i_2], \ldots, \mathbf{y}[i_N]).\end{aligned}$$

Denoting

$$\boldsymbol{\psi}[i] = \sum_{j=i}^{R} \nabla_{\mathbf{y}[i]}\mathbf{F}_{ji}(\mathbf{y}[i]) \cdot \nabla_{\mathbf{y}[j]}Q(\mathbf{y}[i_1], \mathbf{y}[i_2], \ldots, \mathbf{y}[i_N]) \tag{2.12.10}$$

we get

$$\nabla_{\mathbf{u}[i]}\mathcal{I}(\mathbf{u}) = \left(\nabla_{\mathbf{u}[i]}\boldsymbol{\varphi}_i(\mathbf{y}[i-1], \mathbf{u}[i])\right) \cdot \boldsymbol{\psi}[i]$$

which is (2.12.6).

It remains to derive equations for $\boldsymbol{\psi}[i]$. We begin with formula (2.12.9):

$$\mathbf{F}_{ji}(\mathbf{y}[i]) = \mathbf{F}_{j\ i+1}(\mathbf{y}[i+1]).$$

Applying the gradient by $\mathbf{y}[i]$ to both sides we get

$$\nabla_{\mathbf{y}[i]}\mathbf{F}_{ji}(\mathbf{y}[i]) = \nabla_{\mathbf{y}[i]}\boldsymbol{\varphi}_{i+1}(\mathbf{y}[i], \mathbf{u}[i+1]) \cdot \nabla_{\mathbf{y}[i+1]}\mathbf{F}_{j\ i+1}\left(\mathbf{y}[i+1]\right).$$

Substituting this into (2.12.10) we get

$$\begin{aligned}\psi[i] &= \sum_{j=i}^{R} \nabla_{\mathbf{y}[i]}\mathbf{F}_{ji}(\mathbf{y}[i]) \cdot \nabla_{\mathbf{y}[j]} Q(\mathbf{y}[i_1], \mathbf{y}[i_2], \dots, \mathbf{y}[i_N]) \\ &= \sum_{j=i+1}^{R} \nabla_{\mathbf{y}[i]}\varphi_{i+1}(\mathbf{y}[i], \mathbf{u}[i+1]) \cdot \nabla_{\mathbf{y}[i+1]}\mathbf{F}_{j\ i+1}(\mathbf{y}[i+1]) \\ &\qquad \cdot \nabla_{\mathbf{y}[j]} Q(\mathbf{y}[i_1], \mathbf{y}[i_2], \dots, \mathbf{y}[i_N]) \\ &\qquad + \nabla_{\mathbf{y}[i]}\mathbf{F}_{ii}(\mathbf{y}[i]) \cdot \nabla_{\mathbf{y}[i]} Q(\mathbf{y}[i_1], \mathbf{y}[i_2], \dots, \mathbf{y}[i_N]) \\ &= \nabla_{\mathbf{y}[i]}\varphi_{i+1}(\mathbf{y}[i], \mathbf{u}[i+1]) \cdot \psi[i+1] + \nabla_{\mathbf{y}[i]} Q(\mathbf{y}[i_1], \mathbf{y}[i_2], \dots, \mathbf{y}[i_N])\end{aligned}$$

where we have used the fact that $\nabla_{\mathbf{y}[i]}\mathbf{F}_{ii}(\mathbf{y}[i]) = \mathbf{E}$. So we have obtained the first of (2.12.7). From the intermediate result of this equality chain the second of (2.12.7) follows. □

We now turn to another class of problems.

2.13 Optimal Time Control Problems

We recall that the problems of this type are as follows. The object is described by a dynamical system

$$\mathbf{y}'(t) = \mathbf{f}(\mathbf{y}(t), \mathbf{u}(t)) \tag{2.13.1}$$

for which we must find an admissible control function $\mathbf{u}(t)$ in such a way that the parameters of the system must be changed from the initial state

$$\mathbf{y}(0) = \mathbf{y}_0 \tag{2.13.2}$$

to the final state

$$\mathbf{y}(T) = \mathbf{y}_1 \tag{2.13.3}$$

in minimal time T. Unlike the terminal control problem, here the final state of the system is fixed but not the time interval.

Let us note that in this problem the set $\mathcal{U}$ of admissible control functions is limited not only by the external inequality restrictions, but also by the boundary conditions (2.13.2)–(2.13.3) because it may happen so that there are no admissible control vectors such that the system, starting with the initial state $\mathbf{y}_0$, can reach the final state $\mathbf{y}_1$ in finite time T.

Next we recall that for the terminal control problem we obtained a conjugate problem with an initial (i.e., "final") condition at T which was called the condition of transversality. The optimal time problem has both the boundary conditions for $\mathbf{y}$ of the same form as the condition at $t = 0$ of the terminal control problem. Thus we should expect that if Pontryagin's principle of maximum is valid in this or that form for the optimal control problem then any boundary conditions for $\mathbf{\Psi}(s)$ are absent. This means that the uniqueness for finding $\mathbf{\Psi}(s)$ needed for this problem is not provided by some explicit equations. The explicit formula for the increment of the objective functional for the optimal control problem is not obtained. So we formulate without proof the statement of Pontryagin's principle of maximum for the optimal control problem.

Theorem 2.13.1 *Let $\mathbf{u}(t)$ be a control function at which T, the length of the time interval, attains its minimal value among all the admissible control functions, for which* (2.13.1)–(2.13.3) *has a solution $\mathbf{y}(t)$. There is a nontrivial vector function $\mathbf{\Psi}(s)$ that is a solution of the conjugate equation*

$$\frac{d}{ds}\mathbf{\Psi}(s) = -\mathbf{\Psi}(s) \cdot \nabla_{\mathbf{y}}\mathbf{f}(\mathbf{y}(s), \mathbf{u}(s))$$

such that the Pontryagin function $H(\mathbf{y}, \mathbf{\Psi}, \mathbf{u}) = \mathbf{f}(\mathbf{y}, \mathbf{u}) \cdot \mathbf{\Psi}$, with respect to the third argument, takes its maximal value for all points of continuity of $\mathbf{u}(t)$:

$$H(\mathbf{y}(t), \mathbf{\Psi}(t), \mathbf{u}(t)) \geq H(\mathbf{y}(t), \mathbf{\Psi}(t), \mathbf{v}).$$

Let us note that in simple cases when $\mathbf{u}(t)$ comes into the equations linearly this theorem reduces the set of possible control functions to those which take values at boundaries of $\mathcal{U}$ at each time t. Indeed, then $\mathbf{u}(t)$ comes linearly into the presentation of $H(\mathbf{y}, \mathbf{\Psi}, \mathbf{u}) = \mathbf{\Psi} \cdot \mathbf{f}(\mathbf{y}, \mathbf{u})$ and thus its maximal value can be taken only at some extreme points of $\mathbf{u}(t)$.

Example 2.13.1 Consider the simplest optimal time problem. Let a material point of unit mass move along a straight line under the action of a force whose magnitude F cannot exceed unity. How should we vary F so that the point moves from one position to another in the shortest time?

Solution If the velocity of the point at its initial and final states is zero then the solution is clear mechanically: first we need to accelerate the point with maximal force until it comes to the middle point between the initial and final state, and then to switch the force to the opposite direction leaving the maximal magnitude so the point is maximally decelerated. When the appointed initial and final velocities are not zero one must have good mechanical intuition to tell what the law for the force should be. Let us solve this problem using Theorem 2.12.1. The governing equation is

$$x''(t) = F(t), \quad x(0) = a_0, \quad x'(0) = a_1, \quad x(T) = b_0, \quad x'(T) = b_1, \tag{2.13.4}$$

and the restriction for $F(t)$ is

$$|F(t)| \le 1. \tag{2.13.5}$$

Let us rewrite this using the notation we used above:

$$y_1(t) = x(t), \qquad y_2(t) = x_1'(t), \qquad u(t) = F(t).$$

Thus we introduce the phase coordinates of the point. Then equations (2.13.4)–(2.13.5) take the form

$$\begin{aligned} y_1'(t) &= y_2(t), \\ y_2'(t) &= u(t), \end{aligned}$$

the boundary conditions

$$\begin{aligned} y_1(0) &= a_0, \\ y_2(0) &= a_1, \end{aligned} \quad \text{and} \quad \begin{aligned} y_1(T) &= b_0, \\ y_2(T) &= b_1, \end{aligned}$$

and the restriction that defines the set $\mathcal{U}$ of piecewise continuous functions

$$-1 \le u(t) \le 1.$$

Let us first introduce the Pontryagin function $H = y_2\psi_1 + u\psi_2$. Let $\mathbf{y}(t)$ and $\mathbf{\Psi}(t)$ be the needed solutions of the main and conjugate systems of equations. The conjugate equations are

$$\begin{aligned} \psi_1' &= -\partial H/\partial y_1 = 0, \\ \psi_2' &= -\partial H/\partial y_2 = -\psi_1. \end{aligned}$$

The solution of this system results in $\psi_2 = d_1 t + d_2$ and thus may have no more than one point $t_0 \in [0, T]$ at which it changes sign. By Pontryagin's principle, it is the only point at which the control function u must switch sign as H can take its maximum when $\psi_2(t)u(t)$ takes its maximum. Thus t_0 splits $[0, T]$ into two parts having $u = \pm 1$. Thus the solution to our simplest optimal time control problem should be synthesized from trajectories of the two systems

$$\begin{aligned} y_1'(t) &= y_2(t), \\ y_2'(t) &= 1, \end{aligned} \quad \text{and} \quad \begin{aligned} y_1'(t) &= y_2(t), \\ y_2'(t) &= -1. \end{aligned}$$

The particle trajectories on the phase plane (y_1, y_2) are parabolas. For the first system $y_1 = t^2/2 + c_1 t + c_2$ and for the second $y_1 = -t^2/2 + c_3 t + c_4$. Geometrically it is evident that there are no more than two parabolas, one from each family, through the end points which intersect. That is the solution trajectory of the problem. Analytically we must compose five equations for unknown c_i and t_0. The first is that at t_0 the curves intersect, that is

$$t_0^2/2 + c_1 t_0 + c_2 = -t_0^2/2 + c_3 t_0 + c_4.$$

The other four equations (boundary conditions) depend on which of switched values of u goes first. If $u = 1$ on $[0, t_0]$ and thus $u = -1$ on the rest,

$$c_2 = a_0, \quad c_1 = a_1, \quad -T^2/2 + c_3 T + c_4 = b_0, \quad -T + c_3 = b_1.$$

If $u = -1$ on $[0, t_0]$ then

$$c_4 = a_0, \quad c_3 = a_1, \quad T^2/2 + c_1 T + c_2 = b_0, \quad T + c_1 = b_1.$$

Only one of these systems has a solution where real t_0 lies in $[0, T]$ and it is what we have sought.

We would like to note that when the controlled object's equations are simple, the maximum principle of Pontryagin gives a good tool to find an optimal solution. For many industrial problems it is necessary to use other methods. In the same manner as Example 2.13.1, any optimal time problem for a system described by the equation $x'' + ax' + bx = u$ can be solved analytically. Textbooks are full of such problems from various areas of science, their analytical solutions as well as geometrical interpretation of some of their solutions.

Our next remark is the following. The terminal control problems and the optimal time problems are in a certain sense, the extremes of all control problems with respect to boundary conditions. For "intermediate" problems, with other types of boundary conditions at starting and ending moments, the conjugate system is supplemented with some conditions of transversality. The situation is similar to that for natural conditions in the calculus of variations.

2.14 Final Remarks on Control Problems

In this chapter we considered in large part the methods for finding optimal solutions. Of course it was an introductory chapter, and we limited ourselves to a small portion of the theory — that portion which is used in many industrial control processes and other applications. We did not touch on the problem of existence of solutions of control problems, which is extremely important since there are many practical problems that are formulated quite nicely from a common sense standpoint but that lack solutions.

We mention only another important part of control theory that is called dynamical programming. It was developed by R. Bellman and used quite successfully in many problems of optimal control. To give the reader some idea of what this theory is about and to lend vividness to the presentation we consider a very simple problem (in a form that might hold the attention of many undergraduate students):

Example 2.14.1 A racketeer has been drunk for three weeks and has failed to perform his job properly. One morning he receives a phone call from his boss, reminding him of a $32,000 debt he owes the boss in one hour. Along with this reminder comes a suitable threat about one lost tooth for each $1000 he fails to bring in. The racketeer lives quite far from his boss, and wishes to collect as much additional money as possible on the way. He has a street map showing how much money he can collect on each possible route. He is constrained to move ahead only, and cannot turn back.

Solution We draw the map as a graph (Fig. 2.4) that should begin at point O and end at B. To get a more convenient presentation at the final point B we split all routes to B and draw them along the final line B_0-B_1 as shown on the picture. On the lines connecting the nodes we put the amounts of money that the racketeer expects to be able to collect from the peaceful citizenry.

Let us discuss this problem. Of course, for this small map the racketeer could test all the possibilities and find the optimal way quite quickly. There are six levels at each of the way can branch so there are few possibilities. Let us imagine that this map has 1000 such levels; then the number of

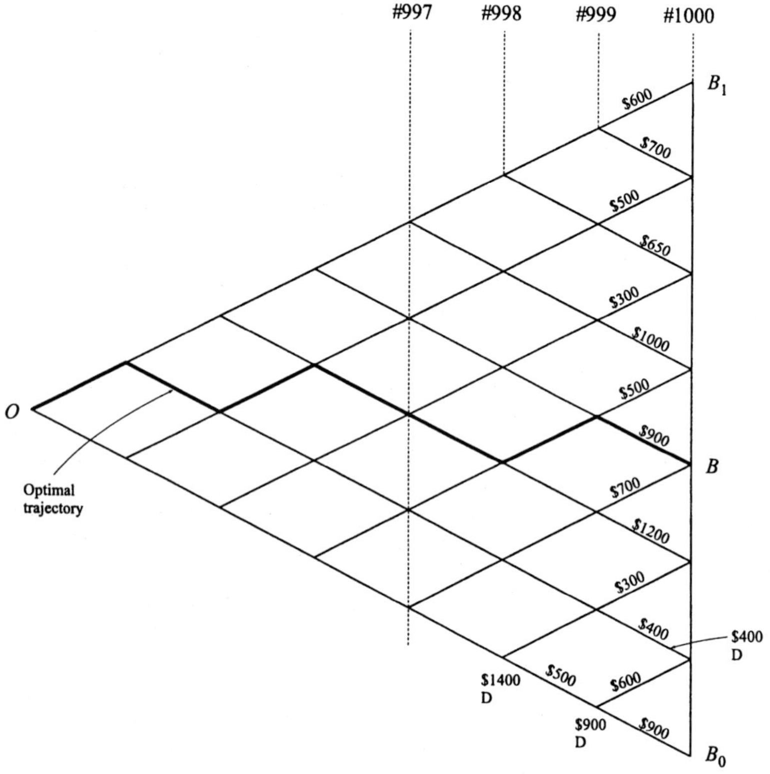

Fig. 2.4 A racketeer's possible routes; optimal trajectory shown as the thick line.

possible ways grows to 2^{1000} and simple experimentation would not bring a quick result. So it becomes necessary to propose a procedure for which the number of operations could be sufficiently small, say several million. Any cross-section of the map would not bring the needed optimal result since the optimal trajectory can be quite strange. The crucial step to the solution is to choose the first step as follows. Suppose that we are at the 999th level of nodes. From each node of this level we exactly know where to move since it is a choice between two possibilities. Near each node of this level we write down where we should move (Down or Up) and the amount. On the 998th level we again should fulfill few operations at each node: moving along the upper street we then add the figure of this street with the price of corresponding 999th node after which we should decide between the two possibilities and to write near the node Up or Down (showing where to go next) and the optimal cost. On the 997th level everything will be repeated: the finding of two sums of two numbers, the choosing of the bigger one, and the placement of the necessary information near the node. This is must be done at each level. In this way we come to the initial point, getting the optimal sum of money as the resulting figure at it, and the optimal trajectory moving along signs Up and Down.

At first glance this seems to be a nice problem for a high school math competition, since it is solved using only "common sense". However, its solution is based on a hard mathematical idea: when we come to some point of the optimal trajectory, the remainder of the optimal trajectory is optimal for the "reduced" problem whose initial point is this one at which we just stopped.

We shall not discuss the many fruitful applications of this principle of Bellman. As the central principle of dynamical programming it has brought many results, both theoretical and practical, in discrete and continuous problems.

We leave it to the reader to explore other books, and thereby to discover other ways to view problems in optimal control and the calculus of variations. These are indeed part of the more general branch of mathematics known as Mathematical Programming.

2.15 Exercises

2.1 Show that the coefficients of the squared gradient

$$\nabla_{\mathbf{y}}^2 = \nabla_{\mathbf{y}} (\nabla_{\mathbf{y}})$$

applied to a scalar valued function $f(\mathbf{y}(t))$ constitute the Hessian matrix of f.

2.2 Establish the formula (2.7.21).

2.3 Formulate the form of the main linear part of the increment of $J(\mathbf{u})$ under the sum of the increments of the control function by the needle-shaped vector function and a small increment as discussed in the current section.

2.4 (A harder problem.) Let the objective functional for the terminal control problem be changed to

$$J^*(\mathbf{u}) = \int_0^T G(\mathbf{y}(t))\, dt.$$

What is the form of the main part of its increment in this case?

2.5 A mechanical oscillator (a mass on a spring) oscillates under force $|F(t)|$ such that $|F(t)| \le 1$. The governing equation is $mx'' + kx = F$, $m = 1$, $k = 1$.

Find the law of the change of the force when the mass goes from state $x(0) = a$, $x'(0) = b$ to the state of equilibrium, $x(T) = 0$, $x'(T) = 0$ in the shortest time T.

Chapter 3

Functional Analysis

A principal tool in the modern analysis of partial differential equations, *functional analysis* allows us to shift our perspective on functions from the viewpoint of ordinary calculus to a viewpoint in which we deal with a function (such as a differential or integral operator) as a whole entity. We accomplish this conceptual shift by extending the notion of an ordinary 3-D vector so that a function can be viewed as an element of a linear vector space. Because this extension involves some subtle points regarding the dimension of a vector space, we devote the present chapter to a suitable introduction for the reader.

As a branch of mathematics, functional analysis is in large part delineated by the tools it offers to the practitioner. Important applications arise in a variety of areas: differential and integral equations, the theory of integration, probability theory, etc. It has been said that functional analysis is not a special branch of mathematics at all, but rather a united point of view on mathematical objects of differing natures. A full presentation of functional analysis would require many volumes. The goal of the present chapter is to offer the reader a relatively brief but still self-contained treatment, and therefore to provide all the tools necessary for the study of boundary value problems.

Before we begin it will be useful to recall two standard theorems from ordinary calculus:

Theorem 3.0.1 *Suppose a sequence $\{f_n(\mathbf{x})\}$ of functions continuous on a compact set $\Omega \subset \mathbb{R}^k$ converges uniformly; that is, for any $\varepsilon > 0$ there is an integer $N = N(\varepsilon)$ such that $|f_n(\mathbf{x}) - f_m(\mathbf{x})| < \varepsilon$ whenever $n, m > N$ and $\mathbf{x} \in \Omega$. Then the limit function*

$$f(\mathbf{x}) = \lim_{n\to\infty} f_n(\mathbf{x})$$

is continuous on Ω.

This is called Weierstrass' theorem. The next one shows the properties of a continuous function on a compact set.

Theorem 3.0.2 *Suppose* $f(\mathbf{x})$ *is continuous on a compact set* $\Omega \subset \mathbb{R}^k$. *Then* $f(\mathbf{x})$ *is uniformly continuous on* Ω; *that is, for any* $\varepsilon > 0$ *there is a* $\delta > 0$ *(independent of* ε*) such that* $|f(\mathbf{x}) - f(\mathbf{y})| < \varepsilon$ *whenever* $\|\mathbf{x} - \mathbf{y}\| < \delta$ *and* $\mathbf{x}, \mathbf{y} \in \Omega$.

3.1 A Normed Space as a Metric Space

Regarding a function as a single object (a viewpoint which functional analysis inherited from the calculus of variations), we must provide a way to quantify the difference between two functions. The simplest and most convenient way to do this is to use the tools of normed spaces. First of all a normed space, consisting of elements of any nature (of functions in particular), must be a *linear space.* This means that we can add or subtract any two elements of the space, or multiply an element of the space by a number, and the result will always be an element of the same space. If complex numbers are used as multipliers then the linear space is called a *complex linear space*; if purely real numbers are used then the space is a *real linear space.* The definition of a linear space can be stated rigorously in terms of axioms and the reader has undoubtedly seen these in a linear algebra course. The main distinction between a general linear space and a normed space is the existence of a norm on the latter. A *norm* is a real-valued functional $\|x\|$ that is determined (which means it takes a unit and a finite value) at each element x of the space and satisfies the following axioms:

(1) $\|x\| \geq 0$ for all x; $\|x\| = 0$ if and only if $x = 0$;
(2) $\|\lambda x\| = |\lambda| \, \|x\|$ for any x and any real number λ;
(3) $\|x + y\| \leq \|x\| + \|y\|$ for all x, y.

The first of these is called the axiom of positiveness, while the second is the axiom of homogeneity and the third is the triangle inequality.

Definition 3.1.1 A *normed linear space* is a linear space X on which a norm $\|\cdot\|$ is defined.

More specifically, $\|\cdot\|$ is "defined" on X if the number $\|x\|$ exists and is finite for every element $x \in X$.

In classical functional analysis one deals with dimensionless quantities. In applications this restriction is not necessary: one can use numbers with dimensional units and get norms having dimensional units. Although this introduces no theoretical complications and is sometimes quite useful, we shall follow the classical procedure and consider all elements to be dimensionless.

Example 3.1.1 Show that if $\|x\|$ is any norm on X and $x, y \in X$, then

$$|\,\|x\| - \|y\|\,| \leq \|x - y\| .$$

We shall find this inequality useful later. For example, in accordance with the definition of continuity it means that the norm is continuous with respect to the norm itself.

Solution Let us begin by replacing x with $x - y$ in norm axiom 3: we get

$$\|x\| - \|y\| \leq \|x - y\| .$$

Interchanging the roles of x and y in this inequality, we get

$$\|y\| - \|x\| \leq \|y - x\| .$$

But the right-hand sides of these two inequalities are the same; indeed, we have $\|y - x\| = \|(-1)(x - y)\| = \|x - y\|$ by norm axiom 2. So the quantity $\|x - y\|$ is greater than or equal to both $\|x\| - \|y\|$ and $\|y\| - \|x\|$. This means it is greater than or equal to $|\,\|x\| - \|y\|\,|$.

We have introduced the normed space $C^{(k)}(\Omega)$ of functions that are k times continuously differentiable on a compact set Ω with the norm

$$\|f\|_{C^{(k)}(\Omega)} = \max_{\mathbf{x}\in\Omega} |f(\mathbf{x})| + \sum_{|\alpha|\leq k} \max_{\mathbf{x}\in\Omega} |D^\alpha f(\mathbf{x})|. \qquad (3.1.1)$$

where

$$D^\alpha f = \frac{\partial^{|\alpha|} f}{\partial x_1^{\alpha_1} \cdots \partial x_n^{\alpha_n}}, \qquad |\alpha| = \alpha_1 + \cdots + \alpha_n.$$

As with any other proposed norm, the reader should verify satisfaction of the axioms.[1] A particular case is the space of all functions continuous on

[1] For example one could take the set of functions continuous on $[0,1]$ and try to introduce a "norm" using the formula $\|f(x)\| = |f(0.5)|$. Which norm axiom would fail?

Ω with the norm

$$\|f\|_{C(\Omega)} = \max_{\mathbf{x}\in\Omega} |f(\mathbf{x})|. \tag{3.1.2}$$

In the space of functions continuous on a compact Ω we can introduce another norm:

$$\|f(\mathbf{x})\| = \left(\int_\Omega |f(\mathbf{x})|^p \, d\Omega\right)^{1/p} \qquad (p \geq 1).$$

The norm axioms can be verified here also (the triangle inequality being known as Minkowski's inequality.).[2] Thus we see that on the same set (linear space) of elements we can introduce one of several norms. On the same compact Ω we can consider the set of all bounded functions and introduce the norm

$$\|f(\mathbf{x})\| = \sup_{\mathbf{x}\in\Omega} |f(\mathbf{x})|. \tag{3.1.3}$$

The resulting space will be called $M(\Omega)$. The space $C(\Omega)$ is a subspace of $M(\Omega)$ (note that for a continuous function the norm (3.1.3) reduces to (3.1.2)). The reader sees that a normed space is defined by the set of elements and the form of the norm imposed on it. So to refer properly a space, we must display a pair consisting of the set of elements X and the form of the norm, something like $(X, \|\cdot\|)$. For the most frequently used spaces it is common to use shorthand notation such as $C(\Omega)$ where the norm is understood. This is especially appropriate when there is a unique norm imposed on a set, and we shall adopt the practice. When it is necessary to distinguish different norms, we shall indicate the space as a subscript on the norm symbol as we have done in (3.1.1) and (3.1.2).

If we define for each pair of elements of a normed space another functional

$$d(x, y) = \|x - y\|, \tag{3.1.4}$$

we see that it satisfies the axioms of a *metric*:

(1) $d(x, y) \geq 0$ for all x, y, and $d(x, y) = 0$ if and only if $x = y$;
(2) $d(x, y) = d(y, x)$ for all x, y;
(3) $d(x, y) \leq d(x, z) + d(z, y)$ for all x, y, z.

[2] We assume Ω is Jordan measurable. This is a safe assumption for our purposes, because we consider only domains occupied by physical bodies having comparatively simple shape.

If such a functional (metric) is defined for any pair of elements of a set X, then we have a *metric space.*

Definition 3.1.2 A *metric space* is a set X on which a metric $d(x, y)$ is defined.

Hence we see that every normed space is a metric space (the metric (3.1.4) is called the *natural metric* and is said to be *induced* by the norm). The notion of metric space is more general than that of normed space. Not all metric spaces can be normed: first of all a metric space need not be a linear space (a fact which is sometimes important, as in applications of the contraction mapping principle). Note that the use of elements with dimensional units would give a metric having dimensions as well; although the metric is a generalization of the notion of distance, this distance can be expressed in units of force, power, etc.

The axioms of a metric replicate the essential properties of distance from ordinary geometry: (1) distance is nonnegative, the distance from a point to itself is zero, and the distance between two distinct points is nonzero; (2) the distance between two points does not depend on the order in which the points are considered; and (3) the triangle inequality holds, meaning that for a triangle the length of any side does not exceed the sum of the lengths of the other two sides. In this way the more general notion of metric preserves many of the terms and concepts of ordinary geometry. For example, we have

Definition 3.1.3 An *open ball* with center x_0 and radius R is the set of points $x \in X$ such that $d(x_0, x) < R$. The corresponding *closed ball* is the set of all $x \in X$ such that $d(x_0, x) \leq R$, and the corresponding *sphere* of radius R is the set of all $x \in X$ such that $d(x_0, x) = R$.

Note that the term "ball" can denote various objects depending on the metric chosen: if we introduce the metric

$$d(x, y) = \max_{1 \leq i \leq 3} |x_i - y_i|$$

in ordinary 3-D space where $x = (x_1, x_2, x_3)$ and $y = (y_1, y_2, y_3)$, then a ball is really shaped like a cube. The other abstract space structures also provide notions that correspond to those of ordinary geometry. In a linear space of vectors we can determine a straight line through the points x_1 and x_2 by the equation

$$tx_1 + (1 - t)x_2, \qquad t \in (-\infty, \infty),$$

and can obtain the segment having x_1 and x_2 as endpoints by restricting t to the interval $[0, 1]$. It is especially important that we can use the notion of metric to introduce the tools of calculus in such a way that functions can be dealt with as whole objects. (We should note that metric spaces are not linear in general, so they include spaces that cannot be normed. However, even some linear metric spaces cannot be normed.)

Armed with a notion of distance in a normed space, we can introduce any of the notions from calculus that are connected with the notion of distance. First is the notion of convergence.

Definition 3.1.4 We say that a sequence $\{x_n\}$ is *convergent* to an element x if to each positive number ε there corresponds a number $N = N(\varepsilon)$ such that $d(x_k, x) < \varepsilon$ whenever $k > N$.

The reader can easily reformulate this definition in terms of the norm.

Just as in calculus we call x the *limit* of $\{x_k\}$ and write $\lim_{k\to\infty} x_k = x$ or $x_k \to x$ as $k \to \infty$.

Example 3.1.2 (a) Show that every convergent sequence in a metric space has a unique limit. (b) Show that if $x_n \to x$ and $y_n \to y$, then $d(x_n, y_n) \to d(x, y)$ as $n \to \infty$.

Solution (a) Our approach will be to suppose that $x_n \to x$ and $x_n \to x'$, and then to show that $x' = x$ follows. Let ε be an arbitrarily small positive number. By assumption we can choose N so large that the inequalities $d(x_N, x) < \varepsilon/2$ and $d(x_N, x') < \varepsilon/2$ both hold. By the triangle inequality then, we have

$$d(x, x') \le d(x, x_N) + d(x_N, x') < \varepsilon.$$

Since the distance $d(x, x')$ is both nonnegative and smaller than any preassigned positive number, it must equal zero. According to metric axiom 1, we conclude that $x = x'$. (b) The generalized triangle inequality

$$d(x_1, x_n) \le d(x_1, x_2) + d(x_2, x_3) + \cdots + d(x_{n-1}, x_n)$$

is easily established through the use of mathematical induction. We can use this fact as follows. We write

$$d(x, y) \le d(x_n, x) + d(x_n, y_n) + d(y_n, y)$$

and

$$d(x_n, y_n) \le d(x_n, x) + d(x, y) + d(y_n, y),$$

and then combine these two inequalities into the form

$$|d(x_n, y_n) - d(x, y)| \le d(x_n, x) + d(y_n, y).$$

Now given any $\varepsilon > 0$ we can choose N so large that $n > N$ implies both $d(x_n, x) < \varepsilon/2$ and $d(y_n, y) < \varepsilon/2$. This means that $|d(x_n, y_n) - d(x, y)| < \varepsilon$, as desired.

It is clear that a sequence of functions continuous on $[0, 1]$ that converges in the norm (3.1.2) also converges in the norm

$$\|f(x)\| = 2 \max_{x\in[0,1]} |f(x)|.$$

However there are other norms, of $L^p(0, 1)$ say, under which the meaning of convergence is different. It is clear that if two norms $\|\cdot\|_1$ and $\|\cdot\|_2$ satisfy the inequalities

$$m\,\|x\|_1 \le \|x\|_2 \le M\,\|x\|_1 \tag{3.1.5}$$

for some positive constants m and M that do not depend on x, then the two resulting notions of convergence on the set of elements are the same.

Definition 3.1.5 Two norms $\|\cdot\|_1$ and $\|\cdot\|_2$ that satisfy (3.1.5) for all $x \in X$ are said to be *equivalent* on X.

We shall not distinguish between normed spaces consisting of the same elements and having equivalent norms.

3.2 Dimension of a Linear Space and Separability

The reader is aware that the *dimension* of a linear space is the maximal number of linearly independent elements of the space. We recall that the elements x_k, $k = 1, 2, \ldots, n$, are linearly independent if the equation

$$c_1x_1 + c_2x_2 + \cdots + c_nx_n = 0$$

with respect to the unknowns c_k implies that $c_k = 0$ for all $k = 1, 2, \ldots, n$. We shall deal for the most part with infinite dimensional spaces. An important example is the space $C(0, 1)$ of functions $f(x)$ continuous on $[0, 1]$. Indeed, any set of monomials $f_k(x) = x^k$ is linearly independent in this space, since for any integer n the equation

$$c_1x + c_2x^2 + \cdots + c_nx^n = 0$$

cannot hold for any x unless $c_k = 0$ for all $k = 1, 2, \ldots, n$. Therefore the dimension of $C(0, 1)$ cannot be finite.

Let us discuss the problem of the number of elements in an abstract set. We shall say that two sets have *equal power* if we can place their elements in one-to-one correspondence. The simplest known infinite sets are those whose elements can be placed in one-to-one correspondence with the set of natural numbers. Such sets are said to be *countable*. An example is the set of al integers. It is clear that a finite union of countable sets is countable, since we can successively count first the elements standing at the first position of each of the sets, then the elements at standing at the second position, etc. There is a sharper result:

Theorem 3.2.1 *A countable union of countable sets is countable.*

Proof. Let X_n be the nth countable set and denote its kth element by x_{nk}, $k = 1, 2, \ldots$. The union of the X_n is the set of all elements x_{nk}. We need only to show how to recount them; this can be done as follows. The first element is x_{11}. The second and third elements are x_{12} and x_{21}, i.e., the elements whose indices sum to 3. The next three elements are the elements whose indices sum to 4: x_{13}, x_{22} x_{31}. We then enumerate the elements whose indices sum to 5, 6, etc. In this way we can associate any element of the union with an integer. □

A consequence of this is that the set $\mathbb{Q}$ of all rational numbers is countable. Recall that a rational number can be represented as i/j where i and j are integers; denoting $x_{ij} = i/j$, we obtain the proof. Thus a countable set can have extremely many elements. However it can be shown that

Theorem 3.2.2 *The points of the interval* $[0, 1]$ *are not countable.*

We shall not prove this result here, but instead refer the interested reader to any book on real analysis. We say that the points of $[0, 1]$ form a *continuum*. It is a valid question whether there exist any sets intermediate in power between the countable sets and continuum sets. It turns out that the existence or non-existence of such a set is an independent axiom of arithmetic, a fact which points to the interesting (and sometimes rather mysterious) nature of the real numbers.

Example 3.2.1 Show that the set P_r of all polynomials with rational coefficients is countable.

Solution For each fixed nonnegative integer n, denote by P_r^n the set of all polynomials of degree n having rational coefficients. The set P_r^n can be put into one-to-one correspondence with the countable set

$$\underbrace{\mathbb{Q} \times \mathbb{Q} \times \cdots \mathbb{Q}}_{n+1 \text{ times}}.$$

Finally, the set P_r is given by

$$P_r = \bigcup_{n=0}^{\infty} P_r^n,$$

and this is a countable union of countable sets.

Another example of a countable set is the collection of all finite trigonometric polynomials of the form

$$a_0 + \sum_{k=1}^{n} (a_k \cos kx + b_k \sin kx)$$

with rational coefficients a_0, a_k, b_k.

Let us discuss the real numbers further, keeping in mind that many of our remarks apply to the complex numbers as well. Any real number can be obtained as a limit point of some sequence of rational numbers. This fundamental fact is, of course, the reason why a computer can approximate a real number by a rational number. The ability to approximate the elements of a given set by elements from a certain subset is important in general.

Definition 3.2.1 Let S be a set in a metric space X. We say that a set $Y \subset S$ is *dense in* S if to each point $s \in S$ and $\varepsilon > 0$, there corresponds a point $y \in Y$ such that $d(s, y) < \varepsilon$.

As an example let us note that the set of rational numbers is dense in the set of real numbers. Next, it is clear that we can express the definition in other terms: Y is dense in S if for any $s \in S$ there is a sequence $\{y_n\} \subset Y$ that converges to s.

Example 3.2.2 Let A, B, C be sets in a metric space. Show that if A is dense in B, and B is dense in C, then A is dense in C.

Solution Let us assume that A is dense in B and B is dense in C. Let c be a given point of the set C, and let $\varepsilon > 0$ be given. We can find a point $b \in B$ such that $d(c, b) < \varepsilon/2$. Similarly, we can find a point $a \in A$ such that $d(b, a) < \varepsilon/2$. Since

$$d(c, a) \leq d(c, b) + d(b, a) < \varepsilon/2 + \varepsilon/2 = \varepsilon,$$

we have found a point $a \in A$ that lies within distance ε of our given point $c \in C$.

Definition 3.2.2 If a metric space X contains a countable subset that is dense in X, then we say that X is *separable.*

Example 3.2.3 Demonstrate that the set of all complex numbers with the natural metric (induced by the absolute value of a number) is a separable metric space.

Solution Consider the subset of complex numbers having rational real and imaginary parts. This set is clearly countable (it can be placed into one-to-one correspondence with the countable set $\mathbb{Q} \times \mathbb{Q}$). We must still show that it is dense in $\mathbb{C}$. Let $z = u + iv$ be a given point of $\mathbb{C}$, $i = \sqrt{-1}$, and let $\varepsilon > 0$ be given. Since u and v are real numbers, and the rationals are dense in the reals, we can find rational numbers $\bar{u}$ and $\bar{v}$ such that

$$|u - \bar{u}| < \varepsilon/\sqrt{2}, \qquad |v - \bar{v}| < \varepsilon/\sqrt{2}.$$

The number $\bar{z} = \bar{u} + i\bar{v}$ is a complex number with rational real and imaginary parts. Noting that

$$d(z, \bar{z}) = \sqrt{(u - \bar{u})^2 + (v - \bar{v})^2} < \sqrt{(\varepsilon/\sqrt{2})^2 + (\varepsilon/\sqrt{2})^2} = \varepsilon,$$

we are finished.

Theorem 3.2.3 *Every finite dimensional normed space is separable.*

We leave the simple proof as an exercise and proceed to

Theorem 3.2.4 *Every subspace of a separable space is separable.*

Proof. Let E be a subspace of a separable space X. Consider a countable set consisting of $(x_1, x_2, \ldots)$ which is dense in X. Let B_{ki} be a ball of radius $1/k$ about x_i. By Theorem 3.2.1, the set of all B_{ki} is countable.

For any fixed k the union $\cup_i B_{ki}$ covers X and thus E. For every B_{ki}, take an element of E which lies in B_{ki} (if it exists). Denote this element by e_{ki}. For any $e \in B_{ki} \cap E$, the distance $d(e, e_{ki})$ is less than $2/k$. It follows that the set of all e_{ki} is, on the one hand, countable, and, on the other hand, dense in E. □

The reader will recall that a subspace of a linear space X is a subset of X whose elements satisfy the linear space axioms. This simple theorem is important in practice because sometimes it is easy to prove that a space is separable, whereas a direct proof of separability for one of its subspaces can be difficult.

An important theorem from analysis is the *Weierstrass approximation theorem*: if f is continuous on a compact domain in $\mathbb{R}^n$, then there is a sequence of polynomials that can "uniformly approximate" f on that domain. Upon this result rests

Theorem 3.2.5 *If Ω is a compact domain in $\mathbb{R}^n$, then the space $C(\Omega)$ is separable.*

Proof. The set of all polynomials with rational coefficients is dense in the set of all polynomials. We may then apply the Weierstrass theorem to see that the set P_r of all polynomials with rational coefficients is dense in $C(\Omega)$. Since P_r is countable, $C(\Omega)$ is separable. □

The next result can also be established:

Theorem 3.2.6 *The space $C^{(k)}(\Omega)$ is separable for $k = 1, 2, \ldots$.*

3.3 Cauchy Sequences and Banach Spaces

If $x_n \to x$, then by the triangle inequality

$$d(x_{n+m}, x_n) \leq d(x_{n+m}, x) + d(x, x_n)$$

we see that for any $\varepsilon > 0$ there is a number $N = N(\varepsilon)$ such that for any $n > N$ and any positive integer m,

$$d(x_{n+m}, x_n) \leq \varepsilon.$$

In calculus such a sequence is given a special name:

Definition 3.3.1 A sequence $\{x_n\}$ is a *Cauchy sequence* if to each $\varepsilon > 0$ there corresponds $N = N(\varepsilon)$ such that for every pair of numbers m, n the inequalities $m > N$ and $n > N$ together imply that $d(x_m, x_n) < \varepsilon$.

It is easy to see that every convergent sequence is a Cauchy sequence. According to a famous theorem of calculus, any Cauchy sequence of real numbers is necessarily convergent to some real number, so in $\mathbb{R}$ the notions of Cauchy sequence and convergent sequence are equivalent. In a general metric space this is not so, as is demonstrated next.

Example 3.3.1 Show that the sequence of functions

$$f_n(x) = \begin{cases} 0, & 0 \le x \le \frac{1}{2}, \\ nx - \frac{n}{2}, & \frac{1}{2} \le x \le \frac{1}{2} + \frac{1}{n}, \\ 1 & \frac{1}{2} + \frac{1}{n} \le x \le 1, \end{cases} \qquad (n = 2, 3, 4, \ldots)$$

continuous on $[0, 1]$ is a Cauchy sequence in $L(0, 1)$ but has no continuous limit. Note: the norm in the space $L(0, 1)$ is given by $\|f(x)\| = \int_0^1 |f(x)|\, dx$. Is this a Cauchy sequence in the norm of $C[0, 1]$?

Solution Each $f_n(x)$ is continuous on $[0, 1]$. To see that $\{f_n\}$ is a Cauchy sequence, we assume $m > n$ and calculate

$$\begin{aligned} d(f_n, f_m) &= \int_{\frac{1}{2}}^{\frac{1}{2}+\frac{1}{m}} \left| \left(mx - \frac{m}{2}\right) - \left(nx - \frac{n}{2}\right) \right| dx \\ &\quad + \int_{\frac{1}{2}+\frac{1}{m}}^{\frac{1}{2}+\frac{1}{n}} \left|1 - \left(nx - \frac{n}{2}\right)\right| dx \\ &= \frac{1}{2}\left(\frac{1}{n} - \frac{1}{m}\right) \to 0 \quad \text{as } m, n \to \infty. \end{aligned}$$

However, we have $f_n \to f$ where

$$f = \begin{cases} 0, & 0 \le x \le \frac{1}{2}, \\ 1, & \frac{1}{2} < x \le 1, \end{cases}$$

because

$$d(f_n, f) = \int_{\frac{1}{2}}^{\frac{1}{2}+\frac{1}{n}} \left|1 - \left(nx - \frac{n}{2}\right)\right| dx = \frac{1}{2n} \to 0 \quad \text{as } n \to \infty.$$

The function $f(x)$ is clearly not continuous.

The property that any Cauchy sequence of a metric space has a limit element belonging to the space is so important that a metric space having this property is called *complete.* If a normed space is complete, it is called a *Banach space* in honor of the Polish mathematician Stefan Banach who discovered many important properties of normed spaces.

Definition 3.3.2 A metric space X is *complete* if every Cauchy sequence in X converges to a point in X. A *Banach space* is a complete normed space.

In applications we encounter solutions to many problems expressed in the form of functional series. To deal with them as with series of elements in the usual calculus, let us introduce a series in a Banach space. We say that a series of the form

$$\sum_{k=1}^{\infty} x_k \qquad (x_k \in X)$$

converges to an element $s \in X$ if the sequence $\{s_n\}$ of partial sums

$$s_n = \sum_{k=1}^{n} x_k$$

converges to $s \in X$ in the norm of X. The notion of absolute convergence may also be adapted to series in Banach spaces.

Definition 3.3.3 We say that the series $\sum_{k=1}^{\infty} x_k$ *converges absolutely* if the numerical series $\sum_{k=1}^{\infty} \|x_k\|$ converges.

In a Banach space, as in ordinary calculus, absolute convergence implies convergence:

Theorem 3.3.1 *Let $\{x_k\}$ be a sequence of elements in a Banach space X. If the series $\sum_{k=1}^{\infty} x_k$ converges absolutely, then it converges.*

Proof. By the triangle inequality we have, for any n and $p \geq 1$,

$$\left\| \sum_{k=1}^{n+p} x_k - \sum_{k=1}^{n} x_k \right\| \leq \left| \sum_{k=1}^{n+p} \|x_k\| - \sum_{k=1}^{n} \|x_k\| \right| .$$

By hypothesis the sequence $\sum_{k=1}^{n} \|x_k\|$ converges and is therefore a Cauchy sequence. By the inequality above, $\sum_{k=1}^{n} x_k$ is a Cauchy sequence and will converge to an element of X by completeness. □

Example 3.3.2 Show that under the conditions of the previous theorem we have

$$\left\| \sum_{k=1}^{\infty} x_k \right\| \leq \sum_{k=1}^{\infty} \|x_k\| .$$

Solution We have

$$\left\| \sum_{k=1}^{\infty} x_k \right\| = \left\| \lim_{n\to\infty} \sum_{k=1}^{n} x_k \right\| = \lim_{n\to\infty} \left\| \sum_{k=1}^{n} x_k \right\| \leq \lim_{n\to\infty} \sum_{k=1}^{n} \|x_k\| = \sum_{k=1}^{\infty} \|x_k\| .$$

Here we used the continuity of the norm, and then the triangle inequality for finite sums.

Many of the other results from ordinary calculus also carry over to series in Banach spaces. For example, we can add two convergent series and perform the addition term-by-term:

$$\sum_{k=1}^{\infty} x_k + \sum_{k=1}^{\infty} y_k = \sum_{k=1}^{\infty} (x_k + y_k).$$

We can also multiply a series by a scalar constant λ in the usual way:

$$\lambda \sum_{k=1}^{\infty} x_k = \sum_{k=1}^{\infty} \lambda x_k.$$

Definition 3.3.4 An element x of a metric space X is called a *limit point* of a set S if any ball centered at x contains a point of S different from x. We say that S is *closed in X* if it contains all its limit points.

A limit point is sometimes referred to as a *point of accumulation.* The following result provides a useful alternative characterization for a closed subset of a complete metric space.

Theorem 3.3.2 *A subset S of a complete metric space X supplied with the metric of X is a complete metric space if and only if S is closed in X.*

Proof. Assume S is complete. If x is a limit point of S, then there is a sequence $\{x_n\} \subset S$ such that $x_n \to x$. But every convergent sequence is a Cauchy sequence, hence by completeness $\{x_n\}$ converges to a point of S. From this and uniqueness of the limit we conclude that $x \in S$. Hence S contains all its limit points and is therefore a closed set by definition.

Now assume S is closed. If $\{x_n\}$ is any Cauchy sequence in S, then $\{x_n\}$ is also a Cauchy sequence in X and converges to a point $x \in X$. This point x is also a limit of S however, hence $x \in S$. So every Cauchy sequence in S converges to a point of S, and S is complete by definition. □

We now turn to some examples of Banach and normed spaces. The simplest kind of Banach space is formed by imposing a norm on the linear space $\mathbb{R}^n$ of n-dimensional vectors $\mathbf{x} = (x_1, x_2, \ldots, x_n)$. A standard norm defined on this space is the Euclidean norm

$$\|\mathbf{x}\|_e = \left(\sum_{i=1}^{n} x_i^2 \right)^{1/2}.$$

The resulting Banach space $(\mathbb{R}^n, \|\cdot\|_e)$ is finite dimensional. The following result allows us to ignore the distinction between different normed spaces that are formed from the same underlying finite dimensional vector space by imposing different norms:

Theorem 3.3.3 *On a finite dimensional vector space all norms are equivalent.*

Proof. It is enough to prove that any norm is equivalent to the Euclidean norm $\|\cdot\|_e$. Take any basis i_k that is orthonormal in the Euclidean inner product. We can express any x as $x = \sum_{k=1}^{n} c_k i_k$. Then

$$\|x\|_e = \left(\sum_{k=1}^{n} c_k^2 \right)^{1/2}.$$

For an arbitrary norm $\|\cdot\|$,

$$\|x\| = \left\| \sum_{k=1}^{n} c_k i_k \right\| \leq \sum_{k=1}^{n} |c_k| \, \|i_k\| \leq \sum_{k=1}^{n} \left(\sum_{j=1}^{n} |c_j|^2 \right)^{1/2} \|i_k\| = m \, \|x\|_e$$

where $m = \sum_{k=1}^{n} \|i_k\|$ is finite. So one side is proved. For the other side, consider $\|x\|$ as a function of the n variables c_k. Because of the above inequality it is a continuous function in the usual sense. Indeed

$$| \, \|x_1\| - \|x_2\| \, | \leq \|x_1 - x_2\| \leq m \, \|x_1 - x_2\|_e \, .$$

It is enough to show that on the sphere $\|x\|_e = 1$ we have $\inf \|x\| = a > 0$ (because of homogeneity of norms). Being a continuous function, $\|x\|$ achieves its minimum on the compact set $\|x\|_e = 1$ at a point x_0. So $\|x_0\| = a$. If $a = 0$ then $x_0 = 0$ and thus x_0 does not belong to the unit sphere (in the Euclidean norm). Thus $a > 0$ and for any x we have $\|x\| / \|x\|_e \geq a$. □

We expect that the reader can deal with finite dimensional vectors. Now we would like to understand how to work with elements in infinitely dimensional spaces. The notion of an infinite dimensional vector $\mathbf{x} = (x_1, x_2, x_3, \ldots)$ with a countable number of components is, of course, a straightforward generalization of the notion of a finite dimensional vector $\mathbf{x} = (x_1, x_2, \ldots, x_n)$. But let us also note that such a vector can be encountered by considering a numerical sequence $\{x_i\}$ as a whole entity. In this case the individual terms x_i of the sequence become the components of a vector $\mathbf{x}$. In fact, we shall use the terms "infinite dimensional vector" and "sequence" interchangeably. Another way to introduce vectors with infinitely many components is to consider expansions of functions into series of different types, say Fourier or Taylor expansions. Combining coefficients of such an expansion, we get something like a vector with infinitely many components.

Our first infinite dimensional Banach spaces can be formed by imposing suitable norms on spaces of infinite dimensional vectors. Such spaces are called *sequence spaces.* For example, we may place under consideration the set c of all convergent numerical sequences and impose the norm

$$\|\mathbf{x}\| = \sup_i |x_i|.$$

An interesting family of sequence spaces can be defined, one for each integer $p \geq 1$. The space ℓ^p is the set of all vectors $\mathbf{x}$ such that $\sum_{i=1}^{\infty} |x_i|^p < \infty$, and its norm is taken to be

$$\|\mathbf{x}\| = \left(\sum_{i=1}^{\infty} |x_i|^p \right)^{1/p} . \tag{3.3.1}$$

The fact that (3.3.1) is a norm is a consequence of the *Minkowski inequality*

$$\left(\sum_{i=1}^{\infty} |x_i + y_i|^p \right)^{1/p} \leq \left(\sum_{i=1}^{\infty} |x_i|^p \right)^{1/p} + \left(\sum_{i=1}^{\infty} |y_i|^p \right)^{1/p}$$

since satisfaction of the other norm axioms for (3.3.1) is evident. An important special case is the space ℓ^2 of *square-summable* sequences $\mathbf{x}$ with $\sum_{i=1}^{\infty} |x_i|^2 < \infty$ and norm

$$\|\mathbf{x}\| = \left(\sum_{i=1}^{\infty} |x_i|^2 \right)^{1/2} .$$

Looking ahead, we mention that any element in a separable Hilbert space H (it is a complete space with an inner product that is similar to the dot product in a Euclidean space) can be represented as a Fourier expansion with respect to an orthonormal basis of H, and there is an one-to-one correspondence between the elements of H and ℓ^2. So all the general properties which we could establish for the elements of ℓ^2 can be reformulated for a separable Hilbert space H and vice versa. We can add that ℓ^2 was the first space introduced by D. Hilbert that initiated functional analysis as a branch of mathematics.

We emphasize that the normed spaces c and ℓ^p are not defined on the same underlying set of vectors. For example, the vector $\mathbf{x} = (1, 1, 1, \ldots)$

obviously belongs to c but not to ℓ^p for any $p \geq 1$. Moreover, there is no analog to Theorem 3.3.3 for infinite dimensional spaces.

There is a subspace of c denoted by c_0 that consists of vectors (sequences) having zero limit. Note that a set of sequences converging to some fixed nonzero limit could not be a linear space. If we wish to consider the set of all convergent sequences with some nonzero limit, we call it a *cone.* We can restrict a cone to some of its subsets by placing additional conditions on the components of vectors.

It is also possible to introduce weighted spaces of sequences with norms of the form

$$\|\mathbf{x}\| = \left(\sum_{i=1}^{\infty} k_i |x_i|^2\right)^{1/2}$$

where the $k_i \geq 0$ are constants used to weight the terms of the sequence.

We can show that all of the spaces mentioned above are Banach spaces. Let us show this, for example, for c. To do this we use the fact that the normed space consisting of the set $\mathbb{R}$ of real numbers under the usual norm $|x|$ is a Banach space. Let $\{\mathbf{x}^{(k)}\}$ be a Cauchy sequence in c. The kth term of this sequence is a numerical sequence:

$$\mathbf{x}^{(k)} = (x_1^{(k)}, x_2^{(k)}, x_3^{(k)}, \ldots).$$

To each $\varepsilon > 0$ there corresponds $N = N(\varepsilon)$ such that

$$\|\mathbf{x}^{(n+m)} - \mathbf{x}^{(n)}\|_c = \sup_i |x_i^{(n+m)} - x_i^{(n)}| \leq \varepsilon$$

whenever $n > N$ and $m > 0$. This implies that

$$|x_i^{(n+m)} - x_i^{(n)}| \leq \varepsilon \quad \text{for each } i \tag{3.3.2}$$

whenever $n > N$ and $m > 0$. Hence $\{x_i^{(j)}\}$ is a Cauchy sequence of real numbers for any fixed i. By the completeness of the normed space $(\mathbb{R}, \|\cdot\|)$ we know that $\{x_i^{(j)}\}$ converges (as $j \to \infty$) to a limit, say x_i^*, in $\mathbb{R}$. Now let

$$\mathbf{x}^* = (x_1^*, x_2^*, x_3^*, \ldots).$$

We will show that

$$\mathbf{x}^{(k)} \to \mathbf{x}^*. \tag{3.3.3}$$

Fix $n > N$; by (3.3.2) and continuity

$$\lim_{m\to\infty} |x_i^{(n+m)} - x_i^{(n)}| \le \varepsilon$$

which gives

$$|x_i^* - x_i^{(n)}| \le \varepsilon \quad \text{for each } i.$$

Hence

$$\sup_i |x_i^* - x_i^{(n)}| = \|\mathbf{x}^* - \mathbf{x}^{(n)}\|_c \le \varepsilon$$

for $n > N$, so we have established (3.3.3). Finally we must show that $\mathbf{x}^* \in c$ by showing that $\{x_i^*\}$ converges. Since every Cauchy sequence of real numbers converges, it suffices to show that $\{x_i^*\}$ is a Cauchy sequence. Let us consider the difference

$$|x_n^* - x_m^*| \le |x_n^* - x_n^{(k)}| + |x_n^{(k)} - x_m^{(k)}| + |x_m^{(k)} - x_m^*|$$

and use an $\varepsilon/3$ argument. Let $\varepsilon > 0$ be given. We can make the first and third terms on the right less than $\varepsilon/3$ for any n, m by fixing k sufficiently large. For this k, $\{x_j^{(k)}\}$ is a Cauchy sequence; hence we can make the second term on the right less than $\varepsilon/3$ by taking n and m sufficiently large.

We see here a general pattern common to many completeness proofs. We take an arbitrary Cauchy sequence $\{x_n\}$ in (X, d), construct an element x that appears to be the limit of $\{x_n\}$, prove that $x \in X$, and prove that $x_n \to x$ under d.

Example 3.3.3 Show that c_0 is a Banach space.

Solution Let $\{\mathbf{x}^{(k)}\}$ be a Cauchy sequence in c_0. The kth term of this sequence is a numerical sequence

$$\mathbf{x}^{(k)} = (x_1^{(k)}, x_2^{(k)}, x_3^{(k)}, \ldots)$$

that converges to 0. As we did with a Cauchy sequence in the space c, we can show that

$$\mathbf{x}^{(k)} \to \mathbf{x}^* = (x_1^*, x_2^*, x_3^*, \ldots) \qquad \text{where} \qquad x_i^* = \lim_{j\to\infty} x_i^{(j)}.$$

(As before, in the process we find that by fixing n sufficiently large we can get the inequality $|x_i^* - x_i^{(n)}| \le \varepsilon$ to hold for all i.) To complete the proof we must show that $\mathbf{x}^* \in c_0$, i.e., that $x_i^* \to 0$ as $i \to \infty$. Let $\varepsilon > 0$ be given. We have

$$|x_i^*| \le |x_i^* - x_i^{(k)}| + |x_i^{(k)}|.$$

We can fix k large enough that the first term on the right is less than $\varepsilon/2$ for all i. For this k, we can choose i large enough that the second term on the right is less than $\varepsilon/2$.

Now let us turn to some spaces of functions. We have introduced the space $C(\Omega)$. If Ω is a compact set in $\mathbb{R}^n$, then $C(\Omega)$ is a Banach space. Indeed, the Weierstrass theorem states that a uniformly convergent sequence of functions defined on a compact set has as a limit a continuous function. A sequence of functions $\{f_k(\mathbf{x})\}$ is a Cauchy sequence in $C(\Omega)$ if to each $\varepsilon > 0$ there corresponds $N = N(\varepsilon)$ such that

$$\max_{\mathbf{x}\in\Omega} |f_{n+m}(\mathbf{x}) - f_n(\mathbf{x})| \le \varepsilon$$

for any $n > N$ and any positive integer m. This definition means that $\{f_n(\mathbf{x})\}$ converges uniformly on Ω and thus its limit point exists and belongs to $C(\Omega)$. (The reader sees that the uniform convergence of a sequence of functions in calculus and convergence with respect to the norm of $C(\Omega)$, Ω being compact, are the same.) That is, by definition, $C(\Omega)$ is a Banach space. Similarly we can show that $C^{(k)}(\Omega)$ is a Banach space.

We mentioned earlier that on the set of functions continuous on a compact set Ω we can introduce

$$\|f(\mathbf{x})\|_{L^p(\Omega)} = \left(\int_\Omega |f(\mathbf{x})|^p \, d\Omega\right)^{1/p}$$

for $p \ge 1$. Writing out the corresponding Riemann sums for the integral and then using the limit passage, we may show that the triangle inequality is fulfilled for this. Fulfillment of the remaining norm axioms is evident. Exercise ?? shows that the set of continuous functions under this norm, for the case $p = 1$, is not a Banach space. The situation is the same for any $p > 1$.

On the set of differentiable functions we can introduce an important class of norms called *Sobolev's norms.* The simplest is one that is called the norm of $W^{1,2}[0,1]$:

$$\|f\|_{W^{1,2}[0,1]} = \left(\int_0^1 \left(|f(x)|^2 + |f'(x)|^2\right) dx\right)^{1/2}$$

This was first studied by S. Banach. The general form of a Sobolev norm is

$$\|f\|_{W^{l,p}(\Omega)} = \left(\int_\Omega \sum_{|\alpha| \le l} |D^\alpha f|^p \, d\Omega \right)^{1/p}, \qquad p \ge 1. \tag{3.3.4}$$

The set of l-times continuously differentiable functions on Ω is not complete in the norm (3.3.4). In these norms as well in the L^p norms the difference of close functions can be very big on subdomains of small area. Later we shall introduce other spaces with these norms that will turn out to be Banach spaces.

Example 3.3.4 (a) Show that if a sequence converges, then any of its subsequences also converges and has the same limit. (b) Show that if some subsequence of a Cauchy sequence has a limit, then the entire sequence must converge to the same limit. (c) A set S in a normed space X is *bounded* if there exists $R > 0$ such that $\|x\| \le R$ whenever $x \in S$. Show that every Cauchy sequence is bounded.

Solution (a) Let $\{x_{n_k}\}$ be a subsequence of $\{x_n\}$ where $x_n \to x$. Given $\varepsilon > 0$, we can find N such that $n \ge N$ implies $d(x_n, x) < \varepsilon$. Since $n_k \ge k$ for all k, we have $d(x_{n_k}, x) < \varepsilon$ whenever $k \ge N$. (b) Suppose that $\{x_{n_k}\}$ is a convergent subsequence of a Cauchy sequence $\{x_n\}$. We show that if $x_{n_k} \to x$, then $x_n \to x$. Let $\varepsilon > 0$ be given and choose N such that $d(x_n, x_m) < \varepsilon/2$ for $n, m > N$. Since $x_{n_k} \to x$, there exists $n_k > N$ such that $d(x_{n_k}, x) < \varepsilon/2$. So for $n > N$ we have $d(x_n, x) \le d(x_n, x_{n_k}) + d(x_{n_k}, x) < \varepsilon/2 + \varepsilon/2 = \varepsilon$. (c) Let $\{x_n\}$ be a Cauchy sequence. There exists N such that

$$\|x_n - x_{N+1}\| < 1$$

whenever $n > N$. For all $n > N$ we have

$$\|x_n\| \le \|x_n - x_{N+1}\| + \|x_{N+1}\| < \|x_{N+1}\| + 1.$$

Hence an upper bound for $\|x_n\|$ is given by

$$B = \max\{\|x_1\|, \ldots, \|x_N\|, \|x_{N+1}\| + 1\}.$$

Example 3.3.5 Show that $\mathbb{R}^n$ is complete.

Solution Let $\{\mathbf{x}^{(k)}\}$ be a Cauchy sequence in $\mathbb{R}^n$. The kth term of this sequence is an n-tuple

$$\mathbf{x}^{(k)} = (x_1^{(k)}, \ldots, x_n^{(k)}).$$

Since $\{\mathbf{x}^{(k)}\}$ is a Cauchy sequence, for each $\varepsilon > 0$ there exists N such that $m > N$ and $p > 0$ imply

$$d(\mathbf{x}^{(m+p)}, \mathbf{x}^{(m)}) = \left[\sum_{i=1}^{n} \left|x_i^{(m+p)} - x_i^{(m)}\right|^2\right]^{1/2} \leq \varepsilon.$$

Since all terms in the sum are non-negative, we have

$$\left|x_i^{(m+p)} - x_i^{(m)}\right| < \varepsilon \quad \text{for each } i = 1, \ldots, n \tag{3.3.5}$$

whenever $m > N$ and $p > 0$. Hence $x_i^{(j)}$ is a Cauchy sequence of reals for any $i = 1, \ldots, n$. By the completeness of $\mathbb{R}$ we know that $x_i^{(j)}$ converges (as $j \to \infty$) to a limit, say x_i^*, in $\mathbb{R}$. Now let

$$\mathbf{x}^* = (x_1^*, \ldots, x_n^*).$$

We will show that

$$\mathbf{x}^{(k)} \to \mathbf{x}^* \tag{3.3.6}$$

where convergence is understood in the sense of the Euclidean metric on $\mathbb{R}^n$. Fix $m > N$; by (3.3.5) we get

$$\lim_{p\to\infty} \left|x_i^{(m+p)} - x_i^{(m)}\right| \leq \varepsilon,$$

hence

$$|x_i^* - x_i^{(m)}| \leq \varepsilon \quad \text{for each } i = 1, \ldots, n.$$

So

$$\left(\sum_{i=1}^{n} \left|x_i^* - x_i^{(m)}\right|^2\right)^{1/2} = d(\mathbf{x}^*, \mathbf{x}^{(m)}) \leq \sqrt{n}\varepsilon$$

for $m > N$, and we have established (3.3.6). We conclude that every Cauchy sequence in $\mathbb{R}^n$ converges to a point of $\mathbb{R}^n$, hence $\mathbb{R}^n$ is complete.

Example 3.3.6 The Cartesian product $X \times Y$ of two linear spaces X and Y can form a linear space under suitable definitions of vector addition and scalar multiplication. If X and Y are also normed spaces with norms $\|\cdot\|_X$, $\|\cdot\|_Y$, respectively, then $X \times Y$ is a normed space under the norm

$$\|(x, y)\| = \|x\|_X + \|y\|_Y .$$

Show that if X and Y are Banach spaces, then so is $X \times Y$.

Solution Choose any Cauchy sequence $\{(x_k, y_k)\} \subset X \times Y$. Then

$$\begin{aligned}\|(x_m, y_m) - (x_n, y_n)\|_{X\times Y} &= \|(x_m - x_n, y_m - y_n)\|_{X\times Y} \\ &= \|x_m - x_n\|_X + \|y_m - y_n\|_Y \to 0\end{aligned}$$

as $m, n \to \infty$, hence

$$\|x_m - x_n\|_X \to 0 \quad \text{and} \quad \|y_m - y_n\|_Y \to 0 \quad \text{as } m, n \to \infty.$$

So $\{x_k\}$ and $\{y_k\}$ are each Cauchy sequences in their respective spaces X, Y; since these are Banach spaces there exist $x \in X$ and $y \in Y$ such that $x_k \to x$ and $y_k \to y$. Finally, we see that $(x_k, y_k) \to (x, y)$ in the norm of $X \times Y$:

$$\begin{aligned}\|(x_k, y_k) - (x, y)\|_{X\times Y} &= \|(x_k - x, y_k - y)\|_{X\times Y} \\ &= \|x_k - x\|_X + \|y_k - y\|_Y \to 0 \quad \text{as } k \to \infty.\end{aligned}$$

3.4 The Completion Theorem

It is inconvenient to deal with an incomplete space. For example, using only rational numbers we leave out such numbers as $\sqrt{2}$ and π, and so cannot obtain exact solutions for many quadratic equations or geometry problems. Various approaches can be used to introduce irrational numbers. To define an irrational number π, we can define a sequence of approximations such as 3, 3.1, 3.14, 3.141, and so on. The limit of this sequence is what we call π. But the approximating sequence $4, 3.2, 3.142, \ldots$ also consists of rational numbers and can be used to define the same number π. There are infinitely many sequences having this same limit, and we can associate this set of Cauchy sequences together as an entity that defines π. We call such sequences *equivalent*. The same can be done with any irrational number. If we then regard a real number as something defined by a set of all equivalent sequences, a rational number can be represented as a set of all equivalent sequences one of which is a *stationary* sequence all of whose terms are this rational number. We shall use this idea to "extend" an incomplete space to one that is complete. In advance we shall introduce several notions.

Definition 3.4.1 Two sequences $\{x_n\}, \{y_n\}$ in a metric space (M, d) are said to be *equivalent* if $d(x_n, y_n) \to 0$ as $n \to \infty$. If $\{x_n\}$ is a Cauchy sequence in M, we can collect into an equivalence class X all Cauchy sequences in M that are equivalent to $\{x_n\}$. Any Cauchy sequence from X is called a *representative* of X. To any $x \in M$ there corresponds a *stationary* equivalence class containing the Cauchy sequence $x, x, x, \ldots$.

Definition 3.4.2 A mapping $F\colon M_1 \to M_2$ is an *isometry* between (M_1, d_1) and (M_2, d_2) if $d_1(x, y) = d_2(F(x), F(y))$ for all $x, y \in M_1$. Distances are obviously preserved under such a mapping. If F is also a one-to-one correspondence between M_1 and M_2, then it is a *one-to-one isometry* and the two metric spaces are said to be *isometric.* Isometric spaces are essentially the same, the isometry amounting to a mere relabeling of the points in each space.

With this terminology in place we can state the important metric space *completion theorem*:

Theorem 3.4.1 *For a metric space M, there is a one-to-one isometry between M and a set $\tilde{M}$ which is dense in a complete metric space M^*. We call M^* the* completion *of M.*

Proof. As we said, we shall use the same idea as above for introducing the needed space. The proof consists of four steps: (1) introduction of the elements of the space M^*; (2) introduction of a metric on this space and verification of the axioms; (3) demonstration that the new space is complete; (4) verification of the remaining statements of the theorem.

1. As indicated in Definition 3.4.1, we collect into an equivalence class X all Cauchy sequences in M that are equivalent to a given Cauchy sequence $\{x_n\}$. We denote the set of all the equivalence classes by M^*, and the set of all stationary equivalence classes by $\tilde{M}$.

2. We impose a metric on M^*. Given $X, Y \in M^*$, we choose any representatives $\{x_n\} \in X$ and $\{y_n\} \in Y$ and define

$$d(X, Y) = \lim_{n \to \infty} d(x_n, y_n). \tag{3.4.1}$$

This same metric is applied to the subspace $\tilde{M}$ of M^*. To see that $d(X, Y)$ is actually a metric, we must first check that the limit in (3.4.1) exists and is independent of the choice of representatives. By metric axiom D4 we have

$$d(x_n, y_n) \le d(x_n, x_m) + d(x_m, y_m) + d(y_m, y_n)$$

so that

$$d(x_n, y_n) - d(x_m, y_m) \le d(x_n, x_m) + d(y_m, y_n).$$

Interchanging m and n we obtain a similar inequality; combining the two, we obtain

$$|d(x_n, y_n) - d(x_m, y_m)| \le d(x_n, x_m) + d(y_n, y_m).$$

But $d(x_n, x_m) \to 0$ and $d(y_n, y_m) \to 0$ as $m, n \to \infty$ because $\{x_n\}$ and $\{y_n\}$ are Cauchy sequences. Thus

$$|d(x_n, y_n) - d(x_m, y_m)| \to 0 \quad \text{as } m, n \to \infty$$

and we see that $\{d(x_n, y_n)\}$ is a Cauchy sequence in $\mathbb{R}$. By completeness of $\mathbb{R}$, the limit in (3.4.1) exists. To show that it does not depend on the choice of representatives, we take any $\{x'_n\} \in X$ and $\{y'_n\} \in Y$ and show that

$$\lim_{n\to\infty} d(x'_n, y'_n) = \lim_{n\to\infty} d(x_n, y_n). \tag{3.4.2}$$

Because $\lim_{n\to\infty} d(x_n, x'_n) = 0 = \lim_{n\to\infty} d(y_n, y'_n)$, the inequality

$$|d(x_n, y_n) - d(x'_n, y'_n)| \le d(x_n, x'_n) + d(y_n, y'_n)$$

gives

$$\lim_{n\to\infty} |d(x_n, y_n) - d(x'_n, y'_n)| = 0$$

which implies (3.4.2). We now check that the metric axioms are satisfied by $d(X, Y)$:

D1: Since $d(x_n, y_n) \ge 0$ for all n, it follows that

$$d(X, Y) = \lim_{n\to\infty} d(x_n, y_n) \ge 0.$$

D2: If $X = Y$ then $d(X, Y) = 0$ (we can choose the same Cauchy sequence $\{x_n\}$ from both X and Y, and since the limit is unique we get the needed conclusion). Conversely, if $d(X, Y) = 0$ then any two Cauchy sequences $\{x_n\} \in X$ and $\{y_n\} \in Y$ satisfy $\lim_{n\to\infty} d(x_n, y_n) = 0$. By definition they are equivalent, hence $X = Y$.

D3: We have

$$d(X,Y) = \lim_{n\to\infty} d(x_n,y_n) = \lim_{n\to\infty} d(y_n,x_n) = d(Y,X).$$

D4: For $x_n, y_n, z_n \in M$ the triangle inequality gives

$$d(x_n,y_n) \le d(x_n,z_n) + d(z_n,y_n);$$

as $n \to \infty$ we have

$$d(X,Y) \le d(X,Z) + d(Z,Y)$$

for the equivalence classes X, Y, Z containing $\{x_n\}$, $\{y_n\}$, $\{z_n\}$, respectively.

3. To see that M^* is complete, we must show that for any Cauchy sequence $\{X^i\} \subset M^*$, there exists

$$X = \lim_{i\to\infty} X^i \in M^*. \tag{3.4.3}$$

Indeed, from each X^i we choose a Cauchy sequence $\{x_j^{(i)}\}$ and from this an element denoted x_i such that $d(x_i, x_j^{(i)}) < 1/i$ whenever $j > i$. To see that $\{x_i\}$ is a Cauchy sequence, we denote by X_i the equivalence class containing the stationary sequence $(x_i, x_i, \ldots)$ and write

$$\begin{aligned} d(x_i,x_j) &= d(X_i,X_j) \\ &\le d(X_i,X^i) + d(X^i,X^j) + d(X^j,X_j) \\ &\le \frac{1}{i} + d(X^i,X^j) + \frac{1}{j}. \end{aligned}$$

As $i,j \to \infty$, $d(x_i,x_j) \to 0$ as required. Finally, denote by X the equivalence class containing $\{x_i\}$. Because $\{x_i\}$ is a Cauchy sequence,

$$\begin{aligned} d(X^i,X) &\le d(X^i,X_i) + d(X_i,X) \\ &\le \frac{1}{i} + d(X_i,X) \\ &= \frac{1}{i} + \lim_{j\to\infty} d(x_i,x_j) \to 0 \quad \text{as } i \to \infty. \end{aligned}$$

This proves (3.4.3).

4. $\tilde{M}$ is dense in M^*. To see this, choose $X \in M^*$. Selecting a representative $\{x_n\}$ from X, we denote by X_n the stationary equivalence class

containing the stationary sequence $(x_n, x_n, \ldots)$. Then

$$d(X_n, X) = \lim_{m\to\infty} d(x_n, x_m) \to 0 \quad \text{as } n \to \infty$$

since $\{x_n\}$ is a Cauchy sequence.

The equality

$$d(X, Y) = d(x, y)$$

if X and Y are stationary classes corresponding to x and y, respectively, demonstrates the one-to-one isometry between M and $\tilde{M}$. □

Corollary 3.4.1 *If M is a linear space, the isometry preserves algebraic operations.*

Since a normed space is a linear metric space we immediately have

Theorem 3.4.2 *Any normed space X can be completed in its natural metric $d(x, y) = \|x - y\|$, resulting in a Banach space X^*.*

We will also make use of the following result:

Theorem 3.4.3 *The completion of a separable metric space is separable.*

Proof. Suppose X is a separable metric space, containing a countable, dense subset S. The completion theorem places X into one-to-one correspondence with a set $\tilde{X}$ that is dense in the completion X^*. Let $\tilde{S}$ be the image of S under this correspondence. Since the correspondence is also an isometry, $\tilde{S}$ is dense in $\tilde{X}$. So we have $\tilde{S} \subseteq \tilde{X} \subseteq X^*$, where each set is dense in the next; therefore $\tilde{S}$ is dense in X^*. Since $\tilde{S}$ is evidently countable, the proof is complete. □

We have lingered over the completion theorem because it is the background for many important notions we will introduce. These include the Lebesgue integral, and the Sobolev and energy spaces.

3.5 Contraction Mapping Principle

We know that the iterative Newton method of tangents for finding zeroes of a differentiable function $g(x)$ demonstrates fast convergence and is widely used in practice. In this method we reduce a given problem to a problem of the form

$$x = f(x) \tag{3.5.1}$$

and the procedure for finding zeroes of $g(x)$ is

$$x_{n+1} = f(x_n).$$

A solution x^* of (3.5.1) is such that the value of $f(x)$ at x^* is x^*, so a solution is a *fixed point* of the mapping f. There are different ways in which an equation $g(x) = 0$ may be reduced to the form (3.5.1), the simplest but not the best of which is to represent the equation as $x = x - g(x)$. Such a transformation is good only when the iterative procedure of solution converges fast enough. It turns out that we can reduce various equations of different natures, from systems of equations to boundary value problems and integral equations, to forms of the type (3.5.1) so that the iterative procedure gives us a good approximation to a solution with few iterations required. The methods of reduction of a general equation $G(x) = 0$ extend those known for the simple equation $g(x) = 0$. In this section we discuss a class of problems of the general form

$$x = F(x) \qquad (3.5.2)$$

where $F(x)$ is a mapping on a metric space M, i.e.,

$$F\colon M \to M,$$

and $x \in M$ is the desired unknown. We see that if x is to satisfy (3.5.2) then the image of x under F must be x itself, so we continue to use the term "fixed point" in this more general case.

We would like to use an iterative process to solve equation (3.5.2). The iteration begins with an initial value $x_0 \in M$ (sometimes called the *seed* element) and proceeds via use of the recursion

$$x_{k+1} = F(x_k) \qquad k = 0, 1, 2, \ldots. \qquad (3.5.3)$$

Under suitable conditions the resulting values $x_0, x_1, x_2, \ldots$ will form a sequence of *successive approximations* to the desired solution. That is, if the approach works we will have

$$\lim_{k\to\infty} x_k = x^*$$

where x^* is a fixed point of F. With this background, let us formulate conditions providing the applicability of the method.

Definition 3.5.1 Let $F(x)$ be a mapping on M. We say that $F(x)$ is a *contraction mapping* if there exists a number $\alpha \in [0, 1)$ such that

$$d(F(x), F(y)) \le \alpha d(x, y) \tag{3.5.4}$$

for every pair of elements $x, y \in M$.

We see that repeated application of (3.5.4) yields

$$d(x_{k+1}, x_k) \le \alpha^k d(x_1, x_0), \qquad k = 0, 1, 2, \ldots,$$

and with $0 \le \alpha < 1$ the successive iterates will land closer and closer together in M. We might expect these iterates to converge to a solution; rigorous confirmation that they do is provided by the following celebrated result due to Banach. It is known as the *contraction mapping theorem.*

Theorem 3.5.1 *A contraction mapping F with constant α, $0 \le \alpha < 1$, on a complete metric space M has a unique fixed point. Convergence of successive approximations to the fixed point occurs independently of the choice of seed element.*

Proof. Let us choose an arbitrary seed element $x_0 \in M$ for the recursion (3.5.3). Using the triangle inequality for several elements, for $m > n$ we have

$$d(x_m, x_n) \le d(x_m, x_{m-1}) + d(x_{m-1}, x_{m-2}) + \cdots + d(x_{n+2}, x_{n+1}) + d(x_{n+1}, x_n)$$

hence

$$\begin{aligned} d(x_m, x_n) &\le (\alpha^{m-1} + \alpha^{m-2} + \cdots + \alpha^{n+1} + \alpha^n) d(x_1, x_0) \\ &= \alpha^n (1 + \alpha + \cdots + \alpha^{m-n-2} + \alpha^{m-n-1}) d(x_1, x_0) \\ &\le \alpha^n (1 - \alpha)^{-1} d(x_1, x_0) \\ &\to 0 \quad \text{as } n \to \infty. \end{aligned}$$

In this, we summed up the geometrical progression. So $\{x_k\}$ is a Cauchy sequence, and by completeness of M there is a point $x^* \in M$ such that $x_k \to x^*$ as $k \to \infty$. From the contraction condition for F it follows that $F(x)$ is continuous on M, hence

$$x^* = \lim_{k\to\infty} F(x_k) = F\left(\lim_{k\to\infty} x_k\right) = F(x^*).$$

We have therefore established the existence of a fixed point of $F(x)$. Uniqueness is proved by assuming the existence of another such point y^*. Then

$$d(x^*, y^*) = d(F(x^*), F(y^*)) \leq \alpha\, d(x^*, y^*)$$

so that

$$(1 - \alpha) d(x^*, y^*) = 0.$$

But $\alpha < 1$, so we must have $d(x^*, y^*) = 0$. Hence $x^* = y^*$ and the proof is complete. □

The proof of the contraction mapping theorem also provides information concerning the rate of convergence of the iterates x_k to x^*. Specifically, we have

Corollary 3.5.1 *Let $F(x)$ be a contraction mapping on a complete metric space M. Then the estimates*

$$d(x_n, x^*) \leq \frac{\alpha^n}{1 - \alpha} d(x_1, x_0) \tag{3.5.5}$$

and

$$d(x_n, x^*) \leq \frac{\alpha}{1 - \alpha} d(x_n, x_{n-1}) \tag{3.5.6}$$

both hold for $n = 0, 1, 2, \ldots$, where α is the contraction constant for $F(x)$ and x^* is the fixed point of F.

Proof. In the inequality

$$d(x_m, x_n) \leq \frac{\alpha^n}{1 - \alpha} d(x_1, x_0)$$

we can pass to the limit as $m \to \infty$ and obtain (3.5.5). If on the right-hand side of (3.5.5) we take x_0 to be x_{n-1}, then x_1 becomes x_n and we obtain (3.5.6). □

The inequality (3.5.5) is known as an *a priori* error estimate, since it provides an upper bound on $d(x_n, x^*)$ in terms of quantities known at the start of the iteration procedure. Inequality (3.5.6) is known as an *a posteriori* error estimate, and can be used to monitor convergence as the iteration proceeds.

The contraction mapping principle can be applied to a variety of problems. Consider a (possibly finite dimensional) system of linear equations

$$x_i = \sum_{j=1}^{\infty} a_{ij} x_j + c_i \qquad (i = 1, 2, 3, \ldots).$$

To solve this problem by iteration we can write

$$\mathbf{x}^{(k+1)} = F(\mathbf{x}^{(k)}) = A(\mathbf{x}^{(k)}) + \mathbf{c}$$

where $\mathbf{c} = (c_1, c_2, c_3, \ldots)$ is a given vector, $\{\mathbf{x}^{(k)}\}$ is a sequence of vector iterates

$$\begin{aligned} \mathbf{x}^{(0)} &= (x_1^{(0)}, x_2^{(0)}, x_3^{(0)}, \ldots), \\ \mathbf{x}^{(1)} &= (x_1^{(1)}, x_2^{(1)}, x_3^{(1)}, \ldots), \\ \mathbf{x}^{(2)} &= (x_1^{(2)}, x_2^{(2)}, x_3^{(2)}, \ldots), \\ &\vdots \end{aligned}$$

and A is the mapping given by

$$A(\mathbf{x}^{(k)}) = \left(\sum_{j=1}^{\infty} a_{1j} x_j^{(k)}, \sum_{j=1}^{\infty} a_{2j} x_j^{(k)}, \sum_{j=1}^{\infty} a_{3j} x_j^{(k)}, \ldots \right).$$

We should note that the possibility to employ iteration (and even simply to solve the system) depends on the space in which we wish to find a solution. Here we shall suppose that $\mathbf{c}$ belongs to the space ℓ^∞, which is the space of all bounded sequences under the norm

$$\|\mathbf{x}\|_\infty = \sup_{i \geq 1} |x_i|.$$

For the operator A to act in ℓ^∞ it is sufficient that the quantity

$$K = \sup_{i \geq 1} \sum_{j=1}^{\infty} |a_{ij}|$$

is finite. This follows from the fact that $\mathbf{c} \in \ell^\infty$ and the next chain of inequalities, with which we will determine when F is a contraction on ℓ^∞.

We have

$$\begin{aligned}\|F(\mathbf{x}) - F(\mathbf{x}')\|_\infty &= \sup_{i\geq 1}\left|\left(\sum_{j=1}^{\infty} a_{ij}x_j + c_i\right) - \left(\sum_{j=1}^{\infty} a_{ij}x'_j + c_i\right)\right| \\ &= \sup_{i\geq 1}\left|\sum_{j=1}^{\infty} a_{ij}(x_j - x'_j)\right| \leq \sup_{i\geq 1}\sum_{j=1}^{\infty} |a_{ij}||x_j - x'_j| \\ &\leq \sup_{i\geq 1}\left[\left(\sup_{j\geq 1}|x_j - x'_j|\right)\left(\sum_{j=1}^{\infty}|a_{ij}|\right)\right] \\ &= \left(\sup_{i\geq 1}\sum_{j=1}^{\infty}|a_{ij}|\right)\left(\sup_{j\geq 1}|x_j - x'_j|\right)\end{aligned}$$

hence

$$\|F(\mathbf{x}) - F(\mathbf{x}')\|_\infty \leq K\,\|\mathbf{x} - \mathbf{x}'\|_\infty .$$

With $K < 1$ we have a contraction and Banach's theorem applies.

In other sequence spaces the appropriate conditions for a_{ij} are different. It is left as an exercise to treat the problem for iterations and a solution in the space ℓ^2.

Before treating the next example let us state another corollary to the contraction mapping theorem. By F^k we denote the k-fold composition of the mapping F: that is, we have

$$F^{n+1}(x) = F(F^n(x)), \qquad n = 1, 2, 3, \ldots,$$

where it is understood that $F^1 = F$.

Corollary 3.5.2 *If F^k is a contraction mapping on a complete metric space for some integer $k \geq 1$, then F has a unique fixed point. Convergence of successive approximations occurs independently of the choice of seed element.*

Proof. F^k has a unique fixed point x^* by Theorem 3.5.1; moreover,

$$\lim_{n\to\infty}(F^k)^n(x) = x^*$$

for any $x \in M$. Putting $x = F(x^*)$ we obtain

$$x^* = \lim_{n\to\infty}(F^k)^n F(x^*) = \lim_{n\to\infty} F(F^k)^n(x^*) = \lim_{n\to\infty} F(x^*) = F(x^*),$$

hence x^* is also a fixed point of F. (Here we have used the assumption that x^* is a fixed point of F^k, hence it is a fixed point of $(F^k)^n$, hence $(F^k)^n(x^*) = x^*$.) If y^* is another fixed point of F, then y^* is also fixed point of F^k, hence $y^* = x^*$. □

We now proceed to our second example. An integral equation of the form

$$\psi(x) = g(x) + \lambda \int_a^x K(x,t)\psi(t)\,dt, \qquad x \in [a,b], \tag{3.5.7}$$

where $\psi(x)$ is unknown, is said to be a *Volterra integral equation.* We suppose that $g(x)$ is continuous on $[a,b]$, and that the kernel $K(x,t)$ is continuous on the closed, triangular region $a \le t \le x$, $a \le x \le b$. Let us show that the mapping F given by

$$F[\psi(x)] = g(x) + \lambda \int_a^x K(x,t)\psi(t)\,dt$$

will generate convergent successive approximations by iteration in $C(a,b)$; our approach to this will be to prove that F^n is a contraction mapping for some integer $n \ge 1$. First, let $u(x)$ and $v(x)$ be any two elements of $C(a,b)$ and observe that

$$|F[v(x)] - F[u(x)]| \le |\lambda| \int_a^x |K(x,t)||v(t) - u(t)|\,dt.$$

Now $K(x,t)$, being continuous on a compact set, is bounded by some number M. So

$$\begin{aligned}|F[v(x)] - F[u(x)]| &\le |\lambda| M \int_a^x |v(t) - u(t)|\,dt \\ &\le |\lambda| M \max_{t\in[a,b]} |v(t) - u(t)| \int_a^x dt \\ &= |\lambda| M (x-a)\, d(v,u). \end{aligned} \tag{3.5.8}$$

We now show by induction that

$$|F^n[v(x)] - F^n[u(x)]| \le |\lambda|^n M^n \frac{(x-a)^n}{n!}\, d(v,u), \quad n = 1,2,3,\ldots. \tag{3.5.9}$$

The case $n = 1$ was established in (3.5.8). Assuming (3.5.9) holds for $n = k$, we have

$$\begin{aligned}\left|F^{k+1}[v(x)] - F^{k+1}[u(x)]\right| &\le |\lambda| \int_a^x |K(x,t)| \left|F^k[v(t)] - F^k[u(t)]\right| dt \\ &\le |\lambda| M \int_a^x |\lambda|^k M^k \frac{(t-a)^k}{k!} d(v,u)\, dt \\ &= |\lambda|^{k+1} M^{k+1} \frac{(x-a)^{k+1}}{(k+1)!} d(v,u),\end{aligned}$$

which is the corresponding statement for $n = k+1$. Taking the maximum of (3.5.9) over $x \in [a,b]$ we get

$$d(F^n[v], F^n[u]) \le |\lambda|^n M^n \frac{(b-a)^n}{n!} d(v,u).$$

For any λ we can choose n so large that

$$|\lambda|^n M^n \frac{(b-a)^n}{n!} < 1,$$

so F^n is a contraction mapping for sufficiently large n. By Corollary 3.5.2 then, (3.5.7) has a unique solution that can be found by successive approximations starting with any seed element. The usual choice for seed element is $\psi(x) = g(x)$.

Example 3.5.1 An integral equation of the form

$$\psi(x) = g(x) + \lambda \int_a^b K(x,t)\psi(t)\, dt \qquad (a \le x \le b),$$

is called a *Fredholm equation of the second kind.* Suppose that $g(x)$ is continuous on $[a,b]$, and that $K(x,t)$ is continuous on the square $[a,b] \times [a,b]$. Find a condition on λ for the equation to be uniquely solvable by iteration in the space $C(a,b)$.

Solution We need the integral operator

$$F(\psi(x)) = g(x) + \lambda \int_a^b K(x,t)\psi(t)\, dt$$

to be a contraction mapping on $C[a,b]$. Now $K(x,t)$, being continuous on a compact set, is bounded by some number B say. Hence if $u(x)$ and $v(x)$

be arbitrary elements of $C[a,b]$, we have

$$\begin{aligned} d(F(u),F(v)) &= \max_{x\in[a,b]} \left| \lambda \int_a^b K(x,t)[u(t)-v(t)]\,dt \right| \\ &\le \max_{x\in[a,b]} |\lambda| \int_a^b |K(x,t)||u(t)-v(t)|\,dt \\ &\le B|\lambda| \max_{x\in[a,b]} \int_a^b |u(t)-v(t)|\,dt \\ &\le B|\lambda|(b-a) \max_{x\in[a,b]} |u(x)-v(x)| \\ &= B|\lambda|(b-a)\,d(u(x),v(x)). \end{aligned}$$

So F will be a contraction on $C[a,b]$ if we require that $|\lambda| < 1/B(b-a)$.

Note that for application of the Banach principle we do not need the space to be linear. This fact is used in the solution of nonlinear problems which can have several solutions. The principle is applicable in cases when it is possible to find a domain M_1 in the original space M such that M_1 is a complete metric space, the operator A acts in M_1, and is a contraction on it.

3.6 L^p Spaces and the Lebesgue Integral

To introduce the Lebesgue integral and the corresponding $L^p(\Omega)$ spaces, we will apply the completion theorem to the set of functions that are continuous on Ω.

Let Ω be a closed and bounded (i.e., compact) subset of $\mathbb{R}^n$, and fix $p \ge 1$. The set S of functions $f(\mathbf{x})$ that are continuous on Ω becomes a normed space under the norm

$$\|f(\mathbf{x})\|_p = \left(\int_\Omega |f(\mathbf{x})|^p \, d\Omega \right)^{1/p}. \tag{3.6.1}$$

It is therefore also a metric space under the natural metric

$$d_p(f(\mathbf{x}), g(\mathbf{x})) = \|f(\mathbf{x}) - g(\mathbf{x})\|_p .$$

In these equations the integral is an ordinary Riemann integral. We saw in Example 3.3.1 that a sequence of continuous functions on $[0,1]$ can be a

Cauchy sequence with respect to the metric

$$\|f - g\| = \int_0^1 |f(x) - g(x)|\, dx$$

and yet lack a continuous limit. More generally, the metric space formed using S and the metric $d_p(f, g)$ for $p \geq 1$ is incomplete. The completion of this space is called $L^p(\Omega)$. The elements of $L^p(\Omega)$ can be integrated in a certain sense; although we have used Riemann integration in the definition, on the resulting space we shall end up introducing a more general type of integration. Our approach to the *Lebesgue integral* will be different from, but equivalent to, the classical one due to Lebesgue. The Lebesgue integral extends the notion of the Riemann integral in the sense that for an element corresponding to a usual continuous function the Lebesgue integral equals the Riemann integral.

In this section we shall denote an element of $L^p(\Omega)$ using uppercase notation such as $F(\mathbf{x})$. An element $F(\mathbf{x}) \in L^p(\Omega)$ is, of course, an equivalence class of Cauchy sequences of continuous functions. In this case "Cauchy" means Cauchy in the norm $\|\cdot\|_p$, and two sequences $\{f_n(\mathbf{x})\}$ and $\{g_n(\mathbf{x})\}$ are equivalent if

$$\|f_n(\mathbf{x}) - g_n(\mathbf{x})\|_p \to 0 \quad \text{as } n \to \infty.$$

Linear space operations may be carried out in the space $L^p(\Omega)$. If $F(\mathbf{x}) \in L^p(\Omega)$ and λ is a scalar, we take $\lambda F(\mathbf{x}) \in L^p(\Omega)$ to be the element for which $\{\lambda f_n(\mathbf{x})\}$ is a representative whenever $\{f_n(\mathbf{x})\}$ is a representative of $F(\mathbf{x})$. A sum such as $F(\mathbf{x}) + G(\mathbf{x})$ is interpreted similarly, in terms of representative Cauchy sequences.

The main goal of this section is to define the Lebesgue integral $\int_\Omega F(\mathbf{x})\, d\Omega$ for $F(\mathbf{x}) \in L^p(\Omega)$. We will do this in such a way that if $F(\mathbf{x})$ belongs to the dense set in $L^p(\Omega)$ that corresponds to the initial set of continuous functions, then the value of this new integral is equal to the Riemann integral of the continuous preimage. In the process we shall make use of *Hölder's inequality*

$$\int_\Omega |f(\mathbf{x})g(\mathbf{x})|\, d\Omega \leq \left(\int_\Omega |f(\mathbf{x})|^p\, d\Omega\right)^{1/p} \left(\int_\Omega |g(\mathbf{x})|^q\, d\Omega\right)^{1/q} \qquad (3.6.2)$$

which holds under the conditions $\frac{1}{p} + \frac{1}{q} = 1$, $p > 1$. This is a consequence of the corresponding inequality

$$\sum_{n=1}^{\infty} |f_n g_n| \leq \left(\sum_{n=1}^{\infty} |f_n|^p \right)^{1/p} \left(\sum_{n=1}^{\infty} |g_n|^q \right)^{1/q}$$

written for the Riemann sums. See Hardy [Hardy, Littlewood, and Pólya (1952)] for further details. Let us mention that for nontrivial $f(\mathbf{x})$ and $g(\mathbf{x})$ the sign of equality in (3.6.2) holds if and only if there is a positive constant λ such that $|f(\mathbf{x})| = \lambda |g(\mathbf{x})|$ almost everywhere. A consequence of (3.6.2) is *Minkowski's inequality*

$$\|f(\mathbf{x}) + g(\mathbf{x})\|_p \leq \|f(\mathbf{x})\|_p + \|g(\mathbf{x})\|_p ,$$

from which the useful result

$$\left| \|f(\mathbf{x})\|_p - \|g(\mathbf{x})\|_p \right| \leq \|f(\mathbf{x}) - g(\mathbf{x})\|_p$$

is easily obtained.

We begin by defining the integral $\int_\Omega |F(\mathbf{x})|^p \, d\Omega$ for $F(\mathbf{x}) \in L^p(\Omega)$. We take a representative Cauchy sequence $\{f_n(\mathbf{x})\}$ from $F(\mathbf{x})$ and consider the sequence $\{K_n\}$ given by

$$K_n = \|f_n(\mathbf{x})\|_p .$$

This is a Cauchy sequence of numbers; indeed

$$\begin{aligned} |K_m - K_n| &= \left| \|f_m(\mathbf{x})\|_p - \|f_n(\mathbf{x})\|_p \right| \\ &\leq \|f_m(\mathbf{x}) - f_n(\mathbf{x})\|_p \to 0 \quad \text{as } m, n \to \infty. \end{aligned}$$

Because $\{K_n\}$ is a Cauchy sequence in $\mathbb{R}$ or $\mathbb{C}$, by completeness there exists a number

$$K = \lim_{n \to \infty} K_n = \lim_{n \to \infty} \left(\int_\Omega |f_n(\mathbf{x})|^p \, d\Omega \right)^{1/p} .$$

It can also be shown that K is independent of the choice of representative sequence. If $\{\tilde{f}_n(\mathbf{x})\}$ is another representative of $F(\mathbf{x})$, i.e., if

$$\|f_n(\mathbf{x}) - \tilde{f}_n(\mathbf{x})\|_p \to 0,$$

then we can set

$$\tilde{K} = \lim_{n \to \infty} \tilde{K}_n = \lim_{n \to \infty} \|\tilde{f}_n(\mathbf{x})\|_p$$

but subsequently find that

$$\begin{aligned}|K - \tilde{K}| &= \left| \lim_{n\to\infty} \|f_n(\mathbf{x})\|_p - \lim_{n\to\infty} \|\tilde{f}_n(\mathbf{x})\|_p \right| \\ &= \lim_{n\to\infty} \left| \|f_n(\mathbf{x})\|_p - \|\tilde{f}_n(\mathbf{x})\|_p \right| \\ &\le \lim_{n\to\infty} \|f_n(\mathbf{x}) - \tilde{f}_n(\mathbf{x})\|_p = 0.\end{aligned}$$

The uniquely determined number K^p,

$$K^p = \left[\lim_{n\to\infty} \left(\int_\Omega |f_n(\mathbf{x})|^p \, d\Omega \right)^{1/p} \right]^p = \lim_{n\to\infty} \int_\Omega |f_n(\mathbf{x})|^p \, d\Omega,$$

is defined as the Lebesgue integral of $|F(\mathbf{x})|^p$. That is, we have

$$\int_\Omega |F(\mathbf{x})|^p \, d\Omega = \lim_{n\to\infty} \int_\Omega |f_n(\mathbf{x})|^p \, d\Omega$$

where $\{f_n(\mathbf{x})\}$ is any representative of $F(\mathbf{x})$.

We now show that when Ω is compact the L^p spaces are nested in the sense that

$$L^p(\Omega) \subseteq L^r(\Omega) \quad \text{whenever} \quad 1 \le r \le p. \tag{3.6.3}$$

Let q be such that $\frac{1}{q} + \frac{r}{p} = 1$ and apply Hölder's inequality:

$$\begin{aligned}\left| \int_\Omega 1 \cdot |f(\mathbf{x})|^r \, d\Omega \right| &\le \left(\int_\Omega 1^q \, d\Omega \right)^{1/q} \left(\int_\Omega |f(\mathbf{x})|^p \, d\Omega \right)^{r/p} \\ &= (\operatorname{mes} \Omega)^{1-\frac{r}{p}} \left(\int_\Omega |f(\mathbf{x})|^p \, d\Omega \right)^{r/p},\end{aligned}$$

or

$$\|f(\mathbf{x})\|_r \le (\operatorname{mes} \Omega)^{\frac{1}{r} - \frac{1}{p}} \|f(\mathbf{x})\|_p \tag{3.6.4}$$

where $\operatorname{mes} \Omega = \int_\Omega 1 \, d\Omega$ is the measure[3] of Ω. Putting $f(\mathbf{x}) = f_n(\mathbf{x}) - f_m(\mathbf{x})$ in (3.6.4), we see that $\{f_n(\mathbf{x})\}$ is a Cauchy sequence in the norm $\|\cdot\|_r$ if it is a Cauchy sequence in the norm $\|\cdot\|_p$. Putting $f(\mathbf{x}) = f_n(\mathbf{x}) - g_n(\mathbf{x})$, we see

[3] Because we use the Riemann integral to construct the Lebesgue integral, we must exclude some "exotic" domains Ω that are actually permitted in Lebesgue integration. But physical problems involve relatively simple domains for which Riemann integration generally suffices. In particular we shall assume that the Riemann integral $\int_\Omega 1 \, d\Omega$ exists for all of our purposes, giving the quantity we are calling the "measure" of Ω. The full notion of Lebesgue measure is far too involved to consider here; fortunately, our domains are all simple enough that we can use the notation "mes Ω" without a full chapter of explanation!

that any two Cauchy sequences equivalent in the norm $\|\cdot\|_p$ are equivalent in the norm $\|\cdot\|_r$. Hence

$$F(\mathbf{x}) \in L^p(\Omega) \implies F(\mathbf{x}) \in L^r(\Omega)$$

for $1 \le r \le p$, and we have established (3.6.3). We thus observe that if $F(\mathbf{x}) \in L^p(\Omega)$ then $\int_\Omega |F(\mathbf{x})|^r \, d\Omega$ is defined for any r such that $1 \le r \le p$. Moreover, putting $f(\mathbf{x}) = f_n(\mathbf{x})$ in (3.6.4) we see that passage to the limit as $n \to \infty$ gives

$$\|F(\mathbf{x})\|_r \le (\operatorname{mes}\Omega)^{\frac{1}{r}-\frac{1}{p}} \|F(\mathbf{x})\|_p, \qquad 1 \le r \le p.$$

Subsequently will interpret this by saying that $L^p(\Omega)$ imbeds continuously into $L^r(\Omega)$. That is, the elements of $L^p(\Omega)$ belong to $L^r(\Omega)$ as well, and the inequality means continuity of the correspondence (imbedding operator) between the elements of $L^p(\Omega)$ and the same elements considered as elements of $L^r(\Omega)$. In a similar way we can show that many inequalities satisfied by the Riemann integral are also satisfied by the Lebesgue integral.

It is now time to introduce the Lebesgue integral

$$\int_\Omega F(\mathbf{x}) \, d\Omega$$

for $F(\mathbf{x}) \in L^p(\Omega)$. Taking a representative $\{f_n(\mathbf{x})\}$ from $F(\mathbf{x})$, we use the modulus inequality

$$\left| \int_\Omega f(\mathbf{x}) \, d\Omega \right| \le \int_\Omega |f(\mathbf{x})| \, d\Omega,$$

to show that the numerical sequence $\{\int_\Omega f_n(\mathbf{x}) \, d\Omega\}$ is a Cauchy sequence:

$$\begin{aligned} \left| \int_\Omega f_n(\mathbf{x}) \, d\Omega - \int_\Omega f_m(\mathbf{x}) \, d\Omega \right| &= \left| \int_\Omega [f_n(\mathbf{x}) - f_m(\mathbf{x})] \, d\Omega \right| \\ &\le \int_\Omega |f_n(\mathbf{x}) - f_m(\mathbf{x})| \, d\Omega \\ &\le (\operatorname{mes}\Omega)^{1-\frac{1}{p}} \|f_n(\mathbf{x}) - f_m(\mathbf{x})\|_p \\ &\to 0 \quad \text{as } m, n \to \infty. \end{aligned}$$

The quantity

$$\int_\Omega F(\mathbf{x}) \, d\Omega \equiv \lim_{n\to\infty} \int_\Omega f_n(\mathbf{x}) \, d\Omega$$

is uniquely determined by $F(\mathbf{x})$ and is called the Lebesgue integral of $F(\mathbf{x})$ over Ω. If the element $F(\mathbf{x})$ happens to correspond to a continuous function, then the Lebesgue integral equals the corresponding Riemann integral. Of course, it is important to understand that $F(\mathbf{x})$ is not a function in the ordinary sense: it is an equivalence class of Cauchy sequence of continuous functions. Nevertheless, for manipulative purposes it often does no harm to treat an element like $F(\mathbf{x})$ as if it were an ordinary function; we may justify this by our ability to choose and work with a representative function that is defined uniquely by some limit passage. With proper understanding we can also relax our notational requirements and employ lowercase notation such as $f(\mathbf{x})$ for an element of $L^p(\Omega)$. We shall do this whenever convenient.

The Lebesgue integral satisfies the inequality

$$\left|\int_\Omega F(\mathbf{x})\,d\Omega\right| \le (\operatorname{mes}\Omega)^{1/q}\,\|F(\mathbf{x})\|_p, \qquad \frac{1}{p}+\frac{1}{q}=1.$$

This results directly from passage to the limit $n \to \infty$ in

$$\left|\int_\Omega f_n(\mathbf{x})\,d\Omega\right| \le (\operatorname{mes}\Omega)^{1-\frac{1}{p}}\,\|f_n(\mathbf{x})\|_p.$$

It can also be shown that a sufficient condition for existence of the integral

$$\int_\Omega F(\mathbf{x})G(\mathbf{x})\,d\Omega$$

is that $F(\mathbf{x}) \in L^p(\Omega)$ and $G(\mathbf{x}) \in L^q(\Omega)$ for some p and q such that $\frac{1}{p}+\frac{1}{q} = 1$. In this case Hölder's inequality

$$\left|\int_\Omega F(\mathbf{x})G(\mathbf{x})\,d\Omega\right| \le \left(\int_\Omega |F(\mathbf{x})|^p\,d\Omega\right)^{1/p}\left(\int_\Omega |G(\mathbf{x})|^q\,d\Omega\right)^{1/q}$$

is valid, and equality holds if and only if $F(\mathbf{x}) = \lambda G(\mathbf{x})$ for some λ.

If $p \ge 1$, then $L^p(\Omega)$ is a Banach space under the norm

$$\|F(\mathbf{x})\|_p = \left(\int_\Omega |F(\mathbf{x})|^p\,d\Omega\right)^{1/p}. \tag{3.6.5}$$

Verification of the norm axioms for $\|F(\mathbf{x})\|_p$ is mostly straightforward, depending on limiting operations of the type we have already seen. To verify the triangle inequality

$$\|F(\mathbf{x})+G(\mathbf{x})\|_p \le \|F(\mathbf{x})\|_p + \|G(\mathbf{x})\|_p,$$

for instance, we simply write $\|f_n(\mathbf{x}) + g_n(\mathbf{x})\|_p \le \|f_n(\mathbf{x})\|_p + \|g_n(\mathbf{x})\|_p$ for representatives $\{f_n(\mathbf{x})\}$ and $\{g_n(\mathbf{x})\}$ of $F(\mathbf{x})$ and $G(\mathbf{x})$, and then let $n \to \infty$. In fact the validity of this is a consequence of the completion theorem, but we wished to prove it independently. The only norm axiom that warrants further mention is

$$\|F(\mathbf{x})\|_p = 0 \iff F(\mathbf{x}) = 0.$$

The statement "$F(\mathbf{x}) = 0$" on the right means that the stationary sequence $(0, 0, 0, \ldots)$, where 0 is the zero function on Ω, belongs to the equivalence class $F(\mathbf{x})$. So $L^p(\Omega)$ is indeed a normed linear space. That it is a Banach space follows immediately by its construction via the metric space completion theorem. According to the completion theorem $L^p(\Omega)$ is complete in the metric

$$\begin{aligned} d(F(\mathbf{x}), G(\mathbf{x})) &= \lim_{n\to\infty} \left(\int_\Omega |f_n(\mathbf{x}) - g_n(\mathbf{x})|^p \, d\Omega \right)^{1/p} \\ &= \left(\int_\Omega |F(\mathbf{x}) - G(\mathbf{x})|^p \, d\Omega \right)^{1/p}, \end{aligned}$$

which of course coincides with the metric induced by the norm (3.6.5).

We began our development with the base set S of continuous functions on Ω, and introduced $L^p(\Omega)$ as the completion of S in the norm (3.6.1). We have introduced the Lebesgue integral in such a way that for any element of $L^p(\Omega)$ it is the unique number that coincides with Riemann integral of f if F corresponds to a continuous function f in the base set. In addition to the fact that the Lebesgue integral is defined for a wider set of functions than the Riemann integral, the Lebesgue integral is more convenient for performing operations involving limit passages. These operations include such important manipulations as taking the limit of an integral with respect to a parameter in the integrand (Lebesgue's theorem) and interchanging the order of integration in a repeated integral (Fubini's theorem). The theory of Riemann integration is based on the notion of Jordan measurability of a set in $\mathbb{R}^n$. The classical theory of Lebesgue integration starts with the introduction of a wider notion of measurability of a set of $\mathbb{R}^n$. In particular, under this definition the set of all rational points on the segment $[0, 1]$ is measurable and its Lebesgue measure is zero. These considerations are outside the scope of this book, and the interested reader should consult standard textbooks in real analysis for details. Lebesgue integration is not only useful in itself, but it finds applications in the theory of Sobolev spaces, and to the generalized setup of boundary value problems.

Example 3.6.1 Show that $L^p(\Omega)$ is separable for compact Ω.

Solution First we show that the space of continuous functions with the L^p metric is separable. We know that the set $P_r(\Omega)$ of polynomials defined on Ω and having rational coefficients is dense in $C(\Omega)$, where $C(\Omega)$ is the space of continuous functions under the metric

$$\|f(\mathbf{x}) - g(\mathbf{x})\|_{C(\Omega)} = \max_{\mathbf{x}\in\Omega} |f(\mathbf{x}) - g(\mathbf{x})|.$$

This follows from the classical Weierstrass theorem. Therefore for any $f(\mathbf{x})$ continuous on Ω we can find $p_\varepsilon(\mathbf{x}) \in P_r(\Omega)$ such that

$$\max_{\mathbf{x}\in\Omega} |f(\mathbf{x}) - p_\varepsilon(\mathbf{x})| \le \frac{\varepsilon}{(\operatorname{mes}\Omega)^{1/p}}.$$

(Here we see why the domain Ω was required to be compact.) Therefore we have

$$\|f(\mathbf{x}) - p_\varepsilon(\mathbf{x})\| = \left(\int_\Omega |f(\mathbf{x}) - p_\varepsilon(\mathbf{x})|^p \, d\Omega\right)^{1/p} \le \left(\frac{\varepsilon^p}{\operatorname{mes}\Omega}\int_\Omega d\Omega\right)^{1/p} = \varepsilon.$$

So imposing the L^p metric on the space of functions continuous on Ω, we get a separable metric space. Furthermore, $L^p(\Omega)$ is the completion of this space. Since the completion of a separable metric space is separable, the conclusion follows.

3.7 Sobolev Spaces

We now proceed to some normed spaces that play an important role in the modern treatment of partial differential equations. On the set of l times continuously differentiable functions $f(\mathbf{x})$ given on a compact set Ω, we have introduced the family of norms

$$\|f\| = \left(\int_\Omega \sum_{|\alpha|\le l} |D^\alpha f|^p \, d\Omega\right)^{1/p}, \qquad p \ge 1. \tag{3.7.1}$$

The resulting normed spaces are, however, incomplete in their natural metrics. Applying the completion theorem to this case (in the same way we produced the Lebesgue spaces $L^p(\Omega)$), we obtain a family of Banach spaces known as the *Sobolev spaces* $W^{l,p}(\Omega)$. The form of the norm (3.7.1) suggests that the elements of a Sobolev space possess something like derivatives. We shall discuss these *generalized derivatives* momentarily, but at this point (3.7.1) seems to indicate that they belong to the space $L^p(\Omega)$. Because $W^{l,p}(\Omega)$ is a completion of the separable space $C^{(l)}(\Omega)$, Theorem 3.4.3 gives us

Theorem 3.7.1 *$W^{l,p}(\Omega)$, $p \geq 1$, is a separable normed space.*

We can use the following definition for a generalized derivative. For $u \in L^p(\Omega)$, K.O. Friedrichs called $v \in L^p(\Omega)$ a *strong derivative* $D^\alpha(u)$ if there exists a sequence $\{\varphi_n\}$, $\varphi_n \in C^{(\infty)}(\Omega)$, such that

$$\int_\Omega |u(\mathbf{x}) - \varphi_n(\mathbf{x})|^p \, d\Omega \to 0 \quad \text{and}$$
$$\int_\Omega |v(\mathbf{x}) - D^\alpha \varphi_n(\mathbf{x})|^p \, d\Omega \to 0 \quad \text{as } n \to \infty.$$

Since $C^{(\infty)}(\Omega)$ is dense in any $C^{(k)}(\Omega)$, we see that an element of $W^{m,p}(\Omega)$ has all strong derivatives up to the order m lying in $L^p(\Omega)$. Note that in this definition we need not define intermediate derivatives as is done for standard derivatives. But this definition does not seem too classical or familiar. In his original monograph [Sobolev (1951)] S.L. Sobolev introduced the notion of a generalized derivative using the ideas of the calculus of variations. He introduced this for elements of $L^p(\Omega)$ (not for just any element of course, but for those elements for which it can be done). S.L. Sobolev called $v \in L^p(\Omega)$ a *weak derivative* $D^\alpha u$ of $u \in L^p(\Omega)$ if for every function $\varphi(\mathbf{x}) \in \mathcal{D}$ the relation

$$\int_\Omega u(\mathbf{x}) D^\alpha \varphi(\mathbf{x}) \, d\Omega = (-1)^{|\alpha|} \int_\Omega v(\mathbf{x}) \varphi(\mathbf{x}) \, d\Omega$$

holds. Here $\mathcal{D}$ is the set of functions that are infinitely differentiable on Ω and that vanish in some neighborhood of the boundary of Ω (the neighborhood can vary from function to function). This definition of derivative inherits some ideas from the calculus of variations: in particular, the Fundamental Lemma insures that we are defining the derivative in a unique way. For elements of $W^{l,p}(\Omega)$ it can be demonstrated that the two notions of generalized derivative are equivalent. Of course, the name "generalized derivative" is warranted because classical derivatives (say, of functions continuous on Ω) are also generalized derivatives, but not vice versa.

The most important result obtained by S.L. Sobolev is called the theorem of imbedding. It gives some properties of the elements of Sobolev spaces and, in particular, relates them to continuously differentiable functions. An example of an imbedding can be seen from the estimate

$$\|f(\mathbf{x})\|_{W^{l,q}(\Omega)} \leq m_{qp} \|f(\mathbf{x})\|_{W^{l,p}(\Omega)}, \qquad q < p, \tag{3.7.2}$$

which can be shown for any $f \in W^{l,p}(\Omega)$ to hold with a constant m_{qp} that depends on q, p, and Ω only. Note that for $q < p$ we have

$$f(\mathbf{x}) \in W^{l,p}(\Omega) \implies f(\mathbf{x}) \in W^{l,q}(\Omega);$$

hence the Sobolev space $W^{l,p}(\Omega)$ is a subset of the Sobolev space $W^{l,q}(\Omega)$:

$$W^{l,p}(\Omega) \subseteq W^{l,q}(\Omega), \qquad q < p.$$

But the estimate (3.7.2) gives us more than just this subset inclusion. We met inclusions of this type when considering the $L^p(\Omega)$ spaces. We called them imbeddings. Now we introduce a general definition of this term.

Definition 3.7.1 The *operator of imbedding* from X to Y is the one-to-one correspondence between a space X and a subspace Y of a space Z under which we identify elements $x \in X$ with elements $y \in Y$ in such a way that the correspondence is linear. If, besides, the correspondence is continuous so that

$$\|y\|_Y \leq m \|x\|_X$$

for some constant m that does not depend on x, then we call it the continuous operator of imbedding. We sometimes employ the notation

$$X \hookrightarrow Y,$$

to indicate the existence of an imbedding from X to Y.

Some words of explanation are in order here. The reader should note that the formal definition of a continuous imbedding operator does not differ from that of a continuous linear operator. However, the term "imbedding" is reserved for situations in which we *identify* an element in X with its image in Y, and thereby effectively consider the "same element" as a member of two different spaces. (In this way an imbedding operator acts somewhat like the identity operator that serves to map elements of a space into themselves; the difference is that in the case of an identity operator the domain and range must be the same space.) The degree to which one may take literally the "identification" process between elements of X and their images in Y depends on the specific type of imbedding under consideration. In some instances we shall see that the elements of X and Y are of the same basic nature (e.g., both are ordinary functions); in other instances this may not be the case (e.g., the elements of Y may be functions while the elements of X are equivalence classes of Cauchy sequences of functions). Note, however, that even when the elements of Y and X are of the same nature, the norms associated with the spaces Y and X may be very different. Finally, we remark that there are imbedding operators that are *compact* and not merely *continuous*. We shall state this when it applies, but shall relegate coverage of compact operators to a later section of this chapter.

Returning to our discussion of Sobolev spaces, we see that the space $W^{l,p}(\Omega)$ is continuously imbedded into the space $W^{l,q}(\Omega)$ when $q < p$, and we write

$$W^{l,p}(\Omega) \hookrightarrow W^{l,q}(\Omega), \qquad q < p.$$

We are also interested in continuous imbeddings from Sobolev spaces into the spaces of continuously differentiable functions. To obtain a relevant example of an *imbedding theorem* let us consider the simple Sobolev space $W^{1,1}(0,1)$, the norm of which is

$$\|f(x)\|_{1,1} = \int_0^1 (|f(x)| + |f'(x)|)\, dx. \tag{3.7.3}$$

So $W^{1,1}(0,1)$ is the completion with respect to the norm (3.7.3) of the set of all functions that are continuously differentiable on $[0,1]$. Let $f(x)$ be continuously differentiable on $[0,1]$. Then for any $x, y \in [1,1]$ we have

$$f(x) - f(y) = \int_y^x f'(t)\, dt$$

and so

$$|f(x)| \le |f(y)| + \left| \int_y^x f'(t)\, dt \right| \le |f(y)| + \int_0^1 |f'(t)|\, dt.$$

Integrating this in y over $[0,1]$ we get

$$\int_0^1 |f(x)|\, dy \le \int_0^1 |f(y)|\, dy + \int_0^1 \int_0^1 |f'(t)|\, dt\, dy$$

or

$$\max_{x\in[0,1]} |f(x)| \le \int_0^1 |f(y)|\, dy + \int_0^1 |f'(t)|\, dt = \|f(x)\|_{1,1}. \tag{3.7.4}$$

Now let $F(x)$ be an equivalence class from $W^{1,1}(0,1)$. A representative of $F(x)$ is a Cauchy sequence $\{f_n(x)\}$ of continuously differentiable functions, and we have

$$\max_{x\in[0,1]} |f_{n+m}(x) - f_n(x)| \le \|f_{n+m}(x) - f_n(x)\|_{1,1} ;$$

it follows that $\{f_n(x)\}$ is a Cauchy sequence in $C[0,1]$ as well, and thus has a limit that is continuous on $[0,1]$. From (3.7.4) it also follows that this limiting function does not depend on the choice of representative sequence of the element of $W^{1,1}(0,1)$. Hence we have a correspondence that is clearly linear, under which to an element $F(x) \in W^{1,1}(0,1)$ there corresponds a unique element $f(x) \in C(0,1)$ such that

$$\|f(x)\|_{C(0,1)} \le \|F(x)\|_{1,1} .$$

We identify this limit element with F, and call F by the name of this limit element. (We can really regard F as this element f if f is continuously differentiable on $[0,1]$ so there is a *stationary* representative sequence $(f, f, f, \ldots)$ from F.) In short, we have

$$W^{1,1}(0,1) \hookrightarrow C(0,1).$$

Similar results for $W^{l,p}(\Omega)$, where Ω is a compact subset of $\mathbb{R}^n$, are called *Sobolev's imbedding theorems.* We shall state one such theorem next. We assume that Ω satisfies the *cone condition*: there is a finite circular cone in $\mathbb{R}^n$ that can touch any point of $\partial\Omega$ with its vertex while lying fully inside Ω (i.e., translations and rotations of the cone are allowed, but not changes in cone angle or height).

Theorem 3.7.2 *Let Ω_r be an r-dimensional piecewise smooth hypersurface in Ω. The imbedding*

$$W^{m,p}(\Omega) \hookrightarrow L^q(\Omega_r)$$

is continuous if one of the following conditions holds:

(i) $n > mp$, $r > n - mp$, $q \le pr/(n - mp)$;
(ii) $n = mp$, q is finite with $q \ge 1$.

It is compact if

(i) $n > mp$, $r > n - mp$, $q < pr/(n - mp)$ or
(ii) $n = mp$ and q is finite with $q \ge 1$.

If $n < mp$ then

$$W^{m,p}(\Omega) \hookrightarrow C^{(k)}(\Omega)$$

for integers k such that $k \leq (mp - n)/p$, and the imbedding is continuous. It is compact if $k < (mp - n)/p$.

Although this theorem is appealing because of its generality, we shall make use only of special cases involving $W^{1,2}(\Omega)$ and $W^{2,2}(\Omega)$. The following special case is used for problems of equilibrium of membranes and 2-D elastic bodies:

Theorem 3.7.3 *Let γ be a piecewise differentiable curve in a compact set $\Omega \subset \mathbb{R}^2$. For any finite $q \geq 1$, there are compact (hence continuous) imbeddings*

$$W^{1,2}(\Omega) \hookrightarrow L^q(\Omega), \qquad W^{1,2}(\Omega) \hookrightarrow L^q(\gamma).$$

For use with problems of equilibrium of plates and shells, we have

Theorem 3.7.4 *Let Ω be a compact subset of $\mathbb{R}^2$. Then there is a continuous imbedding*

$$W^{2,2}(\Omega) \hookrightarrow C(\Omega).$$

For the first derivatives, the imbedding operators to $L^q(\Omega)$ and $L^q(\gamma)$ are compact for any finite $q \geq 1$.

The next result is used for problems of equilibrium of 3-D elastic bodies and dynamic problems for membranes and 2-D elastic bodies.

Theorem 3.7.5 *Let γ be a piecewise smooth surface in a compact set $\Omega \subset \mathbb{R}^3$. The imbeddings*

$$\begin{aligned} W^{1,2}(\Omega) &\hookrightarrow L^q(\Omega), \qquad 1 \leq q \leq 6, \\ W^{1,2}(\Omega) &\hookrightarrow L^p(\gamma), \qquad 1 \leq p \leq 4, \end{aligned}$$

are continuous. They are compact if $1 \leq q < 6$ or $1 \leq p < 4$, respectively.

Example 3.7.1 Show that ℓ^q is continuously imbedded into ℓ^p if $p > q$.

Solution The first step is to show that the norms $\|\cdot\|_p$ and $\|\cdot\|_q$ of the spaces ℓ^p and ℓ^q satisfy $\|\mathbf{x}\|_p \leq \|\mathbf{x}\|_q$ whenever $p \geq q$ (Exercise 3.8). This gives the subset inclusion $\ell^q \subseteq \ell^p$ whenever $p \geq q$, and also shows that $\ell^q \hookrightarrow \ell^p$ with a constant $m = 1$ in the imbedding inequality.

3.8 Compactness

Definition 3.8.1 Let S be a subset of a metric space. We say that S is *precompact* if every sequence taken from S contains a Cauchy subsequence.

Any bounded set in $\mathbb{R}^n$ is precompact. We know this from calculus, where the classical Bolzano–Weierstrass theorem asserts that any bounded sequence from $\mathbb{R}^n$ contains a Cauchy subsequence. This is not necessarily the case in other spaces, however (see Theorem 3.8.4). In § 3.3 we introduced c, the space of convergent numerical sequences with norm

$$\|\mathbf{x}\| = \sup_i |x_i|.$$

The sequence of elements

$$\begin{aligned} \mathbf{x}_1 &= (1,0,0,0,\ldots), \\ \mathbf{x}_2 &= (0,1,0,0,\ldots), \\ \mathbf{x}_3 &= (0,0,1,0,\ldots), \\ &\vdots \end{aligned}$$

taken from c has no Cauchy subsequence, since for any pair of distinct elements $\mathbf{x}_i, \mathbf{x}_j$ we have $\|\mathbf{x}_i - \mathbf{x}_j\| = 1$. Nonetheless, this sequence is bounded: we have $\|\mathbf{x}_i\| = 1$ for each i. So we see that the Bolzano–Weierstrass theorem for $\mathbb{R}^n$ does not automatically extend to all other normed spaces.

What is the principal difference between a bounded set in c and a bounded set in $\mathbb{R}^n$? In $\mathbb{R}^n$, using, say, three decimal places, we can approximate all the coordinates of any point of the unit ball up to an accuracy of 0.001. There are a finite number of points lying within the unit ball whose coordinates are the approximated coordinates of the actual points (the reader could calculate the actual number of such points for a space of n dimensions). Increasing accuracy through the use of m decimal places, $m > 3$, we again have a finite number of points with which we can better approximate any point of the unit ball. In c, as is shown by the above example, such an approximation of all the points of the unit ball by a finite number of elements within a prescribed precision is impossible.

Let us introduce the abstract variant of an approximating finite set for some given set of points:

Definition 3.8.2 Let S and E be subsets of a metric space. We call E a *finite ε-net* for S if E is finite and for every $x \in S$ there exists $e \in E$ such that $d(x, e) < \varepsilon$. We say that S is *totally bounded* if there is a finite ε-net for S for every $\varepsilon > 0$.

Note that a set is totally bounded if when we draw a ball of radius ε about each point of an ε-net of the set, then the set is covered by the union of these balls (i.e., any point of the set is a point of one of the balls).

In particular, if a set is totally bounded, it is bounded. Indeed taking a 1-net we get a finite collection of balls that covers the set. It is clear that there exists some ball of finite radius that contains all these balls inside itself, and so all the points of the initial set, and this implies that the initial set is bounded.

We see that total boundedness of a set is exactly the same property that we described for a ball of $\mathbb{R}^n$, on the existence of finite sets of points with which we can approximate the coordinates of any point of the ball within any prescribed accuracy. We declared that this was a crucial property in determining whether a set is compact. This is confirmed by the following *Hausdorff criterion.*

Theorem 3.8.1 *A subset of a metric space is precompact if and only if it is totally bounded.*

Proof. Let S be a precompact subset of a metric space X. To show that S is totally bounded, we prove the contrapositive statement. Suppose that S has no finite ε_0-net for some particular $\varepsilon_0 > 0$. This means that no finite union of balls of radius ε_0 can contain S. Taking $x_1 \in S$ and a ball B_1 of radius ε_0 about x_1, we know that there exists $x_2 \in S$ such that $x_2 \notin B_1$ (otherwise x_1 by itself would generate a finite ε_0-net for S). Constructing the ball B_2 of radius ε_0 about x_2, we know that there exists $x_3 \in S$ such that $x_3 \notin B_1 \cup B_2$. Continuing in this way, we construct a sequence $\{x_n\}$ such that $d(x_n, x_m) \geq \varepsilon_0$ whenever $n \neq m$. Because $\{x_n\}$ cannot contain a Cauchy subsequence, S is not precompact.

Conversely, suppose that S is totally bounded and take any sequence $\{x_n\}$ from S. We begin to select a Cauchy subsequence from $\{x_n\}$ by taking $\varepsilon_1 = 1/2$ and constructing a finite ε_1-net for S. One of the balls, say B_1, must contain infinitely many elements of $\{x_n\}$. Choose one of these elements and call it x_{i_1}. Then construct a finite ε_2-net for S with $\varepsilon_2 = 1/2^2$. One of the balls, say B_2, must contain infinitely many of those elements of $\{x_n\}$ which belong to B_1. Choose one of these elements and call it x_{i_2}. Note that $d(x_{i_1}, x_{i_2}) \leq (2)(1/2) = 1$ since both x_{i_2} and x_{i_1} belong to B_1. Continuing in this way we obtain a subsequence $\{x_{i_k}\} \subset \{x_n\}$ where, by construction, x_{i_k} and $x_{i_{k+1}}$ reside in a ball B_k of radius $\varepsilon_k = 1/2^k$ so that

$$d(x_{i_k}, x_{i_{k+1}}) \le 2\left(\frac{1}{2^k}\right) = \frac{1}{2^{k-1}}.$$

Thus

$$\begin{aligned} d(x_{i_k}, x_{i_{k+m}}) &\le d(x_{i_k}, x_{i_{k+1}}) + d(x_{i_{k+1}}, x_{i_{k+2}}) + \cdots + d(x_{i_{k+m-1}}, x_{i_{k+m}}) \\ &\le \frac{1}{2^{k-1}} + \frac{1}{2^k} + \cdots + \frac{1}{2^{k+m-2}} < \frac{1}{2^{k-2}} \end{aligned}$$

for any $m \ge 1$, and $\{x_{i_k}\}$ is a Cauchy sequence. □

Definition 3.8.3 Let S be a subset of a metric space. We say that S is *compact* if every sequence taken from S contains a Cauchy subsequence that converges to a point of S.

Note that a compact subset of a metric space is closed. But a closed set is not, in general, compact. (In $\mathbb{R}^n$ a closed *and* bounded set is compact according to the present definition.) We now reformulate the Hausdorff criterion for compactness:

Theorem 3.8.2 *A subset of a complete metric space is compact if and only if it is closed and totally bounded.*

The proof is left as an exercise for the reader.

Example 3.8.1 Show that the *Hilbert cube*

$$S = \{\mathbf{x} = (\xi_1, \xi_2, \ldots) \in \ell^2 \colon |\xi_n| \le \tfrac{1}{n} \text{ for } n = 1, 2, \ldots\}$$

is a compact subset of ℓ^2.

Solution We show that S is closed and totally bounded in the complete space ℓ^2. Let $\mathbf{y} = (\eta_1, \eta_2, \ldots)$ be a limit point of S. There is a sequence $\{\mathbf{x}^{(j)}\} \subset S$ such that

$$\|\mathbf{y} - \mathbf{x}^{(j)}\|_{\ell^2}^2 = \sum_{k=1}^{\infty} |\eta_k - \xi_k^{(j)}|^2 \to 0 \quad \text{as } j \to \infty.$$

Hence for each k we have $|\eta_k - \xi_k^{(j)}| \to 0$ as $j \to \infty$. By the triangle inequality

$$|\eta_k| \le |\eta_k - \xi_k^{(j)}| + |\xi_k^{(j)}| \le |\eta_k - \xi_k^{(j)}| + \frac{1}{k},$$

and passage to the limit as $j \to \infty$ gives $|\eta_k| \leq \frac{1}{k}$ for each k. This shows that $\mathbf{y} \in S$, hence S is closed. Next we show that S is totally bounded. Let $\varepsilon > 0$ be given. We begin to construct a finite ε-net by noting that the nth component of any element $\mathbf{z} = (\zeta_1, \zeta_2, \ldots) \in S$ differs from zero by no more than $1/n$. Since the series $\sum 1/n^2$ is convergent we can choose N such that

$$\sum_{n=N+1}^{\infty} |\zeta_n|^2 < \varepsilon^2/2.$$

Now take the first N components and consider the corresponding bounded closed hypercube in $\mathbb{R}^N$. For this there certainly exists a finite $\varepsilon^2/2$-net of N-tuples, and we can select $(\xi_1, \ldots, \xi_N)$ such that

$$\sum_{n=1}^{N} |\zeta_n - \xi_n|^2 < \varepsilon^2/2.$$

We construct a corresponding element $\mathbf{x}_\varepsilon \in \ell^2$ by appending zeros:

$$\mathbf{x}_\varepsilon = (\xi_1, \ldots, \xi_N, 0, 0, \ldots).$$

For this element

$$\|\mathbf{z} - \mathbf{x}_\varepsilon\|_{\ell^2}^2 = \sum_{n=1}^{N} |\zeta_n - \xi_n|^2 + \sum_{n=N+1}^{\infty} |\zeta_n|^2 < \varepsilon^2/2 + \varepsilon^2/2 = \varepsilon^2$$

as desired.

Theorem 3.8.3 *Every precompact metric space is separable.*

Proof. Let X be a precompact metric space. For each k, $k = 1, 2, 3, \ldots$, let $\varepsilon_k = 1/k$ and construct a finite ε_k-net $(x_{k1}, x_{k2}, \ldots, x_{kN})$ for X. (Here N depends on k.) The union of these nets is countable and dense in X. □

Theorem 3.8.4 *Every closed and bounded subset of a Banach space is compact if and only if the Banach space has finite dimension.*

The proof of Theorem 3.8.4 depends on the following result, known as *Riesz's lemma.*

Lemma 3.8.1 *Let M be a proper closed subspace of a normed space X. If $0 < \varepsilon < 1$, then there is an element $x_\varepsilon \notin M$ having unit norm such that*

$$\inf_{y \in M} \|y - x_\varepsilon\| > 1 - \varepsilon.$$

(Here we use the term "proper" to exclude the case $M = X$.)

Proof. Take an element $x_0 \in X$ that does not belong to M and let

$$d = \inf_{y \in M} \|x_0 - y\| .$$

We have $d > 0$; indeed, the assumption $d = 0$ leads to a contradiction because it implies the existence of a sequence $\{y_k\} \subset M$ such that $\|x_0 - y_k\| \to 0$, hence $y_k \to x_0$, hence $x_0 \in M$ because M is closed. By definition of infimum, for any $\varepsilon > 0$ there exists $y_\varepsilon \in M$ such that

$$d \leq \|x_0 - y_\varepsilon\| < \frac{d}{1 - \varepsilon/2}.$$

The normalized element

$$x_\varepsilon = \frac{x_0 - y_\varepsilon}{\|x_0 - y_\varepsilon\|}$$

has the properties specified in the lemma. It clearly has unit norm and does not belong to M. Moreover, for any $y \in M$ we have

$$\begin{aligned} \|x_\varepsilon - y\| &= \left\| \frac{x_0 - y_\varepsilon}{\|x_0 - y_\varepsilon\|} - y \right\| = \frac{\|x_0 - (y_\varepsilon + \|x_0 - y_\varepsilon\| \, y)\|}{\|x_0 - y_\varepsilon\|} \\ &> d / \frac{d}{1 - \varepsilon/2} = 1 - \frac{\varepsilon}{2} \end{aligned}$$

where the intermediate inequality holds since $y_\varepsilon + \|x_0 - y_\varepsilon\| \, y$ belongs to M. □

As an important application of Riesz's lemma, let us show that the unit ball

$$B = \{x \in X \colon \|x\| \leq 1\}$$

is not compact if X is infinite dimensional. (This is the "only if" part of Theorem 3.8.4.) Take $y_1 \in B$. This element generates a proper closed subspace E_1 of X given by $E_1 = \{\alpha y_1 \colon \alpha \in \mathbb{C}\}$. By Riesz's lemma (with $\varepsilon = 1/2$) there exists y_2 such that $y_2 \in B$, $y_2 \notin E_1$, and $\|y_1 - y_2\| > 1/2$. The elements y_1, y_2 generate a proper closed subspace E_2 of X, and by Riesz's lemma there exists y_3 such that $y_3 \in B$, $y_3 \notin E_2$, and $\|y_i - y_3\| > 1/2$ for $i = 1, 2$. Since X is infinite dimensional we can continue this process indefinitely, producing a sequence $\{y_n\} \subset B$ any two distinct points of which are separated by a distance exceeding $1/2$. Since no subsequence of $\{y_n\}$ is a Cauchy sequence, B is not compact.

Definition 3.8.4 Let M be a set of functions continuous on a compact set $\Omega \subset \mathbb{R}^n$. We say that M is

(1) *uniformly bounded* if there is a constant c such that for every $f(\mathbf{x}) \in M$, $|f(\mathbf{x})| \le c$ for all $\mathbf{x} \in \Omega$.
(2) *equicontinuous* if for any $\varepsilon > 0$ there exists $\delta > 0$, dependent on ε, such that whenever $|\mathbf{x} - \mathbf{y}| < \delta$, $\mathbf{x}, \mathbf{y} \in \Omega$, then $|f(\mathbf{x}) - f(\mathbf{y})| < \varepsilon$ holds for every $f(\mathbf{x}) \in M$.

Uniform boundedness simply means that the set of functions lies in a ball of radius c in $C(\Omega)$ (in Arzelà's time the normed space terminology was not yet in full use). Since the space $C(\Omega)$ is infinite dimensional, this cannot be the sole condition for compactness. We also note that any finite set of continuous functions is equicontinuous by Weierstrass's theorem from calculus; given $\varepsilon > 0$ we can find the required δ for each individual function, then take the smallest of these values and use it as δ for the whole set. An infinite set of continuous functions need not be equicontinuous.

The space of continuous functions is one of the main objects of calculus, differential equations, and many other branches of mathematics. It is important to have a set of practical criteria under which a subset of this space must be precompact. This is provided by *Arzelà's theorem.*

Theorem 3.8.5 *Let Ω be a compact set in $\mathbb{R}^n$, and let M be a set of functions continuous on Ω. Then M is precompact in $C(\Omega)$ if and only if it is uniformly bounded and equicontinuous.*

Proof. Suppose that M is precompact in $C(\Omega)$. By Theorem 3.8.1 there is a finite ε-net for M with $\varepsilon = 1$; i.e., there is a finite set of continuous functions $\{g_i(\mathbf{x})\}_{i=1}^k$ such that to any $f(\mathbf{x})$ there corresponds $g_i(\mathbf{x})$ for which

$$\|f(\mathbf{x}) - g_i(\mathbf{x})\| = \max_{\mathbf{x}\in\Omega} |f(\mathbf{x}) - g_i(\mathbf{x})| \le 1.$$

Since the $g_i(\mathbf{x})$ are continuous there is a constant c_1 such that $|g_i(\mathbf{x})| < c_1$ for each i. Using the inequality $\|f(\mathbf{x})\| \le \|g_i(\mathbf{x})\| + \|f(\mathbf{x}) - g_i(\mathbf{x})\|$, we have

$$\max_{\mathbf{x}\in\Omega} |f(\mathbf{x})| \le c_1 + 1.$$

It follows that M is uniformly bounded with $c = c_1 + 1$. We proceed to verify equicontinuity. Let $\varepsilon > 0$ be given, and choose a finite $\varepsilon/3$-net for M, say $\{g_i(\mathbf{x})\}_{i=1}^m$. Since the number of $g_i(\mathbf{x})$ is finite and, by a calculus theorem, each of them is equicontinuous on Ω, there exists $\delta > 0$ such that

$|\mathbf{x} - \mathbf{y}| < \delta$ implies

$$|g_i(\mathbf{x}) - g_i(\mathbf{y})| < \varepsilon/3, \qquad i = 1, \ldots, m.$$

For each $f(\mathbf{x}) \in M$, there exists $g_r(\mathbf{x})$ such that

$$|f(\mathbf{x}) - g_r(\mathbf{x})| < \varepsilon/3 \text{ for all } \mathbf{x} \in \Omega.$$

Whenever $\mathbf{x}, \mathbf{y} \in \Omega$ are such that $|\mathbf{x} - \mathbf{y}| < \delta$ then, we have

$$\begin{aligned}|f(\mathbf{x}) - f(\mathbf{y})| &\le |f(\mathbf{x}) - g_r(\mathbf{x})| + |g_r(\mathbf{x}) - g_r(\mathbf{y})| + |g_r(\mathbf{y}) - f(\mathbf{y})| \\ &< \varepsilon/3 + \varepsilon/3 + \varepsilon/3 = \varepsilon\end{aligned}$$

as desired.

Conversely suppose that M is uniformly bounded and equicontinuous. We must show that from any sequence of functions $\{f_k(\mathbf{x})\} \subset M$ we can choose a Cauchy subsequence. Let $\{\mathbf{x}_k\}$ be the set of all rational points of Ω (enumerated somehow); this set is countable and dense in Ω. Consider the sequence $\{f_k(\mathbf{x}_1)\}$. Because this numerical sequence is bounded, we can choose a Cauchy subsequence $\{f_{k_1}(\mathbf{x}_1)\}$. We have thus chosen a subsequence $\{f_{k_1}(\mathbf{x})\} \subset \{f_k(\mathbf{x})\}$ that is a Cauchy sequence at $\mathbf{x} = \mathbf{x}_1$. From the bounded numerical sequence $\{f_{k_1}(\mathbf{x}_2)\}$ we can choose a Cauchy subsequence $\{f_{k_2}(\mathbf{x}_2)\}$. The subsequence $\{f_{k_2}(\mathbf{x})\}$ is thus a Cauchy sequence at both $\mathbf{x} = \mathbf{x}_1$ and $\mathbf{x} = \mathbf{x}_2$. We continue in this way, taking subsequences of previously constructed subsequences, so that on the nth step the subsequence $\{f_{k_n}(\mathbf{x}_n)\}$ is a Cauchy sequence and, since it is a subsequence of any previous subsequence, the sequences obtained by evaluating $\{f_{k_n}(\mathbf{x})\}$ at $\mathbf{x}_1, \ldots, \mathbf{x}_{n-1}$ are Cauchy sequences as well.

The diagonal sequence $\{f_{n_n}(\mathbf{x})\}$ is a Cauchy sequence at $\mathbf{x} = \mathbf{x}_i$ for all i. We now show that it is a Cauchy sequence in the norm of $C(\Omega)$. Let $\varepsilon > 0$ be given. According to equicontinuity we can find $\delta > 0$ such that $|\mathbf{x} - \mathbf{y}| < \delta$ gives for every n

$$|f_{n_n}(\mathbf{x}) - f_{n_n}(\mathbf{y})| < \varepsilon/3.$$

Take $\delta' < \delta$ and construct a finite δ'-net for Ω with nodes $\{\mathbf{z}_i\}_{i=1}^r \subset \{\mathbf{x}_i\}$. Since r is finite we can find N such that whenever $n, m > N$ we have

$$|f_{n_n}(\mathbf{z}_i) - f_{m_m}(\mathbf{z}_i)| < \varepsilon/3, \quad i = 1, \ldots, r.$$

Choose any $\mathbf{x} \in \Omega$ and let $\mathbf{z}_k$ be the point of the δ'-net nearest $\mathbf{x}$ so that $|\mathbf{x} - \mathbf{z}_k| < \delta'$. Then $n, m > N$ implies

$$|f_{n_n}(\mathbf{x}) - f_{m_m}(\mathbf{x})| \le |f_{n_n}(\mathbf{x}) - f_{n_n}(\mathbf{z}_k)| + |f_{n_n}(\mathbf{z}_k) - f_{m_m}(\mathbf{z}_k)| \\ + |f_{m_m}(\mathbf{z}_k) - f_{m_m}(\mathbf{x})| < \varepsilon/3 + \varepsilon/3 + \varepsilon/3 = \varepsilon,$$

hence

$$\max_{\mathbf{x}\in\Omega} |f_{n_n}(\mathbf{x}) - f_{m_m}(\mathbf{x})| = \|f_{n_n}(\mathbf{x}) - f_{m_m}(\mathbf{x})\| < \varepsilon$$

for all $n, m > N$. □

Remark 3.8.1 In the proof we made use of the *diagonal sequence* idea. Since this is a standard technique in analysis and will be used again in this chapter, we take a moment to clarify the ideas involved.

Suppose we start with a sequence $\{x_n\}$ and want to extract a subsequence that satisfies some set of convergence-related criteria p_k ($k = 1, 2, 3, \ldots$). Let us agree to write $\{x_{nk}\}$ for the subsequence we select at the kth step of our process ($k = 1, 2, 3, \ldots$), and x_{nk} for the nth element of that subsequence ($n = 1, 2, 3, \ldots$).

Our process begins with the selection of successive subsequences, as follows:

1. From $\{x_n\}$ we select $\{x_{n1}\}$ that satisfies p_1. It is clear that the whole sequence $\{x_{n1}\}$ as well as each of its subsequences satisfies p_1.
2. Then from $\{x_{n1}\}$ we take $\{x_{n2}\}$ that satisfies p_2. The whole sequence as well as each of its subsequences satisfies p_2. Being a subsequence of $\{x_{n1}\}$, it and all of its subsequences satisfy p_1 as well.
3. The same is done with $\{x_{n2}\}$: choose $\{x_{n3}\}$ that satisfies p_3, so all of its subsequences satisfy p_3 and, simultaneously, p_1 and p_2.

⋮

k. Choose $\{x_{nk}\}$ that satisfies p_k and $p_1, \ldots, p_{k-1}$.

⋮

We now form the sequence

$$\{x_{nn}\}_{n=1}^{\infty} = x_{11},\ x_{22},\ x_{33},\ \ldots. \tag{3.8.1}$$

This is the desired diagonal sequence.

The sequence (3.8.1) is automatically contained in $\{x_{n1}\}$. Except possibly for the first term, it is also contained in $\{x_{n2}\}$; the first term is a non-issue because the behavior of a finite number of terms has no impact on the satisfaction of p_2. Except possibly for the first two terms, (3.8.1) is also contained in $\{x_{n3}\}$, and so on. So the diagonal sequence, except for finite numbers of terms, is contained in $\{x_{nk}\}$ for each k. It therefore satisfies p_k for $k = 1, 2, 3, \ldots$.

Example 3.8.2 Let Ω be a compact subset of $\mathbb{R}^n$, and suppose S is a collection of functions $\{f_k(\mathbf{x})\}$ continuous on Ω. Further, suppose that S is bounded in $C(\Omega)$ and that $K(\mathbf{x}, \mathbf{y})$ is a function continuous on $\Omega \times \Omega$. Show that the set

$$A = \left\{ \int_\Omega K(\mathbf{x}, \mathbf{y}) f_k(\mathbf{y})\, d\Omega_{\mathbf{y}} \right\}$$

is precompact in $C(\Omega)$.

Solution The members of A clearly belong to $C(\Omega)$. Uniform boundedness of A is shown by the inequality

$$\max_{\mathbf{x}\in\Omega} \left| \int_\Omega K(\mathbf{x}, \mathbf{y}) f_k(\mathbf{y})\, d\Omega_{\mathbf{y}} \right| \le \max_{\mathbf{x}\in\Omega} |f_k(\mathbf{x})| \cdot \max_{(\mathbf{x},\mathbf{y})\in\Omega\times\Omega} |K(\mathbf{x}, \mathbf{y})| \cdot \operatorname{mes}\Omega,$$

since the set $\{f_k(\mathbf{x})\}$ is itself uniformly bounded so that $\max_{\mathbf{x}\in\Omega} |f_k(\mathbf{x})| \le c$ where c is some constant that does not depend on k. Equicontinuity of A follows from the inequality

$$\left| \int_\Omega K(\mathbf{x}, \mathbf{y}) f_k(\mathbf{y})\, d\Omega_{\mathbf{y}} - \int_\Omega K(\mathbf{x}', \mathbf{y}) f_k(\mathbf{y})\, d\Omega_{\mathbf{y}} \right|$$
$$\le c \cdot \int_\Omega |K(\mathbf{x}, \mathbf{y}) - K(\mathbf{x}', \mathbf{y})|\, d\Omega_{\mathbf{y}}.$$

Indeed, for any $\varepsilon > 0$ there exists $\delta = \delta(\varepsilon)$ such that

$$|K(\mathbf{x}, \mathbf{y}) - K(\mathbf{x}', \mathbf{y})| \le \frac{\varepsilon}{c \operatorname{mes}\Omega}$$

whenever $|\mathbf{x} - \mathbf{x}'| < \delta$ (independent of $\mathbf{y} \in \Omega$). Because A is a uniformly bounded and equicontinuous subset of $C(\Omega)$, it is precompact in $C(\Omega)$ by Arzelà's theorem.

People working in application areas often prefer to have crude but convenient sufficient conditions for the fulfillment of some properties. In the case of $C(a, b)$, the space of functions continuous on $[a, b]$, a sufficient condition is given by

Theorem 3.8.6 *A set of continuously differentiable functions bounded in the space $C^{(1)}(a, b)$ is precompact in the space $C(a, b)$.*

Proof. The proof follows from the classical Lagrange theorem which for any continuously differentiable function $f(x)$ and arbitrary x, y guarantees the existence of $z \in [x, y]$ such that $f(x) - f(y) = f'(z)(x - y)$. Equicontinuity of a bounded subset of $C^{(1)}(a, b)$ is a consequence of this. Uniform boundedness of the set is evident. □

The reader can formulate and prove the similar statement for the more general space $C^{(1)}(\Omega)$. Indeed there is an analogue of the mean value theorem for multivariable functions belonging to $C^{(1)}(\Omega)$ where Ω is compact. Let $\mathbf{x}, \mathbf{y}$ be any two points of Ω such that the connecting segment $A = t\mathbf{y} + (1 - t)\mathbf{x}$, $t \in [0, 1]$, lies in Ω. Consider a function $f(\mathbf{x}) \in C^{(1)}(\Omega)$. For fixed $\mathbf{x}, \mathbf{y}$, the function

$$F(t) = f(t\mathbf{y} + (1 - t)\mathbf{x})$$

belongs to $C^{(1)}(0, 1)$, hence we can apply the one-dimensional form of Lagrange's formula and write

$$F(1) - F(0) = F_t(t)|_{t=\xi}(1 - 0) \quad \text{for some } \xi \in [0, 1].$$

Rewriting this in terms of f we get

$$f(\mathbf{y}) - f(\mathbf{x}) = \nabla f(\mathbf{z})\big|_{\mathbf{z}=\xi\mathbf{y}+(1-\xi)\mathbf{x}} \cdot (\mathbf{y} - \mathbf{x}),$$

which is also called Lagrange's formula. The estimate

$$|f(\mathbf{y}) - f(\mathbf{x})| \le \max_{\mathbf{z} \in A} |\nabla f(\mathbf{z})| |\mathbf{y} - \mathbf{x})|$$

follows immediately. In the same way, beginning with the Newton–Leibniz formula

$$F(1) - F(0) = \int_0^1 F_t(t)\, dt$$

it is easy to prove the integral formula

$$f(\mathbf{y}) - f(\mathbf{x}) = \int_0^1 \nabla f(\mathbf{z})\big|_{\mathbf{z}=\xi\mathbf{y}+(1-\xi)\mathbf{x}} \cdot (\mathbf{y} - \mathbf{x})\, dt.$$

From this we can derive the above estimate as well.

Note that now we consider the same continuously differentiable functions as elements of different spaces, $C^{(1)}(\Omega)$ and $C(\Omega)$. When we consider the correspondence between an element in $C^{(1)}(\Omega)$ and the same element in $C(\Omega)$, it is not an identity mapping since the spaces are different and the properties of the operator are defined not only by the elements but also by the properties of the spaces. This a typical example of an operator of imbedding (we imbed a set of $C^{(1)}(\Omega)$ into $C(\Omega)$). Using this term and the notion of compact operator given later, we can reformulate the last theorem as follows:

Theorem 3.8.7 *Let Ω be a compact set in $\mathbb{R}^n$. The imbedding operator from $C^{(1)}(\Omega)$ into $C(\Omega)$ is compact.*

3.9 Inner Product Spaces, Hilbert Spaces

The existence of the dot product in Euclidean space offers many advantages with respect to the operations that may be performed in the space. The dot product also generates the norm in Euclidean space. In order that there might exist a functional defined on each pair of elements of a normed space and possessing the properties of the dot product, a linear space X should have quite special properties. Let us define what we call an *inner product.* This is a functional (x, y) defined (i.e., always giving a uniquely defined finite result) for any pair of elements x, y of the space X, and having the following properties:

(1) $(x, x) \geq 0$ for all $x \in X$, with $(x, x) = 0$ if and only if $x = 0$.
(2) $(y, x) = \overline{(x, y)}$ for all $x, y \in X$.
(3) $(\lambda x + \mu y, z) = \lambda(x, z) + \mu(y, z)$ for all $x, y, z \in X$ and any complex scalars λ, μ.

We have defined this for a complex space. If X is a real space instead, then property 2 must be changed to

2. $(y, x) = (x, y)$ for all $x, y \in X$

and in property 3 we must use only real scalars λ, μ. Note that the inner product is linear in the first argument and *conjugate linear* in the second argument:

$$(\alpha_1 x_1 + \alpha_2 x_2, y) = \alpha_1(x_1, y) + \alpha_2(x_2, y),$$
$$(x, \alpha_1 y_1 + \alpha_2 y_2) = \overline{\alpha}_1(x, y_1) + \overline{\alpha}_2(x, y_2).$$

Example 3.9.1 Let X be any inner product space under the inner product $(\cdot, \cdot)$. Show that $(x, z) = (y, z)$ holds for arbitrary $z \in X$ if and only if $x = y$.

Solution The "if" part of the proposition is trivial. To prove the "only if" part, we begin by assuming that $(x, z) = (y, z)$ for all $z \in X$. Rearranging this as

$$(x, z) - (y, z) = 0,$$

we can use property 3 to get $(x - y, z) = 0$. Since this holds for all $z \in X$, it holds in particular for $z = x - y$:

$$(x - y, x - y) = 0.$$

By property 1 we conclude that $x - y = 0$. This, of course, implies that $x = y$.

Since this functional, the inner product, is defined by copying the main properties of the dot product, we preserve the terminology connected with the dot product in Euclidean space. In particular there is the notion of orthogonality. We say that two elements x, y are mutually orthogonal if $(x, y) = 0$. We say that x orthogonal to Y, a subspace of X, if x is orthogonal to each element of Y.

Definition 3.9.1 A linear space with an inner product possessing the properties listed above is called an *inner product space* or a *pre-Hilbert space.*

First we demonstrate

Theorem 3.9.1 *A pre-Hilbert space is a normed space.*

Proof. By similarity to Euclidean space let us introduce a functional denoted as a norm

$$\|x\| = (x, x)^{1/2}.$$

This functional is defined for any element of X. Let us demonstrate that it satisfies all the axioms of the norm. Norm axiom 1 is fulfilled by virtue of inner product axiom 1. We verify norm axiom 2 by noting that

$$\begin{aligned}\|\lambda x\| &= [(\lambda x, \lambda x)]^{1/2} = [\lambda(x, \lambda x)]^{1/2} = [\lambda\overline{(\lambda x, x)}]^{1/2} \\ &= [(\lambda\overline{\lambda})\overline{(x, x)}]^{1/2} = [|\lambda|^2(x, x)]^{1/2} \\ &= |\lambda|(x, x)^{1/2}.\end{aligned}$$

Verification of norm axiom 3 requires us to use the *Schwarz inequality*

$$|(x, y)| \le \|x\| \, \|y\|, \tag{3.9.1}$$

in which for nonzero x and y the equality holds if and only if there is a number λ such that $x = \lambda y$. Using it we have

$$\begin{aligned}\|x+y\|^2 &= (x+y, x+y)\\ &= (x,x) + (x,y) + (y,x) + (y,y)\\ &\le \|x\|^2 + \|x\|\,\|y\| + \|x\|\,\|y\| + \|y\|^2\\ &= (\|x\| + \|y\|)^2\end{aligned}$$

as required. □

It remains to establish (3.9.1). We start by noting that if $x = 0$ or $y = 0$ then (3.9.1) is evidently valid. So let $y \neq 0$. If λ is any scalar, then $(x + \lambda y, x + \lambda y) \geq 0$ and expansion gives

$$(x+\lambda y, x+\lambda y) = (x,x) + \lambda(y,x) + \overline{\lambda}(x,y) + \lambda\overline{\lambda}(y,y).$$

The particular choice $\lambda = -(x,y)/(y,y)$ reduces this to

$$\|x\|^2 - 2\frac{|(x,y)|^2}{\|y\|^2} + \frac{|(x,y)|^2\,\|y\|^2}{\|y\|^4} \geq 0,$$

and (3.9.1) follows directly.

Example 3.9.2 Show that

$$\|x+y\|^2 + \|x-y\|^2 = 2\|x\|^2 + 2\|y\|^2.$$

This is known as the *parallelogram equality.*

Solution We write

$$\begin{aligned}\|x+y\|^2 + \|x-y\|^2 &= (x+y, x+y) + (x-y, x-y)\\ &= (x, x+y) + (y, x+y) + (x, x-y) - (y, x-y)\\ &= \overline{(x+y, x)} + \overline{(x+y, y)} + \overline{(x-y, x)} - \overline{(x-y, y)}\\ &= \overline{(x,x)} + \overline{(y,x)} + \overline{(x,y)} + \overline{(y,y)}\\ &\quad + \overline{(x,x)} - \overline{(y,x)} - \overline{(x,y)} + \overline{(y,y)}\\ &= 2(x,x) + 2(y,y)\\ &= 2\|x\|^2 + 2\|y\|^2\end{aligned}$$

and have the desired result.

Example 3.9.3 Show that if x and y are orthogonal vectors in an inner product space, then

$$\|x+y\|^2 = \|x\|^2 + \|y\|^2 .$$

This is known as the *Pythagorean theorem.*

Solution We write

$$\begin{aligned}\|x+y\|^2 &= (x+y, x+y) \\ &= (x, x+y) + (y, x+y) \\ &= (x,x) + (x,y) + (y,x) + (y,y)\end{aligned}$$

and simply note that $(x,y) = (y,x) = 0$ for orthogonal vectors.

Example 3.9.4 (a) Assume the norm is induced by the inner product, and suppose that $x_n \to x$ and $y_n \to y$. Show that $(x_n, y_n) \to (x,y)$. That is, any inner product is a continuous functional in each of its arguments. (b) Let M be a dense subset of an inner product space X, and let $v \in X$. Show that if $(v,m) = 0$ for all $m \in M$, then $v = 0$.

Solution (a) Let us write

$$\begin{aligned}|(x_n,y_n) - (x,y)| &= |(x_n,y_n) - (x_n,y) + (x_n,y) - (x,y)| \\ &= |(x_n, y_n - y) + (x_n - x, y)| \\ &\le |(x_n, y_n - y)| + |(x_n - x, y)| \\ &\le \|x_n\| \, \|y_n - y\| + \|x_n - x\| \, \|y\| .\end{aligned}$$

Since $\{x_n\}$ is convergent it is bounded. The other n-dependent quantities can be made as small as desired by choosing n sufficiently large. (b) Use continuity of the inner product. Let $v \in X$ be fixed. Since M is dense in X there is a sequence of elements $m_k \in M$ such that $m_k \to v$ as $k \to \infty$. Since $0 = (v, m_k)$ for all k, we can take the limit as $k \to \infty$ on both sides and use continuity of the inner product to obtain

$$0 = \lim_{k\to\infty} (v, m_k) = \left(v, \lim_{k\to\infty} m_k\right) = (v,v).$$

Hence $v = 0$.

Definition 3.9.2 A complete pre-Hilbert space is called a *Hilbert space.*

Let us consider some Hilbert spaces. The space ℓ^2 is the space of infinite sequences having inner product

$$(\mathbf{x}, \mathbf{y}) = \sum_{i=1}^{\infty} x_i \overline{y_i}$$

in the complex case and

$$(\mathbf{x}, \mathbf{y}) = \sum_{i=1}^{\infty} x_i y_i$$

in the real case. The corresponding generated (induced) norm is

$$\|\mathbf{x}\| = (x, x)^{1/2} = \left(\sum_{i=1}^{\infty} |x_i|^2 \right)^{1/2} .$$

As we noted earlier, the theory of the space ℓ^2 was the predecessor of functional analysis. It plays an extremely important role in the functional analysis of Hilbert spaces because, as we shall see later, for any separable Hilbert space we can introduce a one-to-one isometric correspondence with ℓ^2 that preserves algebraic operations in the spaces. This is done by the use of Fourier expansion of elements of the Hilbert space.

In the space $L^2(\Omega)$ an inner product can be introduced as

$$(f(\mathbf{x}), g(\mathbf{x})) = \left(\int_{\Omega} f(\mathbf{x}) \overline{g(\mathbf{x})} \, d\Omega \right)^{1/2}$$

in the complex case and

$$(f(\mathbf{x}), g(\mathbf{x})) = \left(\int_{\Omega} f(\mathbf{x}) g(\mathbf{x}) \, d\Omega \right)^{1/2}$$

in the real case. We have introduced the inner product in such a way that the induced norm coincides with the norm that we introduced earlier on $L^2(\Omega)$. This brings us to the question of how to introduce an inner product in any Sobolev space $W^{l,2}(\Omega)$: we use

$$(f(\mathbf{x}), g(\mathbf{x})) = \int_{\Omega} \sum_{|\alpha| \le l} D^{\alpha} f(\mathbf{x}) \overline{D^{\alpha} g(\mathbf{x})} \, d\Omega,$$

Of course, the induced norm is the same as the norm we introduced earlier in $W^{l,2}(\Omega)$.

An important class of Hilbert spaces forms the subject of the next section.

3.10 Some Energy Spaces in Mechanics

To distinguish different states of mechanical objects it is possible to use various norms. To characterize the amplitudes of forces, for example, it is appropriate to use norms of the type of the norm of $M(\Omega)$, defined on the set of bounded functions. If the field is continuous then it is more appropriate to use the tools of the space $C(\Omega)$. The same can be said for fields of displacements, strains, and stresses. However, there is one important characteristic of a body: its energy due to deformation. It is sensible to try to use this quantity when we characterize the state of a body. We would like to consider this possibility in more detail. The most convenient fact is that the energy spaces we shall introduce are subspaces of Sobolev spaces, and thus we can use Sobolev's imbedding theorem to characterize the parameters of corresponding boundary value problems. Of course, it is possible to use Sobolev spaces directly for this, but the use of energy spaces has many advantages. First of all, in this way we take into account the nature of the problem more closely, so sometimes we can make better use of what is often called mechanical intuition. Moreover, the energy norms and corresponding inner products permit the proper and direct use of such fundamental properties as mutual orthogonality of eigensolutions of corresponding problems; these properties form the basis for solutions by the Fourier technique.

A stretched rod

We begin with a very simple problem that could be solved by direct integration. It is the problem of equilibrium of a rod when it is stretched by a distributed load (Figure 3.1). The double inner energy of a rod of length l is

$$2\mathcal{E}(u) = \int_0^l ES(x)u'^2(x)\,dx$$

where the constant E is Young's modulus, $S(x)$ is the area of the cross-section with $0 < S_0 \le S(x) \le S_1$, and $u(x)$ is the displacement of the cross-section of the rod at point x in the longitudinal direction. Suppose

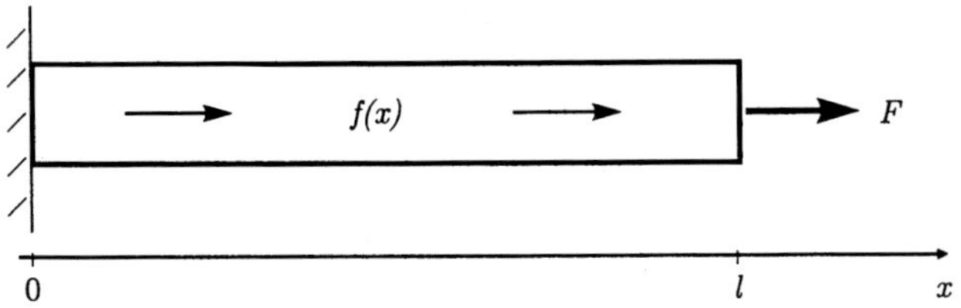

Fig. 3.1 Stretched rod under distributed longitudinal load $f(x)$ and a point force F.

the end at $x = 0$ is fixed:

$$u(0) = 0. \tag{3.10.1}$$

This energy generates a functional in two variables that can be considered as an inner product:

$$(u, v)_{Rc} = \int_0^l ES(x)u'(x)v'(x)\,dx. \tag{3.10.2}$$

(Here the subscripts "*Rc*" are used to remind us that we are dealing with a *clamped rod*: a rod that is fixed in space. Below we will use subscripts "*Rf*" to denote a *free rod*.) The inner product has a clear mechanical meaning: it is the work of internal forces corresponding to the state of the rod $u(x)$ on the admissible displacement field $v(x)$. (We recall that the terms "admissible" and "virtual" are interchangeable.) Considering it on the set C_{Rc} of all continuously differentiable functions on $[0, l]$ that satisfy (3.10.1), we can demonstrate that it really is an inner product (the reader should verify this). Let us demonstrate that on C_{Rc} the norm

$$\|u\|_{Rc} = (u, u)_{Rc}^{1/2} = \left(\int_0^l ES(x)u'^2(x)\,dx\right)^{1/2}$$

induced by the energy inner product is equivalent to the norm of the Sobolev space $W^{1,2}(0, l)$, which is

$$\|u\|_{1,2} = \left(\int_0^l \left(u^2(x) + u'^2(x)\right)\,dx\right)^{1/2}.$$

We must show that there are two positive constants m, M such that for any $u(x) \in C_{Rc}$ we have

$$m\,\|u\|_{Rc} \le \|u\|_{1,2} \le M\,\|u\|_{Rc}.$$

The left-hand inequality is a consequence of

$$\begin{aligned}\|u(x)\|_{Rc}^2 &= \int_0^l ES(x)u'^2(x)\,dx \\ &\le ES_1 \int_0^l \left(u^2(x) + u'^2(x)\right)\,dx \\ &= ES_1\,\|u(x)\|_{1,2}^2\,.\end{aligned}$$

To prove the right-hand inequality we begin with the identity

$$u(x) = \int_0^x u'(t)\,dt.$$

Squaring and then integrating over $[0, l]$ we get

$$\int_0^l u^2(x)\,dx = \int_0^l \left(\int_0^x u'(t)\,dt\right)^2 dx.$$

Applying the Hölder inequality we have

$$\begin{aligned}\int_0^l u^2(x)\,dx &= \int_0^l \left(\int_0^x 1\cdot u'(t)\,dt\right)^2 dx \\ &\le \int_0^l \left(\int_0^x 1^2\,dt \int_0^x u'^2(t)\,dt\right) dx \\ &\le l^2 \int_0^l u'^2(x)\,dx, \qquad (3.10.3)\end{aligned}$$

from which the needed fact follows immediately.

If we now apply the procedure of completion in the set C_{Rc} with respect to the norms $\|\cdot\|_{Rc}$ and $\|\cdot\|_{1,2}$, we get spaces that contain the same elements and have equivalent norms, so they are considered as the same space. Let us denote this energy space by $\mathcal{E}_{Rc}$ and apply the Sobolev imbedding theorem. This space is a subspace of the Sobolev space $W^{1,2}(0,l)$. We said that this space is continuously imbedded into $W^{1,1}(0,l)$, and for the latter we established that to each of its elements there corresponds a continuous function; hence to each element of $\mathcal{E}_{Rc}$ there corresponds a continuous function. It is easy to see that all these continuous functions satisfy (3.10.1). We shall identify these continuous functions with corresponding elements of $\mathcal{E}_{Rc}$, and in this sense say that the elements of $\mathcal{E}_{Rc}$ are continuous functions. The same will be done in other cases.

A free rod

In the same manner we can consider the energy space for a free rod: i.e., a rod with both ends free of geometrical restrictions. In this case longitudinal motions of the rod are unrestricted by boundary conditions, so when we try to use the same inner product (3.10.2) induced by inner energy of the rod, we will meet a situation where there are nontrivial displacements for which the corresponding energy norm is zero. It is easy to see that $u(x) = c$ is such a state of the rod. First we will show that there are no other states with zero inner energy, for which we will derive an inequality that replaces (3.10.3) for the problem of a free rod. We begin with the identity

$$u(x) = u(y) + \int_y^x u'(t)\,dt.$$

First we integrate this with respect to y over $[0, l]$ to get

$$lu(x) = \int_0^l u(y)\,dy + \int_0^l \int_y^x u'(t)\,dt\,dy,$$

then we take the absolute value of both sides of the identity and estimate the right-hand side as in § 3.6:

$$l|u(x)| = \left| \int_0^l u(y)\,dy + \int_0^l \int_y^x u'(t)\,dt\,dy \right| \le \left| \int_0^l u(y)\,dy \right| + l \int_0^l |u'(t)|\,dt. \tag{3.10.4}$$

Let us consider the subset of functions continuously differentiable on $[0, l]$, denoted by C_{Rf}, for which

$$\int_0^l u(y)\,dy = 0. \tag{3.10.5}$$

Note that subtracting a proper constant c from a function, which corresponds to a free motion of the rod through the distance c, we get the displacement field in the rod with property (3.10.5). From (3.10.4) we have three consequences:

$$\int_0^l |u(x)|\,dx \le \left| \int_0^l u(x)\,dx \right| + l \int_0^l |u'(x)|\,dx, \tag{3.10.6}$$

$$l \max_{x \in [0,l]} |u(x)| \le \left| \int_0^l u(x)\,dx \right| + l \int_0^l |u'(x)|\,dx, \tag{3.10.7}$$

and

$$l\int_0^l |u(x)|^2\,dx \le 2\left\{\left|\int_0^l u(t)\,dt\right|^2 + l^3\int_0^l |u'^2(y)|\,dy\right\}. \tag{3.10.8}$$

(cf., Exercise 3.53). From (3.10.6) it follows that the right-hand side can serve as an equivalent norm in the space $W^{1,1}(0,l)$. Result (3.10.7) states that on the subspace of $W^{1,1}(0,l)$ that is the completion of the set C_{Rf} with respect to the norm of $W^{1,1}(0,l)$, we get the continuous imbedding of its elements into $C(0,l)$ and, moreover, the corresponding continuous functions satisfy (3.10.5). Finally, from (3.10.8) it follows that taking the completion of C_{Rf} with respect to the energy norm we get a subspace of $W^{1,2}(0,l)$ whose norm is equivalent to the energy norm $\|\cdot\|_{Rf}$. This was one way in which we could use the energy norm for a free rod to circumvent the difficulty with free motions.

There is another simple way of doing this. We can introduce a factor space of continuously differentiable functions with respect to all constant functions. This means we declare that the union of all the constant functions is the zero element of the new space. Between the elements of this factor set and the set C_{Rf} there is a one-to-one correspondence preserving the energy distances between corresponding elements. So completion in both cases gives the same result from the point of view of isometry, and thus both of the approaches to the introduction of an energy space for a free rod are equivalent.

A bent beam

For a flexible elastic beam (Figure 3.2), equilibrium is governed by the equation

$$(EI\,y''(x))'' = f(x)$$

on $[0,l]$, where E, I are given characteristics of the beam, $y = y(x)$ is the transverse displacement, and $f = f(x)$ is the transverse load. If E and I are piecewise continuous functions of x, then it is natural to assume that

$$0 < c_0 \le EI \le c_1, \tag{3.10.9}$$

where c_0 and c_1 are constants. Let us first consider a cantilever beam:

$$y(0) = 0 = y'(0). \tag{3.10.10}$$

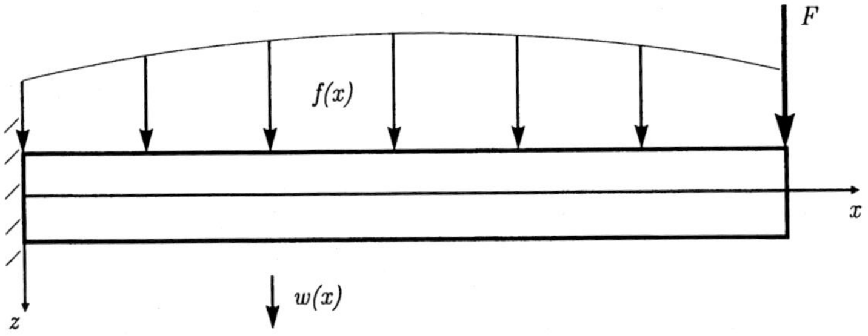

Fig. 3.2 Beam under load $f(x)$ and a point force F acting at the end.

So it is hard-clamped on the left end, and its right end is free from restrictions of geometrical nature. We use dimensionless variables. The elastic energy of the beam is

$$E_B = \frac{1}{2}\int_0^l EI\, y''^2(x)\, dx.$$

On the subset C_B of those $C^{(2)}(0,l)$ functions satisfying the condition (3.10.10), the energy induces a metric

$$d(y,z) = \left(\int_0^l EI\,(y''(x) - z''(x))^2\, dx\right)^{1/2} \tag{3.10.11}$$

(we leave it to the reader to verify that all the metric axioms are valid for this). To this metric there corresponds the energy norm

$$\|y\|_B = \left(\int_0^l EI\, y''^2(x)\, dx\right)^{1/2}$$

and an inner product

$$(y,z)_B = \int_0^l EI\, y''(x) z''(x)\, dx. \tag{3.10.12}$$

This space is not complete. Applying the completion theorem to the space of functions from $C^{(2)}(0,l)$ satisfying (3.10.10) with respect to the metric (3.10.11), we get a complete metric space denoted by $\mathcal{E}_{Bc}$ that is a Hilbert space with inner product (3.10.12). Because of (3.10.9), the norm on $\mathcal{E}_{Bc}$

is equivalent to the auxiliary norm

$$\|y\|_2 = \left(\int_0^l y''^2(x)\, dx \right)^{1/2}, \tag{3.10.13}$$

and thus we will study the properties of elements of $\mathcal{E}_{Bc}$ using the norm (3.10.13). First let us mention that for a function y of the base set C_B from which $\mathcal{E}_{Bc}$ arises, its derivative y' belongs to the base set for the energy space for a stretched rod with clamped edge, and so we can write out inequality (3.10.3) for it,

$$\int_0^l y'^2(x)\, dx \le l^2 \int_0^l y''^2(x)\, dx,$$

in addition to the inequality for y itself,

$$\int_0^l y^2(x)\, dx \le l^2 \int_0^l y'^2(x)\, dx,$$

and thus for any smooth representer of the space $\mathcal{E}_{Bc}$ we have

$$\int_0^l y^2(x)\, dx + \int_0^l y'^2(x)\, dx \le c \int_0^l y''^2(x)\, dx \le c_2 \int_0^l EI\, y''^2(x)\, dx.$$

All together this means that on $\mathcal{E}_{Bc}$ the energy norm is equivalent to the norm of the Sobolev space $W^{2,2}(0,l)$ whose norm is

$$\|y\|_{2,2}^2 = \int_0^l \left(y''^2(x) + y'^2(x) + y^2(x) \right) dx,$$

and so on $\mathcal{E}_{Bc}$ we can use Sobolev's imbedding theorem for $W^{2,2}(0,l)$. In this case each element of $\mathcal{E}_{Bc}$ is identified with a continuously differentiable function; in other words, the space $\mathcal{E}_{Bc}$ is imbedded continuously into the space $C^{(1)}(0,l)$.

A bent beam (free ends)

Now we do not impose any geometric restraints on the left and right ends of the beam. We would like to try the same functional $\|y\|_B$ for the role of a norm. It is easily seen that we are in the same position as for the norm of a free stretched rod, namely, all the axioms of the norm hold except one: when $\|y\|_B = 0$, it follows that there is a non-zero solution to this equation $y = a + bx$. This function has a clear mechanical meaning: it is the motion of the beam in space as a rigid body. We will use this term "rigid motion" very frequently in what follows. In a way how we used inequality (3.10.3), we can subsequently obtain from inequality (3.10.8) that for any function from $C^{(2)}(0,l)$ there holds

$$l\int_0^l y'^2(x)\,dx \le 2\left(\int_0^l y'(x)\,dx\right)^2 + 2l^3\int_0^l y''^2(x)\,dx$$

and

$$l\int_0^l y^2(x)\,dx \le 2\left(\int_0^l y(x)\,dx\right)^2 + 2l^3\int_0^l y'^2(x)\,dx,$$

and thus

$$\int_0^l \left(y^2(x) + y'^2(x)\right)\,dx \le c_3\left[\left(\int_0^l y(x)\,dx\right)^2 + \left(\int_0^l y'(x)\,dx\right)^2 + \int_0^l y''^2(x)\,dx\right]. \qquad (3.10.14)$$

Inequality (3.10.14) means that the expression

$$\|y\|_2 = \left(\left(\int_0^l y(x)\,dx\right)^2 + \left(\int_0^l y'(x)\,dx\right)^2 + \int_0^l EI\,y''^2(x)\,dx\right)^{1/2} \qquad (3.10.15)$$

is a norm that is equivalent to the norm of $W^{2,2}(0,l)$. To introduce the energy space for a free beam we can use this fact in two ways in the same manner as was done for a stretched rod. First we can choose for the base of the energy space only those smooth functions for which

$$\int_0^l y(x)\,dx = 0 = \int_0^l y'(x)\,dx. \qquad (3.10.16)$$

Denote this set C_{Bf}.

To get elements of C_{Bf} we can use the fact that to any smooth function $y = y(x)$ there corresponds the only function satisfying (3.10.16) that is obtained by a proper choice of constants a and b in the expression $y(x) - a - bx$. In this way we do not change the distribution of stresses in the beam, but only fix the beam somehow in space. Thus, because of (3.10.15), on the set of functions from $C^{(2)}(0,l)$ satisfying (3.10.16) the norm $\|y\|_B = \left(\int_0^l EI\,y''^2(x)\,dx\right)^{1/2}$ is equivalent to the norm of $W^{2,2}(0,l)$, and thus after completion we can use the Sobolev imbedding theorem for $W^{2,2}(0,l)$ and hence know that any representative sequence of $\mathcal{E}_{Bf}$ is such that it has a continuous function as a limit and, moreover, the sequence of first derivatives also converges to a continuous function. Moreover, for the limit functions we get that (3.10.16) holds as well.

The second way we can use now is to employ a factor space, declaring that the zero element of the energy space is the set of all linear polynomials that are infinitesimal rigid motions of the beam, $a + bx$. In this case among all the representers of an element there is only one that satisfies (3.10.16), and thus we get an isometric one-to-one correspondence between the elements of the two versions of the energy space and can carry interpretations of results for one version over to the other.

Remark 3.10.1 In order to introduce the energy space for an elastic beam subjected to normal and longitudinal loads, we can consider pairs of displacements (u, w) and combine the energy functionals, norms, and inner products for a rod and a beam.

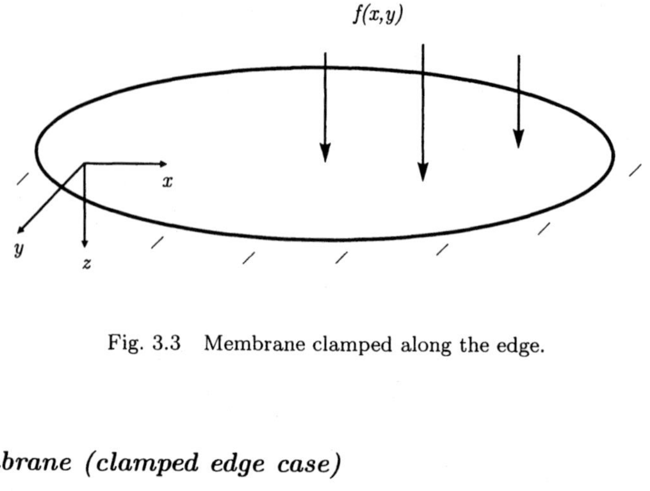

Fig. 3.3 Membrane clamped along the edge.

A membrane (clamped edge case)

The equilibrium of a clamped membrane (Figure 3.3) occupying a domain $\Omega \subset \mathbb{R}^2$ is described by the equations

$$a\,\Delta u = -f, \qquad\qquad u\Big|_{\partial\Omega} = 0,$$

which together make up the Dirichlet problem for Laplace's equation. Here $u = u(x, y)$ is the transverse displacement of the membrane and $f = f(x, y)$

is the external load. The parameter a relates to the tension in the membrane. The potential energy of the membrane is

$$\mathcal{E}_M(u) = \frac{a}{2} \int_\Omega \left[\left(\frac{\partial u}{\partial x} \right)^2 + \left(\frac{\partial u}{\partial y} \right)^2 \right] dx\, dy.$$

By a proper choice of dimensionless variables in what follows, we will put $a = 1$. A metric corresponding to this energy on the set of functions $u(x, y)$ from $C^{(1)}(\Omega)$ that satisfy the boundary condition

$$u(x,y)\Big|_{\partial\Omega} = 0 \tag{3.10.17}$$

is

$$d(u,v) = \left\{ \iint_\Omega \left[\left(\frac{\partial u}{\partial x} - \frac{\partial v}{\partial x} \right)^2 + \left(\frac{\partial u}{\partial y} - \frac{\partial v}{\partial y} \right)^2 \right] dx\, dy \right\}^{1/2}. \tag{3.10.18}$$

The resulting metric space is appropriate as a starting point for investigating the corresponding boundary value problem.

The subset C_{Mc} of $C^{(1)}(\Omega)$ consisting of all functions satisfying (3.10.17) with the metric (3.10.18) is an incomplete metric space. If we introduce an inner product

$$(u,v)_M = \iint_\Omega \left(\frac{\partial u}{\partial x}\frac{\partial v}{\partial x} + \frac{\partial u}{\partial y}\frac{\partial v}{\partial y} \right) dx\, dy$$

consistent with (3.10.18) we get an inner product space. Its completion in the metric (3.10.18) is the energy space for the clamped membrane, denoted $\mathcal{E}_{Mc}$. This is a real Hilbert space.

What can we say about the elements of $\mathcal{E}_{Mc}$? It is obvious that the sequences of first derivatives $\{\partial u_n/\partial x\}$, $\{\partial u_n/\partial y\}$, of a representative sequence $\{u_n\}$ are Cauchy sequences in the norm on $L^2(\Omega)$: i.e., if

$$d(u_m, u_n) = \left\{ \iint_\Omega \left[\left(\frac{\partial u_m}{\partial x} - \frac{\partial u_n}{\partial x} \right)^2 + \left(\frac{\partial u_m}{\partial y} - \frac{\partial u_n}{\partial y} \right)^2 \right] dx\, dy \right\}^{1/2}$$
$$\to 0 \quad \text{as } m, n \to \infty,$$

then

$$\left\{ \iint_{\Omega} \left(\frac{\partial u_m}{\partial x} - \frac{\partial u_n}{\partial x} \right)^2 dx\,dy \right\}^{1/2} = \left\| \frac{\partial u_m}{\partial x} - \frac{\partial u_n}{\partial x} \right\|_{L^2(\Omega)} \to 0 \quad \text{as } m, n \to \infty,$$

and similarly for $\{\partial u_n/\partial y\}$. It takes more work to say something about $\{u_n\}$ itself; we need the *Friedrichs inequality*.

The Friedrichs inequality states that if a continuously differentiable function $u = u(x, y)$ has compact support in Ω, then there is a constant $C > 0$, depending on Ω only, such that

$$\int_{\Omega} |u|^2 \, d\Omega \le C \int_{\Omega} |\nabla u|^2 \, d\Omega.$$

To prove this it is convenient to first suppose Ω is the square $|x| < a$, $|y| < a$. Since

$$u(x, y) = u(-a, y) + \int_{-a}^{x} \frac{\partial u(\xi, y)}{\partial \xi} \, d\xi$$

and $u(-a, y) = 0$, we have

$$\int_{\Omega} |u(x, y)|^2 \, d\Omega = \int_{-a}^{a} \int_{-a}^{a} \left| \int_{-a}^{x} \frac{\partial u(\xi, y)}{\partial \xi} \, d\xi \right|^2 dx\,dy.$$

Then

$$\begin{aligned}
\int_{\Omega} |u(x, y)|^2 \, d\Omega &= \int_{-a}^{a} \int_{-a}^{a} \left| \int_{-a}^{x} 1 \cdot \frac{\partial u(\xi, y)}{\partial \xi} \, d\xi \right|^2 dx\,dy \\
&\le \int_{-a}^{a} \int_{-a}^{a} \int_{-a}^{x} 1^2 \, d\xi \int_{-a}^{x} \left| \frac{\partial u(\xi, y)}{\partial \xi} \right|^2 d\xi\,dx\,dy \\
&\le \int_{-a}^{a} \int_{-a}^{a} \int_{-a}^{a} 1^2 \, d\xi \int_{-a}^{a} \left| \frac{\partial u(\xi, y)}{\partial \xi} \right|^2 d\xi\,dx\,dy \\
&= \int_{-a}^{a} 1^2 \, d\xi \int_{-a}^{a} dx \int_{-a}^{a} \int_{-a}^{a} \left| \frac{\partial u(\xi, y)}{\partial \xi} \right|^2 d\xi\,dy \\
&= 4a^2 \int_{-a}^{a} \int_{-a}^{a} \left| \frac{\partial u(\xi, y)}{\partial \xi} \right|^2 d\xi\,dy,
\end{aligned}$$

hence

$$\int_\Omega |u|^2\,d\Omega \le 4a^2 \int_{-a}^{a}\int_{-a}^{a} \left|\frac{\partial u(x,y)}{\partial x}\right|^2 dx\,dy = 4a^2 \int_\Omega \left|\frac{\partial u}{\partial x}\right|^2 d\Omega.$$

By the same reasoning, an analogous inequality holds with $\partial u/\partial y$ on the right-hand side. Adding these two inequalities we obtain

$$\int_\Omega |u|^2\,d\Omega \le C \int_\Omega \left(\left|\frac{\partial u}{\partial x}\right|^2 + \left|\frac{\partial u}{\partial y}\right|^2\right) d\Omega$$

where $C = 2a^2$. If Ω is not square, we can enclose it in a square $\tilde{\Omega}$ and extend the function u onto the set $\tilde{\Omega}$ by setting $u \equiv 0$ on $\tilde{\Omega} - \Omega$ to obtain a new function $\tilde{u}$; in this case

$$\int_{\tilde{\Omega}} |\tilde{u}|^2\,d\tilde{\Omega} \le C \int_{\tilde{\Omega}} \left(\left|\frac{\partial \tilde{u}}{\partial x}\right|^2 + \left|\frac{\partial \tilde{u}}{\partial y}\right|^2\right) d\tilde{\Omega}$$

follows. (Note that the extension $\tilde{u}$ may have a discontinuous derivative on $\partial\Omega$; however, the presence of such a discontinuity does not invalidate any of the steps above when $\partial\Omega$ is sufficiently smooth.) The constant C depends only on a, hence only on Ω (which dictates the choice of a).

Above we observed that if $\{u_n\}$ is a representative of an element of $\mathcal{E}_{Mc}$, then $\{\partial u_n/\partial x\}$ and $\{\partial u_n/\partial y\}$ are Cauchy sequences in the norm of $L^2(\Omega)$. The Friedrichs inequality applied to $u = u_n(x,y)$ shows that $\{u_n\}$ is also a Cauchy sequence in the norm of $L^2(\Omega)$. Hence to each $U(x,y) \in \mathcal{E}_{Mc}$ having a representative sequence $\{u_n\}$, there correspond elements in $L^2(\Omega)$ having $\{u_n\}$, $\{\partial u_n/\partial x\}$ and $\{\partial u_n/\partial y\}$ as representatives. We denote these elements of $L^2(\Omega)$ by $U(x,y)$, $\partial U(x,y)/\partial x$, and $\partial U(x,y)/\partial y$, respectively. The elements $\partial U/\partial x$ and $\partial U/\partial y$ are assigned interpretations as generalized derivatives of the element U later on. However, we need a result for the elements of the completed energy space. Passage to the limit in the Friedrichs inequality gives

$$\iint_\Omega U^2\,dx\,dy \le C \iint_\Omega \left[\left(\frac{\partial U}{\partial x}\right)^2 + \left(\frac{\partial U}{\partial y}\right)^2\right] dx\,dy \tag{3.10.19}$$

for any $U \in \mathcal{E}_{Mc}$ and a constant C independent of U.

Inequality (3.10.19) also means that in $\mathcal{E}_{Mc}$ the energy norm is equivalent to the norm of $W^{1,2}(\Omega)$, and thus for the space $\mathcal{E}_{Mc}$ there holds an imbedding result in the form of Theorem 3.7.3.

A free membrane

When there are no geometrical restraints on the motion of a membrane, we call it free. As in the case of a free bent beam or a stretched rod, a free membrane has displacements that can be considered as displacements of a rigid body. They are different from the motions of a real rigid body in space because the model of the membrane under consideration reflects only some features of the real object that we consider as a membrane. For characterizing of the state of a free membrane we are choosing the energy functional and so the metric (3.10.18) or, the same, the norm

$$\|u\|_M = \left(\int_\Omega \left(\left(\frac{\partial u}{\partial x}\right)^2 + \left(\frac{\partial u}{\partial y}\right)^2\right) d\Omega\right)^{1/2} \tag{3.10.20}$$

again. It is not a norm on the space of all functions of $C^{(1)}(\Omega)$, where Ω is compact, since the equation $\|u\|_M = 0$ has a solution $u = c =$ constant, so we cannot distinguish two states of the membrane whose difference in position is c. This constant displacement is the only type of rigid motion of a membrane for the model under consideration. The way in which we will circumvent the existence of rigid motions looks similar to the one used above for free rods and beams. It is based on *Poincaré's inequality.* This extends inequality (3.10.8) to a 2-D domain (in fact it is extended for any compact n-dimensional domain with piecewise smooth boundary):

$$\int_\Omega u^2\, d\Omega \le C\left(\left(\int_\Omega u\, d\Omega\right)^2 + \int_\Omega \left(\left(\frac{\partial u}{\partial x}\right)^2 + \left(\frac{\partial u}{\partial y}\right)^2\right) d\Omega\right)$$

with a constant C that does not depend on u. We note that the method of its proof for a rectangle domain is similar to one for (3.10.8). The proof is lengthy and for a general compact domain with piecewise smooth boundary it is even more lengthy, so we leave this inequality without proof.[4] From the Poincaré inequality it follows that on functions from $C^{(1)}(\Omega)$ satisfying the condition

$$\int_\Omega u(x, y)\, d\Omega = 0 \tag{3.10.21}$$

the energy norm $\|u\|_M$ is equivalent to the norm of $W^{1,2}(\Omega)$. Thus defining the energy space $\mathcal{E}_{Mf}$ as the completion of functions from $C^{(1)}(\Omega)$ satisfying (3.10.21) with respect to the norm (3.10.20) we get a subspace of $W^{1,2}(\Omega)$

[4] The interested reader can refer to Courant and Hilbert [Courant and Hilbert (1989)].

and hence we can use the Sobolev imbedding theorem of the elements of this energy space. We can use another way of arrangement of the energy space similarly to another way for free rods and beams. In this we collect all the constants into the one element and declare it to be a zero of the energy space. The energy space in this case is a factor space of $W^{1,2}(\Omega)$ with respect to the set of all the constant functions on Ω. Again, there is one-to-one isometry between elements of the two versions of the energy space, and so we can use any of them in what follows.

An elastic body

The internal energy of an elastic body occupying a 3-D bounded connected volume V is given by

$$\frac{1}{2}\int_V \sum_{ijkl=1}^{3} c^{ijkl} e_{kl} e_{ij} \, dV$$

where c^{ijkl} are the components of the tensor of elastic moduli and e_{ij} are the components of the tensor of small strains. From now on we shall omit the summation symbol when we meet a repeated index in an expression; this is called Einstein's rules of repeated indices. The components of the strain tensor relate to the components of the displacement vector $\mathbf{u} = (u_1, u_2, u_3)$ given in Cartesian coordinates according to

$$e_{ij} = e_{ij}(\mathbf{u}) = \frac{1}{2}\left(u_{i,j} + u_{j,i}\right),$$

where the indices after a comma mean differentiation with respect to the corresponding coordinates:

$$u_{i,j} = \frac{\partial u_i}{\partial x_j}.$$

We suppose that the elastic moduli have the usual properties of symmetry established in the theory of elasticity, and in addition possess the property providing positiveness of the functional of inner energy:

$$c^{ijkl} e_{kl} e_{ij} \geq c_0 e_{mn} e_{mn}$$

for any symmetric tensor with components e_{mn}. Here c_0 is a positive constant.

By symmetry of the components of the tensor of elastic moduli, we can introduce a bilinear functional that has properties of symmetry and so pretends to be an inner product:

$$(\mathbf{u},\mathbf{v})_E = \int_V c^{ijkl} e_{kl}(\mathbf{u}) e_{ij}(\mathbf{v})\, dV.$$

Linearity of this functional in $\mathbf{u}$ and $\mathbf{v}$ is seen, as well as the property of symmetry

$$(\mathbf{u},\mathbf{v})_E = (\mathbf{v},\mathbf{u})_E.$$

It remains to verify that the first axiom of the inner product holds. By the properties of the elastic moduli we get

$$(\mathbf{u},\mathbf{u})_E = \int_V c^{ijkl} e_{kl}(\mathbf{u}) e_{ij}(\mathbf{u})\, dV \geq 0.$$

If $(\mathbf{u},\mathbf{u})_E = \int_V c^{ijkl} e_{kl}(\mathbf{u}) e_{ij}(\mathbf{u})\, dV = 0$ and the components of the vector of displacements are continuously differentiable, we have $e_{ij}(\mathbf{u}) = 0$ for all i, j. From the results of the theory of elasticity it follows that this $\mathbf{u}$ is a vector of infinitesimal motion of the body as a rigid whole that is $\mathbf{u} = \mathbf{a} + \mathbf{b} \times \mathbf{r}$ where $\mathbf{a}$ and $\mathbf{b}$ are constant vectors. If some part of the boundary of the body is fixed, then this provides that $\mathbf{u} = \mathbf{0}$. The needed demonstration is complete.

We consider the case in which the whole boundary of the body is clamped:

$$\mathbf{u}\big|_{\partial\Omega} = \mathbf{0}. \tag{3.10.22}$$

As a base space we take the set C_{Ec} of all vector functions satisfying (3.10.22) whose components belong to $C^{(2)}(V)$. Let us call the completion of C_{Ec} with respect to the induced norm $\|\mathbf{u}\|_E = (\mathbf{u},\mathbf{u})_E^{1/2}$ the energy space of an elastic body with clamped boundary, and denote it as $\mathcal{E}_{Ec}$. We need to study the properties of this Hilbert space.

Theorem 3.10.1 *The space $\mathcal{E}_{Ec}$ is a subspace of the space of 3-D vector functions, each Cartesian component of which belongs to $W^{1,2}(\Omega)$ (the latter space we shall denote by $(W^{1,2}(\Omega))^3$).*

The proof of the theorem is based on an inequality called the *Korn inequality.* In this case it can be written as

$$\int_V \left(|\mathbf{u}(\mathbf{x})|^2 + |\nabla\mathbf{u}(\mathbf{x})|^2\right) dV \leq m \int_V e_{ij}(\mathbf{u}(\mathbf{x})) e_{ij}(\mathbf{u}(\mathbf{x}))\, dV. \tag{3.10.23}$$

We will prove (3.10.23) for the 2-D case in which the functions possess all second continuous derivatives on a compact domain denoted by S and take zero value on the boundary ∂S of S. The spatial variables are x, y. The proof is shorter than that for the 3-D case, but contains all the necessary ideas. We rewrite (3.10.23) for the 2-D case in a modified form:

$$\int_S \left(u^2 + v^2 + u_x^2 + u_y^2 + v_x^2 + v_y^2\right) dx\, dy$$
$$\le m \int_S \left(u_x^2 + \frac{1}{2}(u_y + v_x)^2 + v_y^2\right) dx\, dy. \qquad (3.10.24)$$

Here u, v are the components of vector function $\mathbf{u}$ that are equal to zero on the boundary ∂S:

$$u|_{\partial S} = 0, \qquad v|_{\partial S} = 0, \qquad (3.10.25)$$

and subscripts x, y mean partial derivatives with respect to the corresponding variables. Note the difference between the terms with derivatives of the norm of $\left(W^{1,2}(S)\right)^2$ and the right-hand side of (3.10.24): the latter does not contain the squared derivatives u_y and v_x but their sum.

Let us begin to demonstrate the Korn inequality (3.10.25). Because of the Friedrichs inequality, it is sufficient to demonstrate that there exists a constant $m_1 > 0$ such that

$$\int_S \left(u_x^2 + \frac{1}{2}(u_y + v_x)^2 + v_y^2\right) dx\, dy \ge m_1 \int_S \left(u_x^2 + u_y^2 + v_x^2 + v_y^2\right) dx\, dy. \qquad (3.10.26)$$

Let us transform the intermediate term in the left-hand side of (3.10.26):

$$\int_S (u_y + v_x)^2\, dx\, dy = \int_S \left(u_y^2 + 2u_y v_x + v_x^2\right) dx\, dy$$
$$= \int_S \left(u_y^2 + 2u_x v_y + v_x^2\right) dx\, dy,$$

where we integrated by parts with regard for (3.10.22), so we have

$$\int_S \left(u_x^2 + \frac{1}{2}(u_y + v_x)^2 + v_y^2\right) dx\, dy$$
$$= \int_S \left(u_x^2 + \frac{1}{2}u_y^2 + \frac{1}{2}v_x^2 + v_y^2 + u_x v_y\right) dx\, dy$$
$$\ge \int_S \left(u_x^2 + \frac{1}{2}u_y^2 + \frac{1}{2}v_x^2 + v_y^2 - \frac{1}{2}\left(u_x^2 + v_y^2\right)\right) dx\, dy$$
$$\ge \frac{1}{2} \int_S \left(u_x^2 + u_y^2 + v_x^2 + v_y^2\right) dx\, dy.$$

This completes the proof of the Korn inequality.

We recommend that the reader finish the proof of the theorem for a 3-D body having the above 2-D example. We will not prove Korn's inequality for a body with free boundary (i.e., when there are no boundary conditions for vector functions); the proof is technically much more complex, so we refer the reader to specialized books [Mikhlin (1965); Fichera (1972)]. We note that the form of this inequality is the same if we impose the two conditions

$$\int_V \mathbf{u}(\mathbf{x})\, dV = \mathbf{0}, \qquad \int_V \mathbf{r} \times \mathbf{u}(\mathbf{x})\, dV = \mathbf{0},$$

on each element of the space. These are four scalar conditions in the 2-D case and six conditions in the 3-D case, which coincides with the number of degrees of freedom of a rigid body.

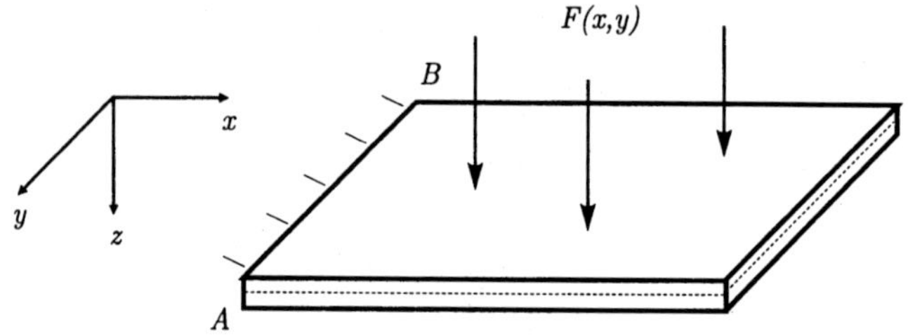

Fig. 3.4 A portion of a plate under a distributed load $F(x, y)$. The plate is clamped along AB.

A plate

Now we begin to consider the energy approach in the theory of a linear plate (Figure 3.4) whose equilibrium was described by the equation

$$D\Delta^2 w = F$$

where $w = w(x, y)$ is the transverse displacement of points of the middle surface of the plate, D the rigidity of the plate, μ is the Poisson ratio, $0 < \mu < 1/2$, and $F = F(x, y)$ is a transverse load. The elastic energy of the plate referred to a compact domain Ω in $\mathbb{R}^2$ is

$$\frac{D}{2} \int_\Omega \left(w_{xx} \left(w_{xx} + \mu w_{yy} \right) + 2(1 - \mu) w_{xy}^2 + w_{yy} \left(w_{yy} + \mu w_{xx} \right) \right) d\Omega$$

where subscripts x and y denote partial derivatives $\partial/\partial x$ and $\partial/\partial y$, respectively. Using dimensionless variables, for the role of a norm we will try the functional

$$\|w\|_P = \left(\int_\Omega \left(w_{xx} \left(w_{xx} + \mu w_{yy} \right) + 2(1-\mu) w_{xy}^2 \right.\right.$$
$$\left.\left. + w_{yy} \left(w_{yy} + \mu w_{xx} \right) \right) d\Omega \right)^{1/2} \qquad (3.10.27)$$

to which there corresponds the inner product

$$(u,v)_P = \int_\Omega \left(u_{xx} \left(v_{xx} + \mu v_{yy} \right) + 2(1-\mu) u_{xy} v_{xy} + u_{yy} \left(v_{yy} + \mu v_{xx} \right) \right) d\Omega.$$

Elementary calculations demonstrate that over functions of $C^{(2)}(\Omega)$ the equation $\|w\|_P = 0$ has a solution $w = a + bx + cy$ with arbitrary constants a, b, c, and no other solution. Considering the case of hard clamping of the edge of the plate, that is

$$w\|_{\partial\Omega} = 0 = \left. \frac{\partial w}{\partial n} \right|_{\partial\Omega}, \qquad (3.10.28)$$

we have $\|w\|_P$ to be a norm on the set C_P of functions of $C^{(2)}(\Omega)$ that satisfy (3.10.28). We will show that the completion of C_P with respect to the energy norm (3.10.27) produces a subspace of $W^{2,2}(\Omega)$ (let us note that for this it is sufficient for the plate to be fixed only at three points that are not on the same straight line: this gives us an energy space being a subspace of $W^{2,2}(\Omega)$).

Let us begin with a simple remark that on C_P the energy norm is equivalent to the norm

$$\|w\|_{2,2} = \left(\int_\Omega \left(w_{xx}^2 + 2w_{xy}^2 + w_{yy}^2 \right) d\Omega \right)^{1/2}$$

and so in the discussion we can use this norm. Next, we see that for functions $w \in C_P$ we have w_x and w_y continuously differentiable on Ω, and it follows from (3.10.28) that on the boundary $w_x = 0 = w_y$. Thus we can apply the Friedrichs inequality, getting

$$\int_\Omega w_x^2 \, d\Omega \le c \int_\Omega \left(w_{xx}^2 + w_{xy}^2 \right) d\Omega$$

and

$$\int_\Omega w_y^2 \, d\Omega \le c \int_\Omega \left(w_{yx}^2 + w_{yy}^2 \right) \, d\Omega.$$

Combining this with the Friedrichs inequality for w we obtain

$$\int_\Omega \left(w^2 + w_x^2 + w_y^2 \right) \, d\Omega \le c_1 \int_\Omega \left(w_{xx}^2 + 2w_{xy}^2 + w_{yy}^2 \right) \, d\Omega$$
$$\le c_2 \int_\Omega \left(w_{xx} \left(w_{xx} + \mu w_{yy} \right) + 2(1-\mu) w_{xy}^2 + w_{yy} \left(w_{yy} + \mu w_{xx} \right) \right) \, d\Omega.$$

Together with a trivial inequality

$$\int_\Omega \left(w_{xx} \left(w_{xx} + \mu w_{yy} \right) + 2(1-\mu) w_{xy}^2 + w_{yy} \left(w_{yy} + \mu w_{xx} \right) \right) \, d\Omega$$
$$\le c_3 \int_\Omega \left(w^2 + w_x^2 + w_y^2 + w_{xx}^2 + 2w_{xy}^2 + w_{yy}^2 \right) \, d\Omega$$

this proves that on C_P the energy norm is equivalent to the norm of $W^{2,2}(\Omega)$, and thus the energy space $\mathcal{E}_{Pc}$ that is the completion of the set C_P with respect to the energy norm (3.10.27) is a subspace of $W^{2,2}(\Omega)$ and we can use in this space the Sobolev imbedding theorem for elements of $W^{2,2}(\Omega)$.

Quite similarly we can consider the case of a plate with the edge free from geometrical restraints. In this case we should circumvent a difficulty of the presence of motions of the plate as a rigid body, which are $w = a + bx + cy$. For this we use the Poincaré inequality for w_x and w_y,

$$\int_\Omega w_x^2 \, d\Omega \le c_4 \left\{ \left(\int_\Omega w_x \, d\Omega \right)^2 + \int_\Omega \left(w_{xx}^2 + w_{xy}^2 \right) \, d\Omega \right\}$$

and

$$\int_\Omega w_y^2 \, d\Omega \le c_4 \left\{ \left(\int_\Omega w_y \, d\Omega \right)^2 + \int_\Omega \left(w_{yx}^2 + w_{yy}^2 \right) \, d\Omega \right\}.$$

Together with the Poincaré inequality for w this gives

$$\begin{aligned}
&\int_\Omega \left(w^2 + w_x^2 + w_y^2\right) d\Omega \\
&\le c_5 \left\{ \left(\int_\Omega w \, d\Omega \right)^2 + \left(\int_\Omega w_x \, d\Omega \right)^2 + \left(\int_\Omega w_y \, d\Omega \right)^2 \right. \\
&\quad \left. + \int_\Omega \left(w_{xx}^2 + 2w_{xy}^2 + w_{yy}^2\right) d\Omega \right\} \\
&\le c_6 \left\{ \left(\int_\Omega w \, d\Omega \right)^2 + \left(\int_\Omega w_x \, d\Omega \right)^2 + \left(\int_\Omega w_y \, d\Omega \right)^2 \right. \\
&\quad \left. + \int_\Omega \left(w_{xx}\left(w_{xx} + \mu w_{yy}\right) + 2(1-\mu) w_{xy}^2 + w_{yy}\left(w_{yy} + \mu w_{xx}\right)\right) d\Omega \right\}.
\end{aligned}$$

For any function from $C^{(2)}(\Omega)$, on proper change by an addendum $a+bx+cy$ we can achieve the equalities

$$\int_\Omega w \, d\Omega = 0, \qquad \int_\Omega w_x \, d\Omega = 0, \qquad \int_\Omega w_y \, d\Omega = 0, \qquad (3.10.29)$$

and for such functions we get the inequality

$$\begin{aligned}
&\int_\Omega \left(w^2 + w_x^2 + w_y^2\right) d\Omega \\
&\qquad \le c_6 \int_\Omega \left(w_{xx}\left(w_{xx} + \mu w_{yy}\right) + 2(1-\mu) w_{xy}^2 + w_{yy}\left(w_{yy} + \mu w_{xx}\right)\right) d\Omega
\end{aligned}$$

that, in the same manner as for hard clamping of the plate gives us that the completion of functions from $C^{(2)}(\Omega)$ satisfying (3.10.29) with respect to the energy norm, denoted by $\mathcal{E}_{Pf}$, is a closed subspace of $W^{2,2}(\Omega)$ whose norm is equivalent to the energy norm (3.10.27). $\mathcal{E}_{Pf}$ is a Hilbert space. Exactly in the same manner as it was done for all "free" cases above, we can introduce the energy space as a factor space of $W^{2,2}(\Omega)$ with respect to the set of all linear polynomials $a + bx + cy$. Using the same energy norm for completion we get another version of the energy space $\mathcal{E}_{Pf}$ whose elements are in one-to-one isometric correspondence with the elements of the previous version, and thus, characterizing their elements, we can use the Sobolev imbedding theorem for $W^{2,2}(\Omega)$.

3.11 Operators and Functionals

We have used the terms "operator" and "functional" frequently, and it may seem strange that we did not introduce these notions carefully long before. However we did not, exploiting instead the synonym "correspondence" for the term "operator".

Definition 3.11.1 A correspondence between two sets (metric spaces) X and Y, when to any element of X there corresponds no more than one element of Y, is called an *operator.* Frequently used synonyms for "operator" include the terms *map*, *mapping*, *function*, and *correspondence.*

The set of those elements x of X at which there is a correspondent element y is called the *domain* of the operator and is denoted $D(A)$. It is not necessarily true that each element $y \in Y$ is the image of some element $x \in X$ under the operator; the set of all elements of Y that *are* images of elements of X is known as the *range* of the operator. The domain and range of an operator A are denoted by $D(A)$ and $R(A)$, respectively.

Definition 3.11.2 If Y is the set of all complex (or real) numbers, then an operator acting from X to Y is called a complex (or real) *functional* defined on X.

An important role in functional analysis is played by linear operators. To introduce this notion we need X and Y to be linear spaces.

Definition 3.11.3 An operator A from a linear space X to a linear space Y is a *linear operator* if for any elements x_1 and x_2 of X and any scalars λ and μ we have

$$A(\lambda x_1 + \mu x_2) = \lambda A(x_1) + \mu A(x_2).$$

For a linear operator A we shall denote $A(x)$ and Ax interchangeably. Linear operators seem to be elementary, but this is not the case. Many physical problems are linear. We will extend the definition of continuity of a function to operators:

Definition 3.11.4 Let A be an operator from a normed space X to a normed space Y. We say that A is continuous at $x_0 \in X$ if to each $\varepsilon > 0$ there corresponds $\delta = \delta(\varepsilon) > 0$ such that $\|Ax - Ax_0\|_Y < \varepsilon$ whenever $\|x - x_0\|_X < \delta$.

Example 3.11.1 Show that any norm is a continuous mapping from X to $\mathbb{R}$. Note, however, that it is not a linear functional.

Solution Using the inequality of Example 3.1.1 we can write

$$| \|x\| - \|x_0\| | \le \|x - x_0\| .$$

Given $\varepsilon > 0$ then, we can choose $\delta = \varepsilon$ in the definition of continuity.

For linear operators there is a convenient theorem:

Theorem 3.11.1 *A linear operator defined on a normed space X is continuous if and only if it is continuous at $x = 0$.*

Proof. Immediate from the relation $Ax - Ax_0 = A(x - x_0)$. □

There is a central theorem that shows us how to check whether a linear operator is continuous:

Theorem 3.11.2 *A linear operator A from a normed space X to normed space Y is continuous if and only if there is a constant c such that for all $x \in D(A)$,*

$$\|Ax\| \le c \|x\| \tag{3.11.1}$$

Proof. Assume (3.11.1) holds. Then with $\delta = \varepsilon/c$ in the definition of continuity we see that A is continuous at $x = 0$. Conversely, suppose A is continuous at $x = 0$. Take $\varepsilon = 1$; by definition there exists $\delta > 0$ such that $\|Ax\| \le 1$ whenever $\|x\| < \delta$. For every nonzero $x \in X$, the norm of $x^* = \delta x / (2 \|x\|)$ is

$$\|x^*\| = \|\delta x/(2\|x\|)\| = \delta/2 < \delta,$$

so $\|Ax^*\| \le 1$. By linearity of A this gives us

$$\|Ax\| \le \frac{2}{\delta} \|x\| ,$$

which is (3.11.1) with $c = 2/\delta$. □

We see why continuous linear operators are often referred to as *bounded* linear operators.

Definition 3.11.5 The least constant c from (3.11.1) is called the *norm* of A and is denoted $\|A\|$.

Note that $\|A\|$ meets all the axioms of a norm:

(1) $\|A\|$ is clearly non-negative. If $\|A\| = 0$ then $\|Ax\| = 0$ for all $x \in X$, i.e., $A = 0$. Conversely, if $A = 0$ then $\|A\| = 0$.
(2) It is obvious that $\|\lambda A\| = |\lambda| \, \|A\|$.

(3) From

$$\|(A+B)x\| = \|Ax+Bx\| \le \|Ax\| + \|Bx\| \le \|A\|\,\|x\| + \|B\|\,\|x\|$$

we see that $\|A+B\| \le \|A\| + \|B\|$.

We denote by $L(X,Y)$ the normed linear space consisting of the set of all continuous linear operators from X to Y under this norm.

There is also a notion of *sequential continuity*, as in ordinary calculus:

Theorem 3.11.3 *The operator A from X to Y is continuous at $x_0 \in X$ if and only if $A(x_n) \to A(x_0)$ whenever $x_n \to x_0$.*

The proof is easily adapted from the corresponding proof that appears in any calculus book. This result justifies manipulations of the form

$$A\left(\lim_{n\to\infty} x_n\right) = \lim_{n\to\infty} Ax_n \tag{3.11.2}$$

for continuous operators A.

Suppose A is a continuous operator acting in a Banach space X. We observed earlier that the series $s = \sum_{k=1}^{\infty} x_k$ may be defined by the limiting operation

$$s = \lim_{n\to\infty} \sum_{k=1}^{n} x_k.$$

But (3.11.2) allows us to write

$$A\left(\sum_{k=1}^{\infty} x_k\right) = \lim_{n\to\infty} A\left(\sum_{k=1}^{n} x_k\right).$$

If A is also linear, then

$$A\left(\sum_{k=1}^{\infty} x_k\right) = \lim_{n\to\infty} \sum_{k=1}^{n} Ax_k = \sum_{k=1}^{\infty} Ax_k.$$

So we see that interchanges of the form

$$A\sum_{k=1}^{\infty} x_k = \sum_{k=1}^{\infty} Ax_k$$

are permissible with convergent series and continuous linear operators.

The most frequent operation in mathematical physics is that of finding a solution x to the equation

$$Ax = y \tag{3.11.3}$$

when y is given. Let us introduce the notion of the inverse to A. If for any $y \in Y$ there is no more than one solution $x \in X$ of (3.11.3), then the correspondence from Y to X defined by the equation is an operator; this operator is called the *inverse* to A and is denoted A^{-1}.

Lemma 3.11.1 *If A and B are each invertible, then the composition BA is invertible with $(BA)^{-1} = A^{-1}B^{-1}$.*

The proof is left to the reader.

Theorem 3.11.4 *Let X, Y be normed spaces. A linear operator A on $D(A) \subseteq X$ admits a continuous inverse on $R(A) \subseteq Y$ if and only if there is a positive constant c such that*

$$\|Ax\| \geq c\,\|x\| \qquad \textit{for all } x \in D(A). \tag{3.11.4}$$

Proof. Assuming the inequality (3.11.4) holds, we see that $Ax = 0$ implies $x = 0$ so the inverse A^{-1} exists. Then the same inequality means that the inverse is bounded (hence continuous) on $R(A)$. The converse is immediate. □

An operator A that satisfies (3.11.4) is said to be *bounded below.*

Example 3.11.2 (a) Show that a bounded linear operator maps Cauchy sequences into Cauchy sequences. (b) Show that every bounded linear operator has a closed null space.

Solution (a) Let $\{x_n\}$ be a Cauchy sequence in X. Let $\varepsilon > 0$ be given and choose N so that $n, m > N$ implies $\|x_n - x_m\| < \varepsilon / \|A\|$. For $n, m > N$ we have

$$\|Ax_n - Ax_m\| = \|A(x_n - x_m)\| \leq \|A\|\,\|x_n - x_m\| < \varepsilon,$$

so $\{Ax_n\}$ is a Cauchy sequence in Y. (b) Let A be a bounded linear operator. The null space of A, often denoted by $N(A)$, is the set of elements x such that $Ax = 0$. Let $\{x_n\}$ be a sequence of points in $N(A)$ with $x_n \to x_0$ as $n \to \infty$. It is easy to see that x_0 belongs to $N(A)$:

$$Ax_0 = A\left(\lim_{n\to\infty} x_n\right) = \lim_{n\to\infty} Ax_n = \lim_{n\to\infty} 0 = 0.$$

This means that $N(A)$ is a closed set.

Example 3.11.3 Show that if $k(x, \xi)$ is a continuous, real-valued function of the real variables x, ξ on $[a, b] \times [a, b]$, then the operator A given by

$$Af = \int_a^b k(x, \xi) f(\xi)\, d\xi$$

is a bounded linear operator from $C(a, b)$ to itself.

Solution The linearity of A is obvious; to see that it is bounded, observe that

$$\begin{aligned} \|Af\| &= \max_{x \in [a,b]} \left| \int_a^b k(x, \xi) f(\xi)\, d\xi \right| \\ &\le \max_{x \in [a,b]} \left[\int_a^b |k(x, \xi)|\, |f(\xi)|\, d\xi \right] \\ &\le \max_{x \in [a,b]} \left[\max_{x \in [a,b]} |f(x)| \int_a^b |k(x, \xi)|\, d\xi \right] \\ &= \|f(x)\| \max_{x \in [a,b]} \int_a^b |k(x, \xi)|\, d\xi. \end{aligned}$$

So $\|Af\| \le \alpha \|f\|$, where

$$\alpha = \max_{x \in [a,b]} \int_a^b |k(x, \xi)|\, d\xi.$$

Example 3.11.4 Show that if a linear operator is invertible, then its inverse is a linear operator.

Solution Suppose A is linear and A^{-1} exists. Let $y_1, y_2 \in R(A)$ where $y_i = Ax_i$ $(i = 1, 2)$ and let a_1, a_2 be scalars. We have

$$a_1 y_1 + a_2 y_2 = a_1 A x_1 + a_2 A x_2 = A(a_1 x_1 + a_2 x_2)$$

so that

$$A^{-1}(a_1 y_1 + a_2 y_2) = a_1 x_1 + a_2 x_2 = a_1 A^{-1} y_1 + a_2 A^{-1} y_2$$

as required.

3.12 Some Approximation Theory

Let X be a normed space. Given $x \in X$ and a set of elements $g_1, \ldots, g_n \in X$, it is reasonable to seek scalars $\lambda_1, \ldots, \lambda_n$ that will minimize the distance between x and the linear combination $\sum_{i=1}^n \lambda_i g_i$. So we would like to find the best approximation of x from among all the linear combinations $\sum_{i=1}^n \lambda_i g_i$. This so called *general problem of approximation* can be rephrased as

$$\phi(\lambda_1, \ldots, \lambda_n) \to \min_{\lambda_1, \ldots, \lambda_n}$$

where ϕ is the functional given by

$$\phi(\lambda_1, \ldots, \lambda_n) = \left\| x - \sum_{i=1}^n \lambda_i g_i \right\| .$$

We take the g_i to be linearly independent. If they are not linearly independent, the solution of the approximation problem will not be unique. Note that $\phi(\lambda_1, \ldots, \lambda_n)$ is a usual function in the n variables λ_i, so we can employ the usual tools of calculus.

Theorem 3.12.1 *For any $x \in X$ there exists $x^* = \sum_{i=1}^n \lambda_i^* g_i$ such that*

$$\|x - x^*\| = \inf_{\lambda_1, \ldots, \lambda_n} \phi(\lambda_1, \ldots, \lambda_n).$$

Proof. An application of the inequality

$$\|x - y\| \geq \big| \|x\| - \|y\| \big| \tag{3.12.1}$$

permits us to show that $\phi(\lambda_1, \ldots, \lambda_n)$ is continuous in the n scalar variables $\lambda_1, \ldots, \lambda_n$:

$$\begin{aligned} &|\phi(\lambda_1 + h_1, \ldots, \lambda_n + h_n) - \phi(\lambda_1, \ldots, \lambda_n)| \\ &\quad = \left| \left\| x - \sum_{i=1}^n (\lambda_i + h_i) g_i \right\| - \left\| x - \sum_{i=1}^n \lambda_i g_i \right\| \right| \\ &\quad \leq \left\| \left[x - \sum_{i=1}^n (\lambda_i + h_i) g_i \right] - \left[x - \sum_{i=1}^n \lambda_i g_i \right] \right\| \\ &\quad = \left\| \sum_{i=1}^n h_i g_i \right\| \leq \sum_{i=1}^n |h_i| \, \|g_i\| . \end{aligned}$$

Continuity of the function

$$\psi(\lambda_1,\ldots,\lambda_n) = \left\| \sum_{i=1}^{n} \lambda_i g_i \right\|$$

is also apparent since it is a particular case of $\phi(\lambda_1,\ldots,\lambda_n)$ at $x = 0$, and $\psi(\lambda_1,\ldots,\lambda_n)$ must therefore reach a minimum on the sphere $\sum_{i=1}^{n} |\lambda_i|^2 = 1$ at some point $(\lambda_{10},\ldots,\lambda_{n0})$. By linear independence of the g_i we have $\psi(\lambda_{10},\ldots,\lambda_{n0}) = d > 0$. Also note that ψ is a homogeneous function,

$$\psi(k\lambda_1,\cdots,k\lambda_n) = |k|\,\psi(\lambda_1,\cdots,\lambda_n),$$

which means that

$$\psi(\lambda_1,\cdots,\lambda_n) \geq Rd \quad \text{when} \quad \left(\sum_{i=1}^{n} |\lambda_i|^2\right)^{1/2} = R,$$

and that $\psi(\lambda_1,\ldots,\lambda_n) > Rd$ for $(\lambda_1,\ldots,\lambda_n)$ outside a sphere of radius R. We wish to show that $\phi(\lambda_1,\ldots,\lambda_n)$ actually attains its minimum value at some finite point.

Since

$$\phi(\lambda_1,\ldots,\lambda_n) \geq \psi(\lambda_1,\ldots,\lambda_n) - \|x\|$$

by (3.12.1), we see that for $(\lambda_1,\ldots,\lambda_n)$ *outside* a ball of radius R we have

$$\phi(\lambda_1,\ldots,\lambda_n) > Rd - \|x\|$$

Outside of the sphere of radius $R = R_0 = 3\|x\|/d$ we have

$$\phi(\lambda_1,\ldots,\lambda_n) > 2\|x\|$$

whereas inside this sphere $\phi(0,\ldots,0) = \|x\|$. Hence when $x \neq 0$ (to the reader: what happens when $x = 0$?) the minimum of ϕ *is* inside the sphere of radius R_0 with the centre at the origin. Thus the corresponding closed ball of radius R_0 contains the minimum point. □

Uniqueness can be addressed with the help of the following concepts.

Definition 3.12.1 A normed space X is said to be *strictly normed* if from the equality

$$\|x+y\| = \|x\| + \|y\|, \quad x \neq 0, \tag{3.12.2}$$

it follows that $y = \lambda x$ for some nonnegative λ.

Not all normed spaces are strictly normed. For example, the space $C(\Omega)$ is not strictly normed. But some important classes of spaces are strictly normed, including $L^p(\Omega)$ and $W^{l,p}(\Omega)$. Later we shall show that every inner product space is strictly normed.

Definition 3.12.2 A subset S of a linear space is said to be *convex* if for any pair $x, y \in S$ it contains the whole segment

$$\lambda x + (1 - \lambda)y, \qquad 0 \leq \lambda \leq 1.$$

Theorem 3.12.2 *Let X be a strictly normed space, and let M be a closed convex subset of X. For any $x \in X$, there is at most one $y \in M$ that minimizes the distance $\|x - y\|$.*

Proof. Suppose that y_1 and y_2 are each minimizers:

$$\|x - y_1\| = \|x - y_2\| = \inf_{y \in M} \|x - y\| \equiv d. \tag{3.12.3}$$

If $x \in M$, we obtain that $y_1 = y_2 = x$. Suppose $x \notin M$. Then $d > 0$. By convexity $(y_1 + y_2)/2 \in M$, hence

$$\left\| x - \frac{y_1 + y_2}{2} \right\| \geq d.$$

But

$$\left\| x - \frac{y_1 + y_2}{2} \right\| = \left\| \frac{x - y_1}{2} + \frac{x - y_2}{2} \right\| \leq \frac{1}{2} \|x - y_1\| + \frac{1}{2} \|x - y_2\| = d,$$

so

$$\left\| \frac{x - y_1}{2} + \frac{x - y_2}{2} \right\| = \left\| \frac{x - y_1}{2} \right\| + \left\| \frac{x - y_2}{2} \right\|.$$

Because X is strictly normed we have $x - y_1 = \lambda(x - y_2)$ for some $\lambda \geq 0$, hence $\|x - y_1\| = \lambda \|x - y_2\|$. From (3.12.3) we deduce that $\lambda = 1$, thus $y_1 = y_2$. □

By this theorem we see that, for a strictly normed space, a solution to the general problem of approximation is unique. A set of spaces important in applications are included here, as shown next.

Lemma 3.12.1 *Every inner product space is strictly normed.*

Proof. Let X be an inner product space, and suppose that $x, y \in X$ are such that (3.12.2) holds. We have $\|x+y\|^2 = (\|x\| + \|y\|)^2$; rewriting this in the form

$$\|x\|^2 + 2\,\mathrm{Re}(x,y) + \|y\|^2 = \|x\|^2 + 2\,\|x\|\,\|y\| + \|y\|^2 ,$$

we obtain

$$\mathrm{Re}(x,y) = \|x\|\,\|y\| .$$

This and the Schwarz inequality show that $\mathrm{Im}(x,y) = 0$ so that

$$(x,y) = \|x\|\,\|y\| .$$

But this last equation represents the case of equality holding in the Schwarz inequality, which can happen only if $y = \lambda x$ for some λ. Making this replacement for y we obtain $(x, \lambda x) = \|x\|\,\|\lambda x\|$, hence $\overline{\lambda}\,\|x\|^2 = |\lambda|\,\|x\|^2$. Since $x \neq 0$ we have $\overline{\lambda} = |\lambda|$, and therefore $\lambda \geq 0$. □

The subspace H_n of an inner product space H that is spanned by g_i, $i = 1, \ldots, n$, is finite dimensional. We know that for any $x \in H$ there is a unique element that minimizes the distance $\|x - y\|$ over $y \in H_n$. In a Euclidean space this element is a projection of the element onto the subspace H_n. Let us show that this result on the unique existence of the projection extends to a Hilbert space. This extension is the basis for an important part of the theory of Hilbert spaces connected with Fourier expansions and many other questions.

Theorem 3.12.3 *Let H be a Hilbert space and let M be closed convex subset of H. For every $x \in H$, there is a unique $y \in M$ that minimizes $\|x - y\|$.*

Proof. Fix $x \in H$. By definition of infimum there is a sequence $\{y_k\} \subset M$ such that

$$\lim_{k\to\infty} \|x - y_k\| = \inf_{y\in M} \|x - y\| .$$

By the parallelogram law

$$\|2x - y_i - y_j\|^2 + \|y_i - y_j\|^2 = 2\left(\|x - y_i\|^2 + \|x - y_j\|^2\right),$$

hence

$$\|y_i - y_j\|^2 = 2\left(\|x - y_i\|^2 + \|x - y_j\|^2\right) - 4\left\|x - \frac{y_i + y_j}{2}\right\|^2 .$$

Since $\|x - y_j\|^2 = d^2 + \varepsilon_j$ where $\varepsilon_j \to 0$ as $j \to \infty$, it follows that

$$\|y_i - y_j\|^2 \leq 2(d^2 + \varepsilon_i + d^2 + \varepsilon_j) - 4d^2 = 2(\varepsilon_i + \varepsilon_j) \to 0 \quad \text{as } i, j \to \infty.$$

Therefore $\{y_k\}$ is a Cauchy sequence, and converges to an element $y \in M$ since M is closed. This minimizer y is unique by Theorem 3.12.2. □

3.13 Orthogonal Decomposition of a Hilbert Space and the Riesz Representation Theorem

Definition 3.13.1 Let M be a subspace of a Hilbert space H. An element $n \in H$ is said to be *orthogonal to* M if n is orthogonal to every element of M.

In $\mathbb{R}^3$ we may imagine a straight line segment inclined with respect to a plane and with one end touching the plane. We may then define the projections of the segment onto the plane and onto the normal, respectively. The length of the normal projection is the shortest distance from the other end of the segment to the surface. The next result is the extension of this fact to inner product spaces.

Lemma 3.13.1 *Let H be a Hilbert space and M a closed linear subspace of H. Given $x \in H$, the unique minimizer $m \in M$ guaranteed by Theorem 3.12.3 is such that $(x - m)$ is orthogonal to M.*

Proof. Let $v \in M$. The function

$$f(\alpha) = \|x - m - \alpha v\|^2$$

of the real variable α takes its minimum value at $\alpha = 0$, hence

$$\left.\frac{df}{d\alpha}\right|_{\alpha=0} = 0.$$

This gives

$$\left.\frac{d}{d\alpha}(x - m - \alpha v, x - m - \alpha v)\right|_{\alpha=0} = -2\,\mathrm{Re}(x - m, v) = 0.$$

Replacing v by iv we get $\mathrm{Im}(x - m, v) = 0$, hence $(x - m, v) = 0$. □

Definition 3.13.2 Two subspaces M and N of H are *mutually orthogonal* if every $n \in N$ is orthogonal to M and every $m \in M$ is orthogonal to N. In this case we write $M \perp N$. If, furthermore, any $x \in H$ can be uniquely represented in the form

$$x = m + n, \qquad m \in M, n \in N, \tag{3.13.1}$$

then we write $H = M \dot{+} N$ and speak of an *orthogonal decomposition* of H into M and N.

Note that mutually orthogonal subspaces have zero as their only point of intersection.

Theorem 3.13.1 *Let M be a closed subspace of a Hilbert space H. There is a closed subspace N of H such that $M \dot{+} N$ is an orthogonal decomposition of H.*

Proof. Let N be the set of all elements of H that are orthogonal to M. We assume $M \neq H$, hence $N \neq \emptyset$. If $n_1, n_2 \in N$ so that $(n_1, m) = (n_2, m) = 0$ for every $m \in M$, then $(\lambda_1 n_1 + \lambda_2 n_2, m) = 0$ for any scalars λ_1, λ_2. Hence N is a subspace of H. To see that N is closed, let $\{n_k\}$ be a Cauchy sequence in N. The limit element $n^* = \lim_{k\to\infty} n_k$ exists; it belongs to N because

$$(n^*, m) = \lim_{k\to\infty} (n_k, m) = 0 \text{ for all } m \in M$$

by continuity of the inner product.

For any element $x \in H$ the representation (3.13.1) exists because we can project x onto M to obtain the element m, then obtain n from $n = x - m$. To show uniqueness, assume that for some x there are two such representations:

$$x = m_1 + n_1, \qquad\qquad x = m_2 + n_2.$$

Equating these, we obtain

$$m_1 - m_2 = n_1 - n_2.$$

Taking the inner product of both sides of this equality with $m_1 - m_2$ and then with $n_1 - n_2$, we get $\|m_1 - m_2\|^2 = 0$ and $\|n_1 - n_2\|^2 = 0$. □

Let us now turn to one of the main facts that we shall need from the theory of Hilbert spaces. We consider a simple case first. Let $\{\mathbf{e}_1, \ldots, \mathbf{e}_n\}$ be an orthonormal basis of $\mathbb{R}^n$ so that any vector $\mathbf{x} \in \mathbb{R}^n$ can be expressed as

$$\mathbf{x} = \sum_{i=1}^{n} x_i \mathbf{e}_i.$$

Now suppose $F(\mathbf{x})$ is a linear functional defined on $\mathbb{R}^n$. It is easy to see that $F(\mathbf{x})$ has a representation of the form

$$F(\mathbf{x}) = \sum_{i=1}^{n} x_i c_i \tag{3.13.2}$$

where the c_i are scalars independent of $\mathbf{x}$; indeed, with $c_i \equiv F(\mathbf{e}_i)$ we have

$$F(\mathbf{x}) = F\left(\sum_{i=1}^{n} x_i \mathbf{e}_i\right) = \sum_{i=1}^{n} x_i F(\mathbf{e}_i) = \sum_{i=1}^{n} x_i c_i$$

by linearity of F. We can write (3.13.2) as

$$F(\mathbf{x}) = (\mathbf{x}, \mathbf{c})$$

where $\mathbf{c}$ is a vector in $\mathbb{R}^n$, independent of $\mathbf{x}$, whose value is uniquely determined by F; in this sense we can say that F has been "represented by an inner product." More generally, we have the following important result known as the *Riesz representation theorem*:

Theorem 3.13.2 *Let $F(x)$ be a continuous linear functional given on a Hilbert space H. There is a unique element $f \in H$ such that*

$$F(x) = (x, f) \quad \textit{for every } x \in H. \tag{3.13.3}$$

Moreover, $\|F\| = \|f\|$.

Hence any bounded linear functional defined on a Hilbert space can be represented by an inner product. The element f is sometimes called the *representer* of $F(x)$.

Proof. Let M be the set of all x for which

$$F(x) = 0. \tag{3.13.4}$$

By linearity of $F(x)$ any finite linear combination of elements of M also belongs to M, hence M is a subspace of H. M is also closed; indeed, a Cauchy sequence $\{m_k\} \subset M$ is convergent in H to some $m^* = \lim_{k\to\infty} m_k$, and by continuity of $F(x)$ we see that m^* satisfies (3.13.4). By Theorem 3.13.1, there is a closed subspace N of H such that $N \perp M$ and such that any $x \in H$ can be uniquely represented as $x = m + n$ for some $m \in M$ and $n \in N$. We can deduce the dimension of N. If n_1 and n_2 are any two elements of N, then so is $n_3 = F(n_1)n_2 - F(n_2)n_1$. Since $F(n_3) = F(n_1)F(n_2) - F(n_2)F(n_1) = 0$ we have $n_3 \in M$. But the only element that belongs to both N and M is the zero vector. This means that n_2 is a scalar multiple of n_1, hence N is one-dimensional.

Now choose $n \in N$ and define $n_0 = n/\|n\|$. Any $x \in H$ can be represented as

$$x = m + \alpha n_0, \qquad m \in M,$$

where $\alpha = (x, n_0)$, and therefore

$$F(x) = F(m) + \alpha F(n_0) = \alpha F(n_0) = F(n_0)(x, n_0) = (x, \overline{F(n_0)} n_0).$$

Denoting $\overline{F(n_0)} n_0$ by f we obtain the representation (3.13.3). To establish its uniqueness, let f_1 and f_2 be two representers:

$$F(x) = (x, f_1) = (x, f_2).$$

So $(x, f_1 - f_2) = 0$ for all x. Setting $x = f_1 - f_2$ we have $\|f_1 - f_2\|^2 = 0$, hence $f_1 = f_2$.

Finally, we must establish $\|F\| = \|f\|$. Since this certainly holds for $F = 0$ we assume $F \neq 0$. Then $f \neq 0$, and

$$\|f\|^2 = (f, f) = F(f) \leq \|F\| \, \|f\|$$

gives $\|f\| \leq \|F\|$. On the other hand

$$\|F\| = \sup_{\|x\| \neq 0} \frac{|F(x)|}{\|x\|} = \sup_{\|x\| \neq 0} \frac{|(x, f)|}{\|x\|} \leq \sup_{\|x\| \neq 0} \frac{\|x\| \, \|f\|}{\|x\|} = \|f\|$$

by the Schwarz inequality. □

The Riesz representation theorem states that a continuous linear functional on a Hilbert space H is identified with an element of H; this correspondence is one-to-one, isometric, and preserves algebraic operations with respect to the elements and functionals. The set of all continuous linear functionals on a normed space X is called the dual space to X and is denoted by X'. In these terms, the Riesz theorem states that X' is isometrically isomorphic to X.

Example 3.13.1 (a) Let a functional in $L^2(0, 2)$ be given by

$$F(f) = \int_0^1 f(x) g(x) \, dx$$

where $g(x) \in L^2(0, 1)$ is given. What is the representer of this functional given by the Riesz representation theorem in $L^2(0, 2)$? (b) Define on $L^2(0,1)$ a linear functional by the formula

$$G(f) = f(0.5).$$

What is the representer of this functional according to the Riesz representation theorem?

Solution (a) We can use

$$G(x) = \begin{cases} g(x), & x \in [0,1], \\ 0, & x \in (1,2], \end{cases}$$

as a representer. (b) The functional G is linear but not continuous in $L^2(0,1)$, so the Riesz representation theorem is not applicable. The functional by its form relates to the δ-function, which is not an element of $L^2(0,1)$.

The Riesz representation theorem will play a key role when we consider the generalized setup of some problems in mechanics.

3.14 Basis, Gram–Schmidt Procedure, Fourier Series in Hilbert Space

If Y is an n-dimensional linear space, then there are n linearly independent elements $g_1, \ldots, g_n \in Y$ such that every $y \in Y$ can be uniquely represented in the form

$$y = \sum_{k=1}^{n} \alpha_k g_k$$

for scalars $\alpha_1, \ldots, \alpha_n$. The scalars are called the components of x. We refer to the finite set $\{g_i\}_{i=1}^{n}$ as a *basis* of Y. A basis of the space is not unique. The concept of basis can be extended to infinite dimensional normed spaces as follows:

Definition 3.14.1 Let X be a normed linear space. A system of elements $\{e_i\}$ is called a basis of X if any $x \in X$ can be represented uniquely as

$$x = \sum_{k=1}^{\infty} \alpha_k e_k \tag{3.14.1}$$

for scalars $\{\alpha_k\}$.

The elements e_i of a basis play the role of coordinate vectors of the space. Every such basis is linearly independent. Indeed, with $x = 0$ equation (3.14.1) holds with $\alpha_k \equiv 0$, and the α_k are unique by assumption.

A normed space X having a basis is separable. To see this, we note that the set of all linear combinations $\sum_{k=1}^{\infty} q_k e_k$ with rational coefficients q_k is countable and dense in X.

In practical calculations we normally use finite approximations of quantities. For this, finite linear combinations of basis elements are appropriate.

Definition 3.14.2 Let X be a normed space. A countable system $\{g_i\} \subset X$ is said to be *complete in* X if for every $x \in X$ and $\varepsilon > 0$ there is a finite linear combination $\sum_{i=1}^{n(\varepsilon)} \alpha_i(\varepsilon) g_i$ such that $\left\| x - \sum_{i=1}^{n(\varepsilon)} \alpha_i(\varepsilon) g_i \right\| < \varepsilon$.

Note that the coefficients α_i of this definition need not be continuous in ε.

The space X is separable if it has a countable complete system: the set of finite linear combinations with rational coefficients is dense in the set of all linear combinations, and thus in the space.

Among all the bases of $\mathbb{R}^n$ an orthonormal basis has some advantages for calculation. The same can be said of an infinite dimensional Hilbert space. A system of elements $\{g_k\} \subset H$ is said to be *orthonormal* if

$$(g_m, g_n) = \begin{cases} 1, & m = n, \\ 0, & m \neq n. \end{cases}$$

If we have an arbitrary basis $\{f_i\}_{i=1}^{\infty}$ of a Hilbert space, we sometimes need to construct an orthonormal basis of the space. An orthonormal basis of a Hilbert space is not unique. One way to produce such a basis is the so-called *Gram–Schmidt procedure.* The process is straightforward. A linearly independent set of elements cannot contain the zero vector, so we may obtain g_1 by normalizing f_1:

$$g_1 = f_1 / \|f_1\| .$$

To obtain g_2, we first generate a vector e_2 by subtracting from f_2 the "component" of f_2 that is the projection of f_2 on the direction of g_1:

$$e_2 = f_2 - (f_2, g_1) g_1$$

(recall that g_1 is a unit vector). We then normalize e_2 to obtain g_2:

$$g_2 = e_2 / \|e_2\| .$$

(Note that $e_2 \neq 0$, otherwise f_1 and f_2 are linearly dependent. The same applies to the rest of the e_i).

We obtain g_3 from f_3 by subtracting the components of f_3 that are the projections of f_3 on both g_1 and g_2:

$$e_3 = f_3 - (f_3, g_1)g_1 - (f_3, g_2)g_2, \qquad g_3 = e_3 / \|e_3\|.$$

In general we set

$$g_i = \frac{e_i}{\|e_i\|} \quad \text{where} \quad e_i = f_i - \sum_{k=1}^{i-1}(f_i, g_k)g_k, \qquad i = 2, 3, 4, \ldots.$$

The reader should verify directly that the Gram–Schmidt procedure actually yields an orthogonal set of elements.

In linear algebra it is shown that a system $\{f_i\}_{i=1}^n$ is linearly independent in $\mathbb{R}^n$ if and only if

$$\begin{vmatrix} (f_1, f_1) & (f_1, f_2) & \cdots & (f_1, f_n) \\ (f_2, f_1) & (f_2, f_2) & \cdots & (f_2, f_n) \\ \vdots & & & \\ (f_n, f_1) & (f_n, f_2) & \cdots & (f_n, f_n) \end{vmatrix} \neq 0.$$

The determinant on the left is called the *Gram determinant*. A finite dimensional inner product space stands in a one-to-one correspondence with $\mathbb{R}^n$, a correspondence in which inner products are preserved. Thus the same Gram criterion is valid for an inner product space as well. It is easy to see that every finite orthonormal system is linearly independent, since the Gram determinant would reduce to $+1$ in that case.

In the space $\mathbb{R}^n$ we find the components of a vector $\mathbf{x}$ with respect to the orthonormal frame vectors $\mathbf{i}_k$ by direct projection of $\mathbf{x}$ onto $\mathbf{i}_k$: $x_k = \mathbf{x} \cdot \mathbf{i}_k$. Similarly we can define the components of an element of a Hilbert space. They are given by

Definition 3.14.3 Let $\{g_i\}$ be an orthonormal system in a complex Hilbert space H. Given $f \in H$, the numbers α_k defined by

$$\alpha_k = (f, g_k), \qquad k = 1, 2, 3, \ldots,$$

are known as the *Fourier coefficients* of f with respect to the system $\{g_i\}$.

We use the same terms as in the classical Fourier theory of expansion of functions, because all the results and even their proofs parallel the results for Fourier expansions established in the space $L^2(a, b)$.

Theorem 3.14.1 *Let H be a Hilbert space. A complete orthonormal system $\{g_i\} \subset H$ is a basis of H; with respect to $\{g_i\}$, any $f \in H$ has the unique representation*

$$f = \sum_{k=1}^{\infty} \alpha_k g_k \tag{3.14.2}$$

where $\alpha_k = (f, g_k)$ is the kth Fourier coefficient of f. The series (3.14.2) is called the Fourier series *of f with respect to $\{g_i\}$.*

Proof. Let $f \in H$ be given, and consider approximating f by a finite linear combination $\sum_{k=1}^{n} c_k g_k$ of the elements $\{g_i\}_{i=1}^{n}$. The approximation error is given by

$$\left\| f - \sum_{k=1}^{n} c_k g_k \right\|^2 = \left(f - \sum_{k=1}^{n} c_k g_k, f - \sum_{k=1}^{n} c_k g_k \right),$$

and manipulation of the right-hand side allows us to put this in the form

$$\left\| f - \sum_{k=1}^{n} c_k g_k \right\|^2 = \|f\|^2 - \sum_{k=1}^{n} |\alpha_k|^2 + \sum_{k=1}^{n} |c_k - \alpha_k|^2.$$

Clearly the error is minimized when $c_k = \alpha_k$ for each k, so the best approximation is the element given by

$$f_n = \sum_{k=1}^{n} (f, g_k) g_k.$$

We call f_n the nth partial sum of the Fourier series for f. Since the error is non-negative we also have

$$\sum_{k=1}^{n} |(f, g_k)|^2 \le \|f\|^2,$$

known as *Bessel's inequality.* This shows that

$$\|f_{n+m} - f_n\|^2 = \left\| \sum_{k=n+1}^{n+m} (f, g_k) g_k \right\|^2 = \sum_{k=n+1}^{n+m} |(f, g_k)|^2 \to 0 \quad \text{as } n \to \infty,$$

hence $\{f_n\}$ is a Cauchy sequence in H. Since H is a Hilbert space the sequence has a limit. We need to show that it coincides with f. Indeed, by completeness of $\{g_i\}$, for any $\varepsilon > 0$ there exists $N = N(\varepsilon)$ and coefficients $c_k(\varepsilon)$ such that

$$\left\| f - \sum_{k=1}^{N} c_k(\varepsilon) g_k \right\|^2 < \varepsilon.$$

But f_N is at least as good an approximation to f, so

$$\|f - f_N\|^2 = \left\| f - \sum_{k=1}^{N} \alpha_k g_k \right\|^2 \le \left\| f - \sum_{k=1}^{N} c_k(\varepsilon) g_k \right\|^2 < \varepsilon$$

and we conclude that $f_N \to f$. From this we see that

$$f = \lim_{n\to\infty} f_n,$$

and the proof is complete. □

Corollary 3.14.1 Parseval's equality

$$\sum_{k=1}^{\infty} |(f, g_k)|^2 = \|f\|^2 \qquad (3.14.3)$$

holds for any $f \in H$ and any complete orthonormal system $\{g_i\}$.

Proof. We established above that

$$\left\| f - \sum_{k=1}^{n} (f, g_k) g_k \right\|^2 = \|f\|^2 - \sum_{k=1}^{n} |(f, g_k)|^2. \qquad (3.14.4)$$

Passage to the limit as $n \to \infty$ yields (3.14.3). □

Proving the theorem, we established that the sequence of partial Fourier sums is a Cauchy sequence and this fact does not depend on whether $\{g_k\}$ is a complete system. We shall use this fact, so we formulate

Corollary 3.14.2 *Let $\{g_k\}$ be an arbitrary orthonormal system in H (not necessarily complete). The sequence of partial Fourier sums f_n of $f \in H$ converges to an element f^* such that $\|f^*\| \le \|f\|$; $f^* = f$ if the system is complete.*

Definition 3.14.4 We say that $\{g_i\} \subset H$ is *closed in H* if the system of equations

$$(f, g_k) = 0 \text{ for all } k = 1, 2, 3, \ldots \qquad (3.14.5)$$

implies that $f = 0$.

Theorem 3.14.2 *An orthonormal system $\{g_i\}$ in a Hilbert space H is complete in H if and only if it is closed in H.*

Proof. If $\{g_i\}$ is a complete orthonormal system in H, then any $f \in H$ can be written as

$$f = \sum_{k=1}^{\infty} (f, g_k) g_k$$

by Theorem 3.14.1. Enforcement of the condition (3.14.5) obviously does yield $f = 0$, hence $\{g_i\}$ is closed. Conversely, assume that $\{g_i\}$ is a closed orthonormal system in H. We established previously (Corollary 3.14.2) that for any $f \in H$ the sequence of partial Fourier sums $f_n = \sum_{k=1}^{n} \alpha_k g_k$ is a Cauchy sequence converging to some $f^* \in H$ since H is a Hilbert space. We have

$$(f - f^*, g_m) = \lim_{n\to\infty} \left(f - \sum_{k=1}^{n} \alpha_k g_k, g_m \right) = \alpha_m - \alpha_m = 0$$

hence

$$(f - f^*, g_m) = 0 \text{ for all } m = 1, 2, 3, \ldots.$$

It follows that $f^* = f$ since $\{g_i\}$ is closed. Because $f_n = \sum_{k=1}^{n} \alpha_k g_k$ converges to f, the system $\{g_i\}$ is complete by Definition 3.14.2. □

The existence of the Gram–Schmidt process implies

Theorem 3.14.3 *Any system of elements $\{g_i\}$ (not necessarily orthonormal) in a Hilbert space H is complete in H if and only if it is closed in H.*

Theorem 3.14.4 *A Hilbert space H has a countable orthonormal basis if and only if H is separable.*

Proof. We saw earlier that the existence of a countable basis in a Hilbert space provides for separability. Conversely, assume H is separable and select a countable set that is dense in H. To this set the Gram–Schmidt procedure can be applied (removing any linearly dependent elements) to produce an orthonormal system. Since the initial set was dense it was complete, hence the Gram–Schmidt procedure yields an orthonormal basis of H. □

One advantage afforded by the tools of functional analysis is that we can discuss many common procedures of numerical analysis in terms to which we are accustomed in finite dimensional spaces. A knowledge of this theory gives us an understanding, without long deliberation, of when we can do so and when we cannot — some nice finite dimensional pictures become invalid or doubtful in spaces of infinite dimension.

The following result will be used later when we cover the Fredholm theory:

Theorem 3.14.5 *Any bounded subset of a Hilbert space H is precompact if and only if H is finite dimensional.*

Proof. If H is finite dimensional then we can place it in one-to-one correspondence with $\mathbb{R}^n$ for some n. Then precompactness of any bounded set follows from calculus.

Next let us suppose that any bounded set of H is precompact but, to the contrary, that H is infinite dimensional. We can construct an infinite Fourier basis $\{e_k\}$. Since $\|e_k - e_n\|^2 = 2$ for $k \neq n$, the sequence $\{e_k\}$ cannot contain a Cauchy subsequence, hence the unit ball of H cannot be precompact. □

Example 3.14.1 Show that every separable, infinite dimensional, complex Hilbert space is isometrically isomorphic to ℓ^2.

Solution Let X be a Hilbert space as described. By separability X has a countable, complete orthonormal set $E = \{e_k\}_{k=1}^{\infty}$. For any $x \in X$, denote the nth Fourier coefficient with respect to E by α_n. Since E is complete we have $\|x\|^2 = \sum_{n=1}^{\infty} |\alpha_n|^2 < \infty$, hence $\alpha = (\alpha_1, \alpha_2, \ldots) \in \ell^2$. Define a transformation A from X to ℓ^2 by $Ax = \alpha$. Because A is clearly linear we can show that it is injective by showing that $N(A) = \{0\}$. But $Ax = 0$ implies $\alpha = 0$, hence each $\alpha_k = 0$, hence $(x, e_k) = 0$ for each k, hence $x = 0$ since the orthonormal set E is closed. Next we show that A is surjective. Choose any $y = (\eta_1, \eta_2, \ldots) \in \ell^2$; since $\sum_{n=1}^{\infty} |\eta_n|^2 < \infty$, the series $\sum_{n=1}^{\infty} \eta_n e_n = x$ for some $x \in X$. Moreover we have $\eta_n = (x, e_n)$ for all n, and from this we see that $\|Ax\|^2 = \|y\|^2 = \sum_{n=1}^{\infty} |\eta_n|^2 = \sum_{n=1}^{\infty} |(x, e_n)|^2 = \|x\|^2$. That is, A is also an isometry.

3.15 Weak Convergence

It is easy to show that $\{\mathbf{x}_k\}$ is a Cauchy sequence in $\mathbb{R}^n$ if and only if each of its component sequences $\{(\mathbf{x}_k, \mathbf{i}_j)\}$, $j = 1, \ldots, n$, is a numerical Cauchy sequence. So in $\mathbb{R}^n$, norm convergence is equivalent to component-wise convergence. Remember that, besides, all the norms in $\mathbb{R}^n$ are equivalent.

Unlike $\mathbb{R}^n$, in an infinite dimensional Hilbert space, where the role of components is played by the Fourier coefficients of an element, the component-wise convergence of a sequence does not guarantee strong convergence of the same sequence. Indeed, consider the sequence composed of the elements of an orthonormal basis $\{g_k\}$. The sequence of the jth Fourier component $(g_k, g_j) \to 0$ as $k \to \infty$ because of the mutual orthogonality of the elements of the basis; hence, by similarity to the case of $\mathbb{R}^n$, we could conclude that the zero element is a limit. But $\{g_k\}$ does not have a strong limit, because $\|g_k - g_m\| = \sqrt{2}$ whenever $k \neq m$. However, component-wise convergence in a Hilbert space is still important, and we need to introduce a suitable notion. A component in Hilbert space is given by the Fourier coefficient, which is found through the use of an inner product. This coefficient is a continuous linear functional on H. So a natural extension of the definition of component-wise convergence is

Definition 3.15.1 Let $\{x_k\} \subset H$ where H is a Hilbert space. We say that $\{x_k\}$ is a *weak Cauchy sequence* if $\{F(x_k)\}$ is a (numerical) Cauchy sequence for every continuous linear functional $F(x)$ defined on H.

In contrast, we know that $\{x_k\}$ is a Cauchy sequence in H if

$$\|x_n - x_m\| \to 0 \quad \text{as } m, n \to \infty.$$

In this latter case we shall refer to $\{x_k\}$ as a *strong Cauchy sequence* whenever there is danger of ambiguity. It is apparent that every strong Cauchy sequence is a weak Cauchy sequence. We also observe that, by the Riesz representation theorem, $\{x_k\}$ is a weak Cauchy sequence if the numerical sequence $\{(x_n, f)\}$ is a Cauchy sequence for every element $f \in H$. But above we showed the existence of a sequence that is a weak Cauchy sequence but not a strong Cauchy sequence. Thus we have defined a new kind of convergence in a Hilbert space. We shall rephrase all the notions of strong continuity for the weak version.

Definition 3.15.2 Let $x_0 \in H$. If $F(x_n) \to F(x_0)$ for every continuous linear functional $F(x)$ defined on H, we write

$$x_n \rightharpoonup x_0$$

and say that $\{x_n\}$ is *weakly convergent* to x_0. Alternatively, by the Riesz representation theorem we have $x_n \rightharpoonup x_0$ if and only if $(x_n, f) \to (x_0, f)$ for every element $f \in H$.

Recalling that the strong limit of a sequence is unique, we might wonder whether weak limits also share this property. The answer is affirmative:

Theorem 3.15.1 *If a sequence in a Hilbert space has a weak limit, the limit is unique.*

Proof. Suppose there are two weak limits x^* and x^{**} of a sequence $\{x_k\}$. An arbitrary continuous linear functional, by the Riesz representation theorem, is $F(x) = (x, f)$. When k tends to infinity the numerical sequence (x_k, f) can have only one limit (by calculus), so $(x^{**}, f) = (x^*, f)$. This holds for any $f \in H$, and thus for $f = x^{**} - x^*$. But then it follows that $\|x^{**} - x^*\|^2 = 0$. □

There is a simple and convenient sufficient condition for a weakly convergent sequence to be strongly convergent:

Theorem 3.15.2 *Suppose $x_k \rightharpoonup x_0$ in a Hilbert space H. Then $\|x_k\| \to \|x_0\|$ implies that $x_k \to x_0$ as $k \to \infty$.*

Proof. For each k we have

$$\|x_k - x_0\|^2 = (x_k - x_0, x_k - x_0) = \|x_k\|^2 - (x_0, x_k) - (x_k, x_0) + \|x_0\|^2 .$$

But as $k \to \infty$ both (x_0, x_k) and (x_k, x_0) approach $\|x_0\|^2$ by definition of weak convergence, and we have $\|x_k\| \to \|x_0\|$ by assumption. So $\|x_k - x_0\| \to 0$ as $k \to \infty$. □

We know that a strong Cauchy sequence is bounded. It is not immediately apparent that a weak Cauchy sequence has this property. However, we have

Theorem 3.15.3 *In a Hilbert space, every weak Cauchy sequence is bounded.*

Proof. Suppose that $\{x_n\}$ is a weak Cauchy sequence in H with $\|x_n\| \to \infty$ as $n \to \infty$. Before seeking a contradiction we establish an auxiliary fact: if $B(y_0, \varepsilon)$ is a closed ball of some radius $\varepsilon > 0$ and arbitrary center $y_0 \in H$, then it is possible to find a sequence $\{y_n\} \subset B(y_0, \varepsilon)$ such that the numerical sequence

$$(x_n, y_n) \to \infty \quad \text{as } n \to \infty. \tag{3.15.1}$$

The sequence $\{y_n\}$ given by

$$y_n = y_0 + \varepsilon \frac{x_n}{2\|x_n\|}$$

is suitable. Indeed

$$\|y_n - y_0\| = \left\| \frac{\varepsilon x_n}{2\|x_n\|} \right\| = \frac{\varepsilon}{2} < \varepsilon$$

shows that $y_n \in B(y_0, \varepsilon)$ for each n. Furthermore,

$$(x_n, y_n) = (x_n, y_0) + \frac{\varepsilon}{2\|x_n\|}(x_n, x_n) = (x_n, y_0) + \frac{\varepsilon}{2}\|x_n\|$$

establishes (3.15.1) since the numerical sequence $\{(x_n, y_0)\}$ is a Cauchy sequence by definition of weak convergence of $\{x_n\}$, and every Cauchy sequence is bounded.

We are now ready to obtain a contradiction. Starting with $\varepsilon_1 = 1$ and $y_0 = 0$, we can find x_{n_1} and $y_1 \in B(y_0, \varepsilon_1)$ such that

$$(x_{n_1}, y_1) > 1. \tag{3.15.2}$$

By continuity of the inner product in the second argument, there is a ball $B(y_1, \varepsilon_2) \subset B(y_0, \varepsilon_1)$ such that (3.15.2) holds not only for y_1 but for all $y \in B(y_1, \varepsilon_2)$:

$$(x_{n_1}, y) > 1 \text{ for all } y \in B(y_1, \varepsilon_2).$$

Similarly, we can find x_{n_2} (with $n_2 > n_1$) and $y_2 \in B(y_1, \varepsilon_2)$ such that

$$(x_{n_2}, y_2) > 2,$$

and, by continuity, a ball $B(y_2, \varepsilon_3) \subset B(y_1, \varepsilon_2)$ such that

$$(x_{n_2}, y) > 2 \text{ for all } y \in B(y_2, \varepsilon_3).$$

Continuing this process we generate a nested sequence of balls $B(y_k, \varepsilon_{k+1})$ and a corresponding subsequence $\{x_{n_k}\}$ of $\{x_n\}$ such that

$$(x_{n_k}, y) > k \text{ for all } y \in B(y_k, \varepsilon_{k+1}).$$

Since H is a Hilbert space the intersection $\bigcap_k B(y_k, \varepsilon_{k+1})$ is nonempty, hence there exists y^* such that $(x_{n_k}, y^*) > k$ for each k. For the continuous linear function $F^*(x) = (x, y^*)$ then, the numerical sequence $\{F^*(x_{n_k})\}$ is not a Cauchy sequence. Because $\{x_{n_k}\}$ is not a weak Cauchy sequence, neither is $\{x_n\}$. This is the contradiction sought. □

As a byproduct of this proof we have

Lemma 3.15.1 *If $\{x_k\}$ is an unbounded sequence in H, i.e., $\|x_k\| \to \infty$ as $k \to \infty$, then there exists $y^* \in H$ and a subsequence $\{x_{n_k}\}$ such that $(x_{n_k}, y^*) \to \infty$ as $k \to \infty$.*

We now present another important theorem with which we can show boundedness of some sets in a Hilbert space. Set boundedness plays an important role in the applications of functional analysis to mathematical physics. The present result is called the *principle of uniform boundedness*:

Theorem 3.15.4 *Let $\{F_k(x)\}_{k=1}^{\infty}$ be a family of continuous linear functionals defined on a Hilbert space H. If $\sup_k |F_k(x)| < \infty$ for each $x \in H$, then $\sup_k \|F_k\| < \infty$.*

Proof. Each $F_k(x)$ has Riesz representation $F_k(x) = (x, f_k)$ for a unique $f_k \in H$ such that $\|f_k\| = \|F_k\|$. So it suffices to show that if $\sup_k |(x, f_k)| < \infty$ for each $x \in H$, then $\sup_k \|f_k\| < \infty$. We prove the contrapositive of this. Assuming $\sup_k \|f_k\| = \infty$, we see that Lemma 3.15.1 guarantees the existence of $x_0 \in H$ and a subsequence $\{f_{k_n}\}$ such that $|(x_0, f_{k_n})| \to \infty$ as $k \to \infty$. This completes the proof. □

Corollary 3.15.1 *Let $\{F_k(x)\}$ be a sequence of continuous linear functionals given on H. If for every $x \in H$ the numerical sequence $\{F_k(x)\}$ is a Cauchy sequence, then there is a continuous linear functional $F(x)$ on H such that*

$$F(x) = \lim_{k\to\infty} F_k(x) \quad \textit{for all } x \in H \tag{3.15.3}$$

and

$$\|F\| \le \liminf_{k\to\infty} \|F_k\| < \infty. \tag{3.15.4}$$

Proof. The limit in (3.15.3) exists by hypothesis and clearly defines a linear functional $F(x)$. By Theorem 3.15.4 we have $\sup_k \|F_k\| < \infty$; from

$$|F(x)| = \lim_{k\to\infty} |F_k(x)| \le \sup_k \|F_k\| \, \|x\|$$

it follows that $F(x)$ is continuous. Writing

$$|F(x)| = \lim_{k\to\infty} |F_k(x)| \le \liminf_{k\to\infty} \|F_k\| \, \|x\| \,,$$

we establish (3.15.4). □

Because of the Riesz representation theorem we can rephrase this as

Theorem 3.15.5 *A weak Cauchy sequence in a Hilbert space has a weak limit belonging to the space. This means that any Hilbert space is* weakly complete.

It is therefore unnecessary for us to introduce the notion of weak completeness of a Hilbert space separately.

Theorem 3.15.6 *A sequence $\{x_n\} \subset H$ is a weak Cauchy sequence if and only if the following two conditions hold:*

(i) $\{x_n\}$ is bounded in H;
(ii) for any element from a complete system $\{f_\alpha\}$ in H, the sequence of numbers $\{(x_n, f_\alpha)\}$ is a Cauchy sequence.

Proof. Since necessity of the two conditions follows from Theorem 3.15.3 and Definition 3.15.2, we proceed to prove sufficiency. Suppose conditions (i) and (ii) hold, and let $\varepsilon > 0$ be given. Condition (i) means that $\|x_n\| \le M$ for all n. Take an arbitrary continuous linear functional defined by its Riesz representer $f \in H$ as (x, f). By (ii) there is a linear combination $f_\varepsilon = \sum_{k=1}^{N} c_k f_k$ such that

$$\|f - f_\varepsilon\| < \varepsilon/3M.$$

We have

$$\begin{aligned} |(x_n - x_m, f)| &= |(x_n - x_m, f_\varepsilon + f - f_\varepsilon)| \\ &\le |(x_n - x_m, f_\varepsilon)| + |(x_n - x_m, f - f_\varepsilon)| \\ &\le \sum_{k=1}^{N} |c_k||(x_n - x_m, f_k)| + (\|x_n\| + \|x_m\|)\, \|f - f_\varepsilon\|\,. \end{aligned}$$

By (ii), $\{(x_n, f_k)\}$ is a Cauchy sequence for each k. Therefore for sufficiently large m, n we have

$$\sum_{k=1}^{N} |c_k||(x_n - x_m, f_k)| < \varepsilon/3.$$

So

$$|(x_n - x_m, f)| \le \varepsilon/3 + 2M\varepsilon/(3M) = \varepsilon$$

for sufficiently large m, n, as required. □

Definition 3.15.3 A set S in an inner product space X is said to be *weakly closed* if $x_n \rightharpoonup x_0 \in X$ implies that $x_0 \in S$.

Lemma 3.15.2 *In a Hilbert space, any closed ball with center at the origin is weakly closed.*

Proof. From the ball $\|x\| \leq M$, choose a sequence $\{x_n\}$ that converges weakly to $x_0 \in H$. We shall show that $\|x_0\| \leq M$. The formula

$$F(y) = \lim_{n\to\infty} (y, x_n)$$

defines a linear functional on H. This functional is bounded (i.e., continuous) because

$$|F(y)| = \lim_{n\to\infty} |(y, x_n)| \leq M \|y\|,$$

and we see that $\|F\| \leq M$. Applying the Riesz representation theorem we obtain $F(y) = (y, f)$ for a unique $f \in H$ such that $\|f\| \leq M$. So we can write

$$\lim_{n\to\infty} (y, x_n) = (y, f)$$

for any $y \in H$, and conclude that $x_n \rightharpoonup f$. □

A result known as *Mazur's theorem* (see, for example, Yosida [Yosida (1965)]) states that every closed convex set in a Hilbert space is weakly closed. This would apply to the previous case, as well as to any closed subspace of a Hilbert space.

Definition 3.15.4 Let S be a subset of an inner product space. We say that S is *weakly precompact* if every sequence taken from S contains a weak Cauchy subsequence. We say that S is *weakly compact* if every sequence taken from S contains a weak Cauchy subsequence that converges weakly to a point of S.

Next, we see that a bounded set in a separable Hilbert space is weakly precompact.

Theorem 3.15.7 *Every bounded sequence in a separable Hilbert space contains a weak Cauchy subsequence.*

Proof. Let $\{x_n\}$ be a bounded sequence in a separable Hilbert space H, and let $\{g_n\}$ be an orthonormal basis of H. By Theorem 3.15.6 it suffices to show that there is a subsequence $\{x_{n_k}\}$ such that, for any fixed g_m, the numerical sequence $\{(x_{n_k}, g_m)\}$ is a Cauchy sequence. Let us demonstrate its existence. From the bounded numerical sequence $\{(x_n, g_1)\}$ we can choose a Cauchy subsequence $\{(x_{n_1}, g_1)\}$. Then, from the bounded numerical sequence $\{(x_{n_1}, g_2)\}$ we can choose a Cauchy subsequence $\{(x_{n_2}, g_2)\}$. We can continue this process, on the kth step obtaining a Cauchy subsequence $\{(x_{n_k}, g_k)\}$. The diagonal sequence $\{x_{n_n}\}$ has the property that for any fixed g_m the numerical sequence $\{(x_{n_n}, g_m)\}$ is a Cauchy sequence. Hence $\{x_{n_n}\}$ is a weak Cauchy sequence. □

A simple but important corollary of this and Lemma 3.15.2 we formulate as

Theorem 3.15.8 *In a Hilbert space, any closed ball with center at the origin is weakly compact.*

That is, a bounded sequence $\{x_n\}$ with $\|x_n\| \le M$ has a subsequence that converges weakly to some x^* with $\|x^*\| \le M$. We shall use this fact in the next chapter.

Example 3.15.1 Prove the following assertions. (a) If $\{x_n\}$ is a (strong) Cauchy sequence, then it is a weak Cauchy sequence. (b) Let $\{x_n\}$ be a weak Cauchy sequence, and suppose that one of its subsequences converges (strongly) to x_0. Then $\{x_n\}$ converges weakly to x_0. (c) If $\{x_n\}$ converges weakly to x_0, so do each of its subsequences. (d) Suppose $x_k \rightharpoonup x$ and $y_k \rightharpoonup y$. Then $x_k + y_k \rightharpoonup x + y$, and $\alpha x_k \rightharpoonup \alpha x$ for any scalar α. (e) Let $x_n \rightharpoonup x_0$ and $y_n \to y_0$. Then $(x_n, y_n) \to (x_0, y_0)$ as $n \to \infty$.

Solution Let F be an arbitrary continuous linear functional. (a) Let $\varepsilon > 0$ be given, and choose N so large that $n, m > N$ imply $\|x_n - x_m\| < \varepsilon / \|F\|$. Then for $n, m > N$ we have

$$|F(x_n) - F(x_m)| = |F(x_n - x_m)| \le \|F\| \, \|x_n - x_m\| < \varepsilon.$$

(b) Since $\{x_n\}$ is weakly Cauchy, the sequence $\{F(x_n)\}$ is Cauchy. Also, $x_{n_k} \to x_0$ implies that $F(x_{n_k}) \to F(x_0)$. Because the Cauchy sequence $\{F(x_n)\}$ has a subsequence $\{F(x_{n_k})\}$ that converges to $F(x_0)$, the whole sequence converges to $F(x_0)$. This shows that x_n converges to x_0 weakly. (c) If $x_n \rightharpoonup x_0$, then $F(x_n) \to F(x_0)$. But then $F(x_{n_k}) \to F(x_0)$ for every subsequence $\{F(x_{n_k})\}$ of $\{F(x_n)\}$. (d) We have $F(x_k) \to F(x)$ and $F(y_k) \to F(y)$. Hence

$$F(x_k + y_k) = F(x_k) + F(y_k) \to F(x) + F(y) = F(x + y)$$

and

$$F(\alpha x_k) = \alpha F(x_k) \to \alpha F(x) = F(\alpha x).$$

(e) We have

$$\begin{aligned}|(x_n, y_n) - (x_0, y_0)| &\le |(x_n, y_n) - (x_n, y_0)| + |(x_n, y_0) - (x_0, y_0)| \\ &= |(x_n, y_n - y_0)| + |(x_n, y_0) - (x_0, y_0)| \\ &\le \|x_n\| \, \|y_n - y_0\| + |(x_n, y_0) - (x_0, y_0)|.\end{aligned}$$

The first term tends to zero as $n \to \infty$ because the weakly convergent sequence $\{x_n\}$ is bounded and $\|y_n - y_0\| \to 0$. The second term tends to zero by weak convergence of $\{x_n\}$ to x_0.

3.16 Adjoint and Self-adjoint Operators

In the theory of matrices, for a matrix A the equality

$$(A\mathbf{x}, \mathbf{y}) = (\mathbf{x}, A^T\mathbf{y})$$

which is valid for any $\mathbf{x}, \mathbf{y}$, introduces a dual (conjugate) matrix A^T. The formula for integration by parts (when $g(0) = 0 = g(1)$),

$$\int_0^1 f'(x)g(x)\,dx = -\int_0^1 f(x)g'(x)\,dx,$$

introduces a correspondence between the operator of differentiation (of the first argument f) and a dual operator, $-d/dx$, for the second argument. For a linear differential operator with constant coefficients, integration by parts can be used to find a corresponding dual operator that plays an important role in the theory of differential equations. An extension of these ideas to the general case brings us to the notion of adjoint operator.

Let H be a Hilbert space and A a continuous linear operator from H to H. For any fixed $y \in H$, we can view the inner product (Ax, y) as a functional with respect to the variable $x \in H$. This functional is linear:

$$(A(\lambda x_1 + \mu x_2), y) = (\lambda A x_1 + \mu A x_2, y) = \lambda\,(Ax_1, y) + \mu\,(Ax_2, y)\,.$$

It is also bounded (i.e., continuous) since

$$|(Ax, y)| \le \|Ax\| \, \|y\| \le \|A\| \, \|y\| \, \|x\|$$

by the Schwarz inequality and the fact that A is bounded. By the Riesz representation theorem we can write

$$(Ax, y) = (x, z)$$

where $z \in H$ is uniquely determined by y and A. The correspondence $y \mapsto z$ defines an operator that we shall denote by A^*.

Definition 3.16.1 Let A be a continuous linear operator acting in H. The operator A^* from H to H given by

$$(Ax, y) = (x, A^*y) \qquad \text{for all } x \in H$$

is called the *adjoint* of A.

Let us verify that A^* is a linear operator. For any $y_1, y_2 \in H$ we have

$$(Ax, y_1) = (x, A^*y_1), \qquad\qquad (Ax, y_2) = (x, A^*y_2),$$

and, if λ and μ are any scalars, $(Ax, \lambda y_1 + \mu y_2) = (x, A^*(\lambda y_1 + \mu y_2))$. Hence

$$\begin{aligned}(x, A^*(\lambda y_1 + \mu y_2)) &= \overline{\lambda}(Ax, y_1) + \overline{\mu}(Ax, y_2) \\ &= \overline{\lambda}(x, A^*y_1) + \overline{\mu}(x, A^*y_2) \\ &= (x, \lambda A^*y_1) + (x, \mu A^*y_2).\end{aligned}$$

Therefore, since $x \in H$ is arbitrary,

$$A^*(\lambda y_1 + \mu y_2) = \lambda A^*y_1 + \mu A^*y_2$$

Let us proceed to some other properties of A^*.

Lemma 3.16.1 *We have*

$$(A + B)^* = A^* + B^*, \qquad\qquad (AB)^* = B^*A^*,$$

for any continuous linear operators A, B acting in H.

Proof. The first property is evident. We write

$$\begin{aligned}(x, (AB)^*y) &= ((AB)x, y) = (A(Bx), y) = (Bx, A^*y) = (x, B^*(A^*y)) \\ &= (x, (B^*A^*)y)\end{aligned}$$

to establish the second property. □

Lemma 3.16.2 *If A is a continuous linear operator, then so is A^*; moreover,*

$$\|A^*\| = \|A\|.$$

Proof. Define[5]

$$M = \sup_{x,y\in H} \frac{|(Ax, y)|}{\|x\|\,\|y\|}.$$

By the Schwarz inequality

$$M \leq \sup_{x,y\in H} \frac{\|A\|\,\|x\|\,\|y\|}{\|x\|\,\|y\|} = \|A\|\,.$$

But we also have

$$M = \sup_{x,y\in H} \frac{|(x, A^*y)|}{\|x\|\,\|y\|}$$

and can put $x = A^*y$ to obtain a new value

$$M_1 = \sup_{y\in H} \frac{|(A^*y, A^*y)|}{\|A^*y\|\,\|y\|} = \sup_{y\in H} \frac{\|A^*y\|}{\|y\|}.$$

Since $M_1 \leq M$ we see that A^* is bounded and

$$M_1 = \|A^*\| \leq M \leq \|A\|\,.$$

So A^* is continuous with $\|A^*\| \leq \|A\|$. The reverse inequality follows from the next lemma. □

Lemma 3.16.3 $(A^*)^* = A$.

Proof. Since A^* is continuous we have

$$(x, (A^*)^*y) = (A^*x, y) = \overline{(y, A^*x)} = \overline{(Ay, x)} = (x, Ay)$$

for any $x, y \in H$. □

We are now ready to consider some specific examples. In preparation for this it will be helpful to have

Definition 3.16.2 An operator A is said to be *self-adjoint* if $A^* = A$.

Let us note that for boundary value problems the equality $A^* = A$ means not only coincidence of the form of the operators, but coincidence of their domains as well. This remark becomes important when in mathematical physics one introduces the notion of the adjoint to an operator having a domain that is only dense in the space. Then one may introduce symmetrical operators (these are such that the form of the adjoint operator

[5]Here it is evident that the sup should be taken only over $x, y \neq 0$, so we suppress this condition to simplify the notation.

remains the same) and self-adjoint operators for which there is complete coincidence with the original operator.

On the space ℓ^2 having elements $\mathbf{x} = (x_1, x_2, \ldots)$, we can define a matrix operator A by

$$(A\mathbf{x})_i = \sum_{j=1}^{\infty} a_{ij} x_j.$$

It follows from

$$\|A\mathbf{x}\|_{\ell^2} = \left[\sum_{i=1}^{\infty} \left(\sum_{j=1}^{\infty} a_{ij} x_j \right)^2 \right]^{1/2} \leq \left[\sum_{i=1}^{\infty} \sum_{j=1}^{\infty} |a_{ij}|^2 \sum_{k=1}^{\infty} |x_k|^2 \right]^{1/2}$$

that

$$\|A\| \leq \left(\sum_{i=1}^{\infty} \sum_{j=1}^{\infty} |a_{ij}|^2 \right)^{1/2}.$$

Suppose

$$\left(\sum_{i=1}^{\infty} \sum_{j=1}^{\infty} |a_{ij}|^2 \right)^{1/2} \leq M$$

so A becomes continuous. From

$$(A\mathbf{x}, \mathbf{y}) = \sum_{i=1}^{\infty} \sum_{j=1}^{\infty} a_{ij} x_j \overline{y_i} = \sum_{j=1}^{\infty} x_j \overline{\left(\sum_{i=1}^{\infty} \overline{a}_{ij} y_i \right)} = (\mathbf{x}, A^* \mathbf{y})$$

we see that A^* is defined by

$$(A^* \mathbf{y})_j = \sum_{i=1}^{\infty} \overline{a}_{ij} y_i.$$

It is evident that A is self-adjoint if $a_{ij} = \overline{a}_{ji}$ for all indices i, j. A continuous analogue of this example is the integral operator B acting in $L^2(0,1)$ defined by

$$(Bf)(x) = \int_0^1 k(x,s) f(s)\, ds.$$

Here $k(x, s)$ is a function known as the kernel of the operator. The inequality

$$\begin{aligned} \|Bf\|_{L^2(0,1)} &= \left(\int_0^1 \left| \int_0^1 k(x,s) f(s)\, ds \right|^2 dx \right)^{1/2} \\ &\leq \left(\int_0^1 \left(\int_0^1 |k(x,s)|^2\, ds \int_0^1 |f(s)|^2\, ds \right) dx \right)^{1/2} \\ &= \left(\int_0^1 \int_0^1 |k(x,s)|^2\, ds\, dx \right)^{1/2} \|f\|_{L^2(0,1)} \end{aligned}$$

shows that B is bounded if $k(x, s) \in L^2([0,1] \times [0,1])$ and that

$$\|B\| \leq \left(\int_0^1 \int_0^1 |k(x,s)|^2\, ds\, dx \right)^{1/2}.$$

Manipulations analogous to those done for the matrix example above show that B^* is given by

$$(B^*g)(s) = \int_0^1 \overline{k(x,s)} g(x)\, dx.$$

Clearly B is self-adjoint if $k(x, s) = \overline{k(s, x)}$ and $k(x, s) \in L^2([0,1] \times [0,1])$.

Definition 3.16.3 An operator acting in a Hilbert space is said to be *weakly continuous* if it maps every weakly convergent sequence into a weakly convergent sequence.

Lemma 3.16.4 *A continuous linear operator acting in a Hilbert space is also weakly continuous.*

Proof. Let A be continuous on H and choose $\{x_n\}$ such that $x_n \rightharpoonup x_0$ in H. An arbitrary continuous linear functional $F(x)$ takes the form $F(x) = (x, f)$ for some $f \in H$, hence we must show that $(Ax_n - Ax_0, f) \to 0$ as $n \to \infty$. But

$$(Ax_n - Ax_0, f) = (x_n - x_0, A^*f) \to 0 \quad \text{as } n \to \infty$$

since $A^*f \in H$ and $\{x_n\}$ converges weakly to x_0. □

We see from the above proof that

$$x_n \rightharpoonup x_0 \implies Ax_n \rightharpoonup Ax_0,$$

analogous to the case with ordinary (strong) continuity.

In the justification of many numerical methods for the solution of boundary value problems, the following simple lemma plays an important role.

Lemma 3.16.5 *Assume that A is a continuous linear operator acting in a Hilbert space H. If $x_n \rightharpoonup x_0$ and $y_n \to y_0$ in H, then $(Ax_n, y_n) \to (Ax_0, y_0)$.*

Proof. We will show that $(Ax_n, y_n) - (Ax_0, y_0) \to 0$. We have

$$\begin{aligned}(Ax_n, y_n) - (Ax_0, y_0) &= (x_n, A^*y_n) - (x_0, A^*y_0) \\ &= (x_n, A^*y_n) - (x_n, A^*y_0) + (x_n, A^*y_0) - (x_0, A^*y_0) \\ &= (x_n, A^*(y_n - y_0)) + (x_n - x_0, A^*y_0).\end{aligned}$$

The first term on the right tends to zero because

$$|(x_n, A^*(y_n - y_0))| \le \|x_n\| \, \|A^*\| \, \|y_n - y_0\|$$

and $y_n \to y_0$ (here $\|x_n\|$ is bounded since $\{x_n\}$ is weakly convergent); the second term tends to zero because $x_n \rightharpoonup x_0$. □

Sometimes it is important to obtain an exact value or accurate bound for the norm of an operator. For a self-adjoint operator this can be done through the use of the following theorem.

Theorem 3.16.1 *If A is a self-adjoint continuous linear operator given on a Hilbert space H, then*

$$\|A\| = \sup_{\|x\| \le 1} |(Ax, x)|. \tag{3.16.1}$$

Proof. Let us denote the right side of (3.16.1) by γ. By the Schwarz inequality

$$\gamma \le \sup_{\|x\| \le 1} \{\|Ax\| \, \|x\|\} \le \sup_{\|x\| \le 1} \{\|A\| \, \|x\|^2\} = \|A\|.$$

The reverse inequality, which completes the proof, takes a bit more effort to establish. First, by definition of γ we have $|(Ax, x)| \le \gamma$ whenever $\|x\| \le 1$. Hence, replacing x by $x / \|x\|$, we can write

$$|(Ax, x)| \le \gamma \, \|x\|^2$$

for any $x \in H$. Setting $x_1 = y + \lambda z$ and $x_2 = y - \lambda z$ where $\lambda \in \mathbb{R}$ and

$y, z \in H$, we have

$$\begin{aligned} C &\equiv |(Ax_1, x_1) - (Ax_2, x_2)| \\ &= |2\lambda|\,|(Ay, z) + (Az, y)| \\ &= |2\lambda|\,|(Ay, z) + (z, Ay)|. \end{aligned}$$

On the other hand

$$\begin{aligned} C &\leq |(Ax_1, x_1)| + |(Ax_2, x_2)| \\ &\leq \gamma(\|x_1\|^2 + \|x_2\|^2) \\ &= 2\gamma(\|y\|^2 + \lambda^2\,\|z\|^2) \end{aligned}$$

by the parallelogram equality, so

$$|2\lambda|\,|(Ay, z) + (z, Ay)| \leq 2\gamma(\|y\|^2 + \lambda^2\,\|z\|^2).$$

Since this holds for all $y, z \in H$ we may set $z = Ay$ to obtain

$$|4\lambda|\,\|Ay\|^2 \leq 2\gamma(\|y\|^2 + \lambda^2\,\|Ay\|^2).$$

With $\lambda = \|y\| / \|Ay\|$ this reduces to $\|Ay\| \leq \gamma\,\|y\|$ and so $\|A\| \leq \gamma$. □

Note that if A satisfies the conditions of the theorem and $(Ax, x) = 0$ for all $x \in H$, then A is the zero operator.

3.17 Compact Operators

Using computers we can successfully solve finite systems of linear algebraic equations. A computer performs a finite number of operations, so if we need to solve a problem with some accuracy it should have a structure close to that of finite algebraic equations. An important class of operators with which problems of this kind arise is the class of compact operators.

In this section we take X to be a normed space and Y to be a Banach space.

Definition 3.17.1 Let A be a linear operator from X to Y. We say that A is *compact* if it maps bounded subsets of X into precompact subsets of Y.

It suffices to show that A maps the *unit ball* of X into a precompact subset of Y. (By "the unit ball" of a space, if nothing is said about its center, we mean a ball of radius 1 centered at the origin of the space.) This follows from the linearity of A. It is also evident that A is compact if and only if every bounded sequence $\{x_n\}$ in X has a subsequence whose image under A is a Cauchy sequence in Y.

In the space $\mathbb{R}^n$ with a fixed basis, a matrix A defines a continuous linear operator that is denoted by A as well. Such an operator A maps a closed and bounded subset of $\mathbb{R}^n$ into a closed and bounded subset of $\mathbb{R}^n$; so the image is compact, and A is a compact operator. In an infinite dimensional space a continuous linear operator is not in general compact. For example, the identity operator I on $C(0,1)$ performs the simple mapping $f(x) \mapsto f(x)$. Therefore I maps the unit ball of $C(0,1)$ into itself, but the unit ball of $C(0,1)$ is not precompact.

Theorem 3.17.1 *Every compact linear operator is bounded, hence continuous.*

Proof. Suppose A is not bounded. Then we can find a bounded sequence $\{x_n\}$ in X such that $\|Ax_n\| \to \infty$. As $\{Ax_n\}$ contains no convergent subsequence, A is not compact. □

It is clear that the zero operator is compact. Let us present a non-trivial example of a compact linear operator. Consider the operator A from $C(0,1)$ to $C(0,1)$ given by

$$(Af)(t) = \int_0^1 h(t,\tau)f(\tau)\,d\tau,$$

where the kernel function $h(t,\tau)$ is continuous on the square $[0,1]\times[0,1]$. Let B_1 be the unit ball of $C(0,1)$, and let $S = A(B_1)$. Because h is continuous there exists $\alpha > 0$ such that $|h(t,\tau)| \le \alpha$, and thus

$$\max_{t\in[0,1]} |(Af)(t)| \le \alpha \max_{t\in[0,1]} |f(t)| \le \alpha$$

whenever $f(t) \in B_1$ (i.e., whenever $|f(t)| \le 1$ on $[0,1]$). We conclude that S is uniformly bounded. S is also equicontinuous: we have

$$\begin{aligned} |(Af)(t_2) - (Af)(t_1)| &\le \int_0^1 |h(t_2,\tau) - h(t_1,\tau)|\,|f(\tau)|\,d\tau \\ &\le \max_{\tau\in[0,1]} |h(t_2,\tau) - h(t_1,\tau)| \end{aligned}$$

for $f(t) \in B_1$, and, given $\varepsilon > 0$, the uniform continuity of $h(t,\tau)$ guarantees that we can find δ such that $|h(t_1,\tau) - h(t_1,\tau)| < \varepsilon$ whenever $|t_2 - t_1| < \delta$ and $\tau \in [0,1]$. So by Arzelà's theorem S is precompact, and we conclude that A is a compact operator.

We now consider a practically important class of compact linear operators. An operator is called *one dimensional* if its image is a one dimensional subspace. The general form of a continuous one dimensional linear operator T is evidently

$$Tx = (F(x))y_0$$

where F is a continuous linear functional and y_0 is some fixed element of the image. A one dimensional linear operator is compact. Indeed, the functional F maps the unit ball B with center at the origin into a bounded numerical set $F(B)$, so it is precompact. Thus the set $F(B)y_0$ is precompact in the space Y as well. A linear operator T_n is called *finite dimensional* if

$$T_n x = \sum_{k=1}^{n} (F_k(x))y_k$$

where the F_k are linear functionals in X and the y_k are some elements of Y. If the F_k are continuous then so is T_n. Because each component of T_n is a compact linear operator, so is T_n; this is a consequence of the following general theorem.

Theorem 3.17.2 *If A_1 and A_2 are compact linear operators from X to Y, then so is each operator of the form $\lambda A_1 + \mu A_2$ where λ, μ are scalars.*

Proof. If $\{x_n\}$ is a bounded sequence in X, it has a subsequence $\{x_{n_1}\}$ for which $\{A_1 x_{n_1}\}$ is a Cauchy sequence in Y. Because this subsequence is itself a bounded sequence, it has a subsequence $\{x_{n_2}\}$ for which $\{A_2 x_{n_2}\}$ is a Cauchy sequence. The image subsequences $\{A_1 x_{n_2}\}$ and $\{A_2 x_{n_2}\}$ are both Cauchy sequences then. Weighting by the scalars λ and μ does not affect whether a sequence is a Cauchy sequence, and the sum of two Cauchy sequences is a Cauchy sequence. Therefore the operator $\lambda A_1 + \mu A_2$ is compact. □

This theorem means that the set of compact linear operators from X to Y is a linear subspace of $L(X, Y)$.

Lemma 3.17.1 *Let A and B be linear operators in X. If A is compact and B is continuous, then the composition operators AB and BA are compact.*

Proof. If M is any bounded subset of X, then $B(M)$ is bounded. But the compact operator A maps bounded sets to precompact sets, so $AB(M)$ is precompact as required. The proof for BA is left to the reader. □

Theorem 3.17.3 *If $A \in L(X, Y)$ is compact, then A maps weak Cauchy sequences from X into strong Cauchy sequences in Y.*

Proof. Let $\{x_n\}$ be a weak Cauchy sequence in X. Then $\{x_n\}$ is bounded and, since A is a compact operator, the sequence $\{Ax_n\}$ contains a strong Cauchy subsequence $\{Ax_{n_1}\}$. This Cauchy subsequence converges to some $y \in Y$ since Y is a Banach space. It is easy to show that $\{Ax_n\}$ is a weak Cauchy sequence in Y; furthermore, because one of its subsequences converges strongly to y, the whole sequence $\{Ax_n\}$ converges weakly to $y \in Y$.

We now show that $\{Ax_n\}$ converges strongly to y. Suppose to the contrary that it does not. Then there is a subsequence $\{Ax_{n_2}\}$ and $\varepsilon > 0$ such that

$$\|Ax_{n_2} - y\| > \varepsilon \qquad (3.17.1)$$

for each n_2. But from $\{Ax_{n_2}\}$ we can select a subsequence $\{Ax_{n_3}\}$ that is a strongly Cauchy sequence in Y and thus has a limit $y_1 \in Y$. This subsequence converges weakly to the same element y_1. By the paragraph above it also converges weakly to y. But we must have $y_1 = y$ by uniqueness of the weak limit; hence $Ax_{n_3} \to y$, and this contradicts (3.17.1). □

In a separable Hilbert space this result can be strengthened:

Theorem 3.17.4 *A linear operator A acting in a separable Hilbert space H is compact if and only if it takes every weak Cauchy sequence $\{x_n\}$ into the strong Cauchy sequence $\{Ax_n\}$ in H.*

Proof. Suppose that A maps every weak Cauchy sequence $\{x_n\} \subset H$ into the strong Cauchy sequence $\{Ax_n\} \subset H$. To show that A is compact, we take a bounded set $M \subset H$ and show that $A(M)$ is precompact. Take a sequence $\{y_n\} \subset A(M)$ and consider its preimage $\{x_n\} \subset M$ (i.e., the sequence for which $Ax_n = y_n$). Because $\{x_n\}$ is bounded it contains a weak Cauchy subsequence $\{x_{n_k}\}$. By hypothesis $\{Ax_{n_k}\}$ is a strong Cauchy sequence in H, hence $A(M)$ is precompact. The converse was proved in Theorem 3.17.3. □

Example 3.17.1 Show that if $x_n \rightharpoonup x_0$, and A from X to Y is compact, then $Ax_n \to Ax_0$ as $n \to \infty$.

Solution If $\{x_n\}$ is weakly convergent then it is weakly Cauchy and by Theorem 3.17.3 we have $Ax_n \to y$ for some $y \in Y$. Since strong convergence implies weak convergence we have $Ax_n \rightharpoonup y$ for some $y \in Y$. On the other hand A is compact, hence continuous, hence weakly continuous, so $x_n \rightharpoonup x_0$ implies $Ax_n \rightharpoonup Ax_0$. Finally, $y = Ax_0$ by uniqueness of the weak limit.

Recall that $L(X,Y)$ is a normed linear space under the operator norm $\|\cdot\|$. If $\{A_n\}$ is a sequence of linear operators such that

$$\lim_{n\to\infty} \|A_n - A\| = 0,$$

then $\{A_n\}$ is said to be *uniformly convergent* and the operator A is known as the *uniform operator limit* of the sequence $\{A_n\}$.

Theorem 3.17.5 *A uniform operator limit of a sequence of compact linear operators is a compact linear operator.*

Proof. Let $\{A_n\} \subset L(X,Y)$ be a sequence of compact linear operators and suppose $\|A_n - A\| \to 0$ as $n \to \infty$. Our approach is to take any bounded sequence $\{x_n\} \subset X$ and show that we can select a subsequence whose image under A is a Cauchy sequence in Y. By compactness of A_1 we can select from $\{x_n\}$ a subsequence $\{x_{n_1}\}$ such that $\{A_1 x_{n_1}\}$ is a Cauchy sequence. Similarly, by compactness of A_2 we can select from $\{x_{n_1}\}$ a subsequence $\{x_{n_2}\}$ such that $\{A_2 x_{n_2}\}$ is a Cauchy sequence. Continuing in this way, after the kth step we have a subsequence $\{x_{n_k}\}$ for which $\{A_k x_{n_k}\}$ is a Cauchy sequence. The diagonal sequence $\xi_n \equiv x_{n_n}$ has the property that $\{A_k \xi_n\}$ is a Cauchy sequence for each fixed k. Then for any $m \geq 1$ we have

$$\begin{aligned}
&\|A\xi_{n+m} - A\xi_n\| \\
&\quad = \|(A\xi_{n+m} - A_k\xi_{n+m}) + (A_k\xi_{n+m} - A_k\xi_n) + (A_k\xi_n - A\xi_n)\| \\
&\quad \leq \|A - A_k\|\,\|\xi_{n+m}\| + \|A_k\xi_{n+m} - A_k\xi_n\| + \|A_k - A\|\,\|\xi_n\| \\
&\quad \leq 2b\,\|A - A_k\| + \|A_k\xi_{n+m} - A_k\xi_n\|
\end{aligned}$$

where $\|\xi_n\| \leq b$ for all n. Given $\varepsilon > 0$ we can choose and fix p so that $\|A - A_p\| < \varepsilon/4b$; then

$$\|A\xi_{n+m} - A\xi_n\| \leq \varepsilon/2 + \|A_p\xi_{n+m} - A_p\xi_n\|,$$

and we can finish the proof by choosing N so large that the second term on the right is less than $\varepsilon/2$ for $n > N$ and any $m \geq 1$. □

Thus the set of all compact linear operators from X to Y is a closed linear subspace of $L(X,Y)$.

Above we introduced the set of finite dimensional linear operators; these, being continuous, are compact. The importance of this class is given by the following theorem, which states that this class is dense in the set of compact linear operators in a Hilbert space.

Theorem 3.17.6 *If A is a compact operator acting in a separable Hilbert space, then there is a sequence of finite dimensional continuous linear operators $\{A_n\}$ having uniform operator limit A.*

Proof. A Hilbert space H has an orthonormal basis $\{g_n\}$, in terms of which any $f \in H$ can be represented as

$$f = \sum_{k=1}^{\infty} (f, g_k) g_k.$$

Since A is a continuous operator we have

$$Af = \sum_{k=1}^{\infty} (f, g_k) A g_k.$$

We define A_n by

$$A_n f = \sum_{k=1}^{n} (f, g_k) A g_k$$

and show that

$$\lim_{n \to \infty} \|A - A_n\| = 0. \tag{3.17.2}$$

By definition

$$\|A - A_n\| = \sup_{\|f\| \le 1} \|(A - A_n) f\| .$$

First we show that there exists f_n^* such that

$$\|f_n^*\| \le 1 \quad \text{and} \quad \|A - A_n\| = \|(A - A_n) f_n^*\| . \tag{3.17.3}$$

By definition of the supremum there is a sequence $\{f_k\}$ such that

$$\|f_k\| \le 1 \quad \text{and} \quad \lim_{k \to \infty} \|(A - A_n) f_k\| = \|A - A_n\| .$$

This bounded sequence in a separable Hilbert space has a weak Cauchy subsequence $\{f_{k_1}\}$, and this subsequence converges weakly to an element f_n^*; moreover, by the proof of Lemma 3.15.2 we have $\|f_n^*\| \le 1$. Because $A - A_n$ is compact the sequence $\{(A - A_n) f_{k_1}\}$ converges strongly to $(A - A_n) f_n^*$, i.e., a subsequence of the convergent sequence $\{\|(A - A_n) f_k\|\}$ converges to the number $\|(A - A_n)f_n^*\|$ as $k \to \infty$. So the second relation in (3.17.3) also holds.

But

$$(A - A_n)f_n^* = A\left(\sum_{k=n+1}^{\infty} (f_n^*, g_k)g_k\right)$$

so taking the norm of both sides we have, by (3.17.3),

$$\|A - A_n\| = \|A\varphi_n\| \qquad \text{where} \qquad \varphi_n = \sum_{k=n+1}^{\infty} (f_n^*, g_k)g_k. \tag{3.17.4}$$

The sequence $\{\varphi_n\} \subset H$ converges weakly to zero. Indeed for any $f \in H$ we can write

$$\begin{aligned}(\varphi_n, f) &= \left(\sum_{k=n+1}^{\infty} (f_n^*, g_k)g_k, \sum_{m=1}^{\infty}(f, g_m)g_m\right)\\ &= \left(\sum_{k=n+1}^{\infty} (f_n^*, g_k)g_k, \sum_{m=n+1}^{\infty} (f, g_m)g_m\right)\\ &= \sum_{k=n+1}^{\infty} (f_n^*, g_k)\overline{(f, g_k)},\end{aligned}$$

hence

$$\begin{aligned}|(\varphi_n, f)| &\le \left(\sum_{k=n+1}^{\infty} |(f_n^*, g_k)|^2\right)^{1/2} \left(\sum_{k=n+1}^{\infty} |(f, g_k)|^2\right)^{1/2}\\ &\le \left(\sum_{k=n+1}^{\infty} |(f, g_k)|^2\right)^{1/2} \|f_n^*\| \to 0 \quad \text{as } n \to \infty\end{aligned}$$

since $\|f_n^*\| \le 1$ and $\sum_{k=1}^{\infty} |(f, g_k)|^2 = \|f\|^2 < \infty$ by Parseval's equality (i.e., the parenthetical quantity represents the tail of a convergent series). Since $\varphi_n \rightharpoonup 0$ and A is compact we have

$$\lim_{n\to\infty} \|A\varphi_n\| = 0.$$

By (3.17.4) this proves (3.17.2). □

We will need the following simple theorem.

Theorem 3.17.7 *If A is a compact linear operator acting in a Hilbert space, then A^* is compact.*

Proof. We take a sequence $\{f_n\}$ such that $f_n \rightharpoonup f_0$ and show that $A^* f_n \to A^* f_0$. We have

$$\begin{aligned}\|A^* f_n - A^* f_0\|^2 &= (A^* f_n - A^* f_0, A^* f_n - A^* f_0) \\ &= (f_n - f_0, AA^*(f_n - f_0)) \\ &\le \|f_n - f_0\| \, \|AA^*(f_n - f_0)\| \\ &\le (\|f_n\| + \|f_0\|) \, \|AA^*(f_n - f_0)\| \, .\end{aligned}$$

But $\|f_n\| \le M$ for some constant M, and the product AA^* is compact since A^* is continuous. Hence

$$\|A^* f_n - A^* f_0\|^2 \to 0 \quad \text{as } n \to \infty$$

which completes the proof. □

Sobolev's imbedding theorem states that some imbedding operators from a Sobolev space are compact. A simple illustration can serve to clarify this idea. Let us consider the mapping under which a continuously differentiable function $f(x)$ (we show this as $f(x) \in C^{(1)}(0,1)$) is regarded as an element of the space $C(0,1)$, the space of functions continuous on $[0,1]$. Although this mapping is an operator, we cannot call it an identity operator since its domain and range are different spaces. Instead, we refer to it as the imbedding operator from $C^{(1)}(0,1)$ to $C(0,1)$.

Theorem 3.17.8 *The imbedding operator from $C^{(1)}(0,1)$ to $C(0,1)$ is compact.*

Proof. We need to check that the image S of the unit ball of the domain is a precompact set in $C(0,1)$. By Arzelà's theorem we need to show that the set of functions S is uniformly bounded and equicontinuous. It is uniformly bounded since a function of the unit ball of $C^{(1)}(0,1)$ satisfies $|f(x)| \le 1$ and thus is inside the unit ball of $C(0,1)$. The Lagrange mean value theorem then states that for any $x_1 < x_2$ from $[0,1]$ where the function is continuously differentiable there exists $\xi \in [x_1, x_2]$ such that

$$f(x_2) - f(x_1) = f'(\xi)(x_2 - x_1).$$

Since $|f'(\xi)| \le 1$ for any $f \in S$, we have

$$|f(x_2) - f(x_1)| \le |x_2 - x_1| \, .$$

This implies the equicontinuity of S. □

3.18 Closed Operators

Thus far we have considered the case of a continuous linear operator whose domain is the whole space. However, in many circumstances we are forced to consider operators whose domains are not the whole space. For example, the operator of differentiation d/dx on the space of functions continuous on $[0,1]$ does not have the entire space $C[0,1]$ as its domain, since there are continuous functions that are nowhere differentiable on $[0,1]$. But this operator, as we shall see below, has some properties that are "better" than the properties of a general operator with an arbitrary domain. We shall show that it resides in a class of operators that is wider than the class of continuous operators, but such that there remains the possibility for us to perform some limit passages with it. The class is given by the following definition.

Definition 3.18.1 Let A be a linear operator mapping elements of a Banach space X into elements of a Banach space Y. We say that A is *closed* if for any sequence $\{x_n\} \subset D(A)$ such that $x_n \to x$ and $Ax_n \to y$ as $n \to \infty$, it follows that $x \in D(A)$ and $y = Ax$.

It is evident that A is closed if A is continuous and $D(A) = X$. There are, however, closed operators that are not continuous. An example is the derivative operator $A = d/dt$ acting from $C(0,1)$ to $C(0,1)$. The domain of A is the subset of $C(0,1)$ consisting of those functions having continuous first derivatives on $[0,1]$. To see that A is closed, we first assume that

$$x_n(t) \to x(t) \quad \text{as } n \to \infty$$

in the norm of $C(0,1)$, where each $x_n'(t)$ is continuous, and that

$$Ax_n(t) = x_n'(t) \to y(t) \quad \text{as } n \to \infty,$$

also in the norm of $C(0,1)$. Realizing that convergence in the max norm is uniform convergence, we recall a theorem from ordinary calculus:

Theorem 3.18.1 *If $f_n(t)$ is continuous for each n and $f_n(t) \to f(t)$ uniformly on $[0,1]$, then*

(1) $f(t)$ is continuous on $[0,1]$, and
(2) uniform convergence of the sequence $\{f_n'(t)\}$ of derivatives that are continuous on $[0,1]$ implies that $f'(t)$ exists, is continuous on $[0,1]$, and that $f_n'(t) \to f'(t)$.

By this theorem $A = d/dx$ on $C(0,1)$ meets the definition of a closed operator. To see that A is not continuous, consider its action on the set of functions $\{t^n\}$. This set is bounded with

$$\|t^n\| = 1 \quad \text{for each } n,$$

but its image under A is unbounded with

$$\left\| \frac{d}{dt} x_n(t) \right\| = \|nt^{n-1}\| = n.$$

So A does not map every bounded set into a bounded set.

If $\Omega \subset \mathbb{R}^n$ is compact, then the more general differential operator A given by

$$Af(\mathbf{x}) = \sum_{|\alpha| \leq n} c_\alpha(\mathbf{x}) D^\alpha f(\mathbf{x}),$$

with continuous coefficients $c_\alpha(\mathbf{x})$ and acting from $C^{(n)}(\Omega)$ to $C(\Omega)$, is a closed operator.

Definition 3.18.2 Let A be an operator from X to Y. Suppose that an operator B, also from X to Y, satisfies the following two conditions:

(1) $D(A) \subseteq D(B)$, and
(2) $B(x) = A(x)$ for all $x \in D(A)$.

Then B is said to be an *extension of* A.

Lemma 3.18.1 *A linear operator A acting from a Banach space X to a Banach space Y has a closed extension if and only if from the condition*

(*) let $\{x_n\} \subset D(A)$ be an arbitrary sequence such that $x_n \to 0$ and $Ax_n \to y$

it follows that $y = 0$.

Proof. Necessity follows from Definition 3.18.1. To prove sufficiency let us explicitly construct a closed extension B of A.

We first define B, then verify its properties. Let $D(B)$ consist of those elements x for which there exists $\{x_n\} \subset D(A)$ such that $x_n \to x$ and $Ax_n \to y$ as $n \to \infty$; for each such x, define $Bx = y$. The condition (*) ensures that y is uniquely defined by x. Indeed, suppose two sequences

$\{x_n\}$ and $\{z_n\}$ in $D(A)$ both converge to x, and $Ax_n \to y$ while $Az_n \to y'$. Then

$$x_n - z_n \to 0, \qquad A(x_n - z_n) = Ax_n - Az_n \to y - y',$$

and from (*) it follows that $y - y' = 0$.

To see that B is linear, we take two elements $x, \tilde{x}$ in $D(B)$ and any two scalars λ, μ. By definition of $D(B)$ there are sequences $\{x_n\}$ and $\{\tilde{x}_n\}$ in $D(A)$ such that

$$x_n \to x, \quad Ax_n \to y, \qquad \tilde{x}_n \to \tilde{x}, \quad A\tilde{x}_n \to \tilde{y},$$

and we define $Bx = y$, $B\tilde{x} = \tilde{y}$. But $\lambda x + \mu\tilde{x} \in D(B)$ because

$$\lambda x_n + \mu\tilde{x}_n \to \lambda x + \mu\tilde{x}, \quad A(\lambda x_n + \mu\tilde{x}_n) = \lambda Ax_n + \mu A\tilde{x}_n \to \lambda y + \mu\tilde{y},$$

and we therefore *define* $B(\lambda x + \mu\tilde{x}) = \lambda y + \mu\tilde{y} = \lambda Bx + \mu B\tilde{x}$.

Finally, let $\{u_n\} \subset D(B)$ be such that $u_n \to u$ and $Bu_n \to v$. According to Definition 3.18.1 we must prove that $u \in D(B)$ and $Bu = v$. Let us construct a sequence $\{x_n\} \subset D(A)$ that is equivalent to $\{u_n\}$, and then verify the desired properties for $\{x_n\}$. Fix u_n. By definition of B there exists $\{w_{nk}\} \subset D(A)$ such that $w_{nk} \to u_n$ and $Aw_{nk} \to Bu_n$ as $k \to \infty$. Hence there exists N such that for all $k > N$ we have both $\|w_{nk} - u_n\| < 1/n$ and $\|Aw_{nk} - Bu_n\| < 1/n$. Choose one of the points w_{nk_0} where $k_0 > N$, and denote this point x_n. Now consider the sequence of points $\{x_n\} \subset D(A)$. The inequalities $\|x_n - u_n\| < 1/n$ and $\|Ax_n - Bu_n\| < 1/n$ show that $x_n \to u$ and $Ax_n \to v$ as $n \to \infty$. By definition of B we have $u \in D(B)$ and $Bu = v$. □

It sometimes happens that we can establish boundedness of an operator directly on a subspace that is everywhere dense in the space. To establish that it is continuous on the whole space, we may employ

Theorem 3.18.2 *Let A be a closed linear operator whose domain is a Banach space X and whose range lies in a Banach space Y. Assume there is a set M which is dense in X and a positive constant c such that*

$$\|Ax\| \le c\,\|x\| \quad \textit{for all } x \in M.$$

Then A is continuous on the whole space X.

Proof. For any $x_0 \in X$, we can find $\{x_n\} \subset M$ such that $\|x_n - x_0\| < 1/n$ for each n. The inequality

$$\|Ax_{k+m} - Ax_k\| \le c\,\|x_{k+m} - x_k\| \le c(\|x_{k+m} - x_0\| + \|x_k - x_0\|) \le 2c/k$$

shows that $\{Ax_k\}$ is a Cauchy sequence in Y. We have $Ax_k \to y$ for some $y \in Y$ since Y is a Banach space; since A is closed, $Ax_0 = y$. Now we can write

$$\|Ax_0\| = \lim_{k\to\infty} \|Ax_k\| \le \lim_{k\to\infty} c\,\|x_k\| = c\,\|x_0\|.$$

Since x_0 is an arbitrary element of X and c does not depend on x_0, the proof is complete. □

Closed operators can be considered from another viewpoint. We begin by noting that if X and Y are Banach spaces over the same scalar field, then the Cartesian product space $X \times Y$ with algebraic operations defined by

$$(x_1, y_1) + (x_2, y_2) = (x_1 + x_2, y_1 + y_2), \qquad \alpha(x, y) = (\alpha x, \alpha y),$$

and norm defined by

$$\|(x, y)\| = (\|x\|_X^2 + \|y\|_Y^2)^{1/2},$$

is also a Banach space.

Definition 3.18.3 Let A be an operator acting from $D(A) \subset X$ to Y. Then the set

$$G(A) = \{(x, Ax) \in X \times Y \mid x \in D(A)\}$$

is called the *graph* of A.

Theorem 3.18.3 *A linear operator A acting from $D(A) \subset X$ to Y is closed if and only if $G(A)$ is a closed linear subspace of $X \times Y$.*

Proof. Suppose A is a closed operator. Let (x, y) be a limit point of $G(A)$. Then there is a sequence $\{(x_n, Ax_n)\} \subset G(A)$ that converges to (x, y) in the norm of $X \times Y$. Evidently this implies that as $n \to \infty$ we have $x_n \to x$ in X and $Ax_n \to y$ in Y. Because A is closed, $x \in D(A)$ and $y = Ax$. Hence $(x, Ax) \in G(A)$ by definition of $G(A)$.

Conversely, suppose $G(A)$ is closed in $X \times Y$. Let $\{x_n\} \subset D(A)$ be such that, as $n \to \infty$, $x_n \to x$ in X and $Ax_n \to y$ in Y. The sequence $\{(x_n, Ax_n)\} \subset G(A)$ converges in the norm of $X \times Y$ to (x, y). Since $G(A)$ is closed, $(x, y) \in G(A)$. By definition of $G(A)$ this means that $x \in D(A)$ and $y = Ax$. □

Theorem 3.18.4 *If A is an invertible closed linear operator, then A^{-1} is also closed.*

Proof. We can obtain $G(A^{-1})$ from the graph of $G(A)$ by the simple rearrangement $(x, Ax) \mapsto (Ax, x)$. Hence $G(A^{-1})$ is closed in $Y \times X$. □

We can now formulate Banach's *closed graph theorem.*

Theorem 3.18.5 *Let X and Y be Banach spaces. If A is a closed linear operator having $D(A) = X$, then A is continuous on X.*

See Yosida [Yosida (1965)] for a proof. In applications the following simple consequence of the theorem can be used to establish continuity of an operator.

Corollary 3.18.1 *Let X and Y be Banach spaces. If a closed linear operator A from X to Y is one-to-one and onto, then A^{-1} is continuous on Y.*

Proof. The operator A^{-1} is closed by Theorem 3.18.4, and is continuous by Theorem 3.18.5. □

3.19 Introduction to Spectral Concepts

We begin this section by recalling that the equation

$$A\mathbf{x} = \lambda\mathbf{x} \tag{3.19.1}$$

plays an important role in the theory of an $n \times n$ matrix A. Any number λ that satisfies (3.19.1) for some nonzero vector $\mathbf{x}$ is called an *eigenvalue* of A, and $\mathbf{x}$ is a corresponding *eigenvector*. An alternative form for (3.19.1) is, of course,

$$(A - \lambda I)\mathbf{x} = 0$$

where I is the $n \times n$ identity matrix. To this equation we can relate another, inhomogeneous equation which corresponds to most mechanical problems involving periodic forced oscillations of a finite number of oscillators:

$$(A - \lambda I)\mathbf{x} = \mathbf{b}. \tag{3.19.2}$$

We know that if λ is not an eigenvalue of A this equation is solvable for any $\mathbf{b}$. The eigenvalues of A correspond to the frequencies of external forces that put the system into the resonance state when the amplitude of vibrations grows without bound.

But equations having the form (3.19.2) also occur outside the realm of matrix theory. Equations of the form

$$(A - \lambda I)x = b, \tag{3.19.3}$$

where A is a more general operator, arise naturally in continuum physics. Usually we get an equation of this form when studying the oscillations of a medium. Then A is a differential or integral operator acting on the set of admissible functions that represent distributions of displacement, strain, stress, heat, etc. This operator is linear. Defining properly the set of admissible functions x and loading terms b (note that b may represent actual mechanical loads in some problems, but may represent sources, say of heat, in other problems) we get an operator equation. If $b = 0$ we then have the problem of finding nontrivial solutions to the homogeneous equation. These are called *eigensolutions.* The terminology of matrix theory is retained in this case. These eigensolutions, as for a finite system of oscillators, represent eigen-oscillations of some elastic bodies or fields. Even when they do not represent oscillations of the system, they still participate in the Fourier method of separation of variables to solve the problem and, in any case, give us an understanding of how the system functions. Note that unlike the situation for a matrix equation where we seek solutions in space $\mathbb{R}^n$ for which all norms are equivalent, the choice of admissible sets for continuum problems brings a new situation: with a proper choice of the space of solution we can gain or lose eigensolutions. To decide which spaces are "correct" spaces we should rely on our understanding of the physics of the corresponding processes.

The simple relation between the existence of solution for an inhomogeneous matrix equation and λ being or not being an eigenvalue may fail for continuum problems. It turns out that there are situations in which λ is not an eigenvalue of the corresponding operator equation, so there are no eigenvectors of the operator A, but we cannot find a solution to (3.19.3) in such a way that it depends continuously on changes in b. The collection of "trouble spots" for λ in the complex plane (including the eigenvalues) is known as the *spectrum* of the operator A. We give a formal definition of this concept next, as well as a classification of the points of the spectrum.

Definition 3.19.1 Let A be a linear operator having domain and range in a complex normed space X. For a complex parameter λ, denote by A_λ the operator

$$A_\lambda = A - \lambda I$$

where I is the identity operator on X. The *resolvent set* of A is the set $\rho(A)$ of all $\lambda \in \mathbb{C}$ for which the range of A_λ is dense in X and for which A_λ has a bounded inverse. For any $\lambda \in \rho(A)$, we call A_λ^{-1} the *resolvent* of A at λ and write

$$R(\lambda; A) = (A - \lambda I)^{-1}.$$

The complement of $\rho(A)$ in $\mathbb{C}$ is a set called the *spectrum* of A, denoted $\sigma(A)$.

Any value $\lambda \in \rho(A)$ is known as a *regular value* of A. Any $\lambda \in \sigma(A)$ is called a *spectral value* of A. The spectrum of any operator A is naturally partitioned into three disjoint subsets:

(1) $P_\sigma(A)$, the *point spectrum* of A, is the set of all spectral values for which the resolvent $R(\lambda; A)$ does not exist. Its elements are called the *eigenvalues* of A.
(2) $C_\sigma(A)$, the *continuous spectrum* of A, is the set of all spectral values for which $R(\lambda; A)$ exists on a dense subset of X but is not a bounded operator.
(3) $R_\sigma(A)$, the *residual spectrum* of A, is the set of all spectral values for which $R(\lambda; A)$ exists but with a domain that is not dense in X.

So

$$\sigma(A) = P_\sigma(A) \cup C_\sigma(A) \cup R_\sigma(A)$$

(we shall see that some of the sets on the right may be empty). The use of the term "eigenvalue" for an element $\lambda \in P_\sigma(A)$ may be justified as follows. We have $\lambda \in P_\sigma(A)$ if and only if the linear operator $A - \lambda I$ is not one-to-one, which is true if and only if its null space does not consist only of the zero vector. In other words, we can have $\lambda \in P_\sigma(A)$ if and only if the equation

$$(A - \lambda I)x = 0$$

has a nontrivial solution x. Such an element x would be, of course, an eigenvector of A corresponding to the eigenvalue λ.

Example 3.19.1 Let $X = \ell^1$, and let A from X to X be given by

$$A\mathbf{x} = \left(\frac{\xi_1}{1}, \frac{\xi_2}{2}, \frac{\xi_3}{3}, \dots\right)$$

for $\mathbf{x} = (\xi_1, \xi_2, \xi_3, \dots) \in \ell^1$. Find $P_\sigma(A)$.

Solution We have

$$(A-\lambda I)\mathbf{x} = \left(\left(\frac{1}{1}-\lambda\right)\xi_1, \left(\frac{1}{2}-\lambda\right)\xi_2, \left(\frac{1}{3}-\lambda\right)\xi_3, \ldots\right).$$

$A - \lambda I$ is not one-to-one if and only if λ is such that $\frac{1}{k} - \lambda = 0$ for some $k = 1, 2, 3, \ldots$. Hence

$$P_\sigma(A) = \left\{1, \frac{1}{2}, \frac{1}{3}, \ldots\right\}.$$

Example 3.19.2 Show that if A is a bounded linear operator and λ is an eigenvalue of A, then $|\lambda| \leq \|A\|$.

Solution For some nonzero vector v we have $Av = \lambda v$, hence $|\lambda|\,\|v\| = \|Av\| \leq \|A\|\,\|v\|$.

For a bounded operator we can show an important part of the resolvent set immediately.

Theorem 3.19.1 *Let A be a bounded linear operator on a Banach space X. All the $\lambda \in \mathbb{C}$ such that $\|A\| < |\lambda|$ are points of the resolvent set of operator A, that is $(A-\lambda I)^{-1}$ is a bounded linear operator on X. Moreover, there holds*

$$(A-\lambda I)^{-1} = -\frac{1}{\lambda}\sum_{k=0}^{\infty}\frac{1}{\lambda^k}A^k.$$

The series on the right is called the Neumann series *for A.*

Proof. Thus A is a bounded linear operator on a Banach space X. Let us take a value $\lambda \in \mathbb{C}$ and consider solving the equation

$$Ax - \lambda x = y \tag{3.19.4}$$

for $x \in X$ when $y \in X$ is given. We rewrite this as

$$x = -\frac{1}{\lambda}y + \frac{1}{\lambda}Ax,$$

define the right member as the mapping $F(x) = -\lambda^{-1}y + \lambda^{-1}Ax$, and check to see whether F can be a contraction. We have

$$\|F(x_1) - F(x_2)\| = |\lambda|^{-1}\|Ax_1 - Ax_2\| \leq |\lambda|^{-1}\|A\|\,\|x_1 - x_2\|,$$

hence F is a contraction whenever

$$|\lambda| > \|A\|.$$

Provided this condition is fulfilled we can employ the iteration scheme

$$x_{j+1} = -\frac{1}{\lambda}y + \frac{1}{\lambda}Ax_j, \quad j = 0, 1, 2, \ldots$$

to solve (3.19.4). Starting with $x_0 = -y/\lambda$, we may generate a sequence of iterates:

$$\begin{aligned}
x_0 &= -\frac{1}{\lambda}y \\
x_1 &= -\frac{1}{\lambda}y + \frac{1}{\lambda}Ax_0 = -\frac{1}{\lambda}y - \frac{1}{\lambda^2}Ay \\
x_2 &= -\frac{1}{\lambda}y + \frac{1}{\lambda}Ax_1 = -\frac{1}{\lambda}y - \frac{1}{\lambda^2}Ay - \frac{1}{\lambda^3}A^2y \\
&\vdots \\
x_n &= -\frac{1}{\lambda}\sum_{k=0}^{n}\frac{1}{\lambda^k}A^k y.
\end{aligned}$$

These iterates converge to the unique solution

$$x = -\frac{1}{\lambda}\sum_{k=0}^{\infty}\frac{1}{\lambda^k}A^k y.$$

It is therefore clear that the operator given by the absolutely convergent series

$$-\frac{1}{\lambda}\sum_{k=0}^{\infty}\frac{1}{\lambda^k}A^k,$$

is the inverse of the operator $A - \lambda I$. We can also check this statement explicitly. To see that it is a right inverse, we write

$$\begin{aligned}
(A - \lambda I)\left(-\frac{1}{\lambda}\sum_{k=0}^{\infty}\frac{1}{\lambda^k}A^k\right) &= \left(I - \frac{1}{\lambda}A\right)\left(I + \sum_{k=1}^{\infty}\frac{1}{\lambda^k}A^k\right) \\
&= I - \left(\frac{1}{\lambda}A + \frac{1}{\lambda}A\sum_{k=1}^{\infty}\frac{1}{\lambda^k}A^k\right) + \sum_{k=1}^{\infty}\frac{1}{\lambda^k}A^k \\
&= I - \sum_{k=1}^{\infty}\frac{1}{\lambda^k}A^k + \sum_{k=1}^{\infty}\frac{1}{\lambda^k}A^k \\
&= I.
\end{aligned}$$

Verification that it is a left inverse is similar. □

By this theorem the set

$$\{\lambda \in \mathbb{C} \colon |\lambda| > \|A\|\}$$

does not contain any points of the spectrum of A, which is another solution of Example 3.19.2.

Certain kinds of operators have simple and convenient spectral properties. In our future work we shall need the results given in the following lemma:

Lemma 3.19.1 *Let A be a self-adjoint continuous linear operator A acting in a Hilbert space H. Then*

(i) the functional (Ax, x) is real valued;
(ii) the eigenvalues of A are real;
(iii) if x_1, x_2 are two eigenvectors corresponding to distinct eigenvalues λ_1, λ_2, then $(x_1, x_2) = 0$ and $(Ax_1, x_2) = 0$.

Proof. To prove item (i) we merely write

$$(Ax, x) = (x, Ax) = \overline{(Ax, x)}.$$

If $Ax = \lambda x$ then $(Ax, x) = \lambda(x, x)$, hence λ is real. This proves (ii). Now suppose $Ax_1 = \lambda_1 x_1$ and $Ax_2 = \lambda_2 x_2$ where $\lambda_2 \neq \lambda_1$. Forming inner products with x_2 and x_1 respectively, we obtain

$$\lambda_1(x_1, x_2) = (Ax_1, x_2), \qquad \lambda_2(x_1, x_2) = (x_1, Ax_2) = (Ax_1, x_2);$$

subtracting these we find $(\lambda_2 - \lambda_1)(x_1, x_2) = 0$, hence $(x_1, x_2) = 0$. Returning to $\lambda_1(x_1, x_2) = (Ax_1, x_2)$, we have $(Ax_1, x_2) = 0$. This proves (iii). □

3.20 The Fredholm Theory in Hilbert Spaces

It is a quite common problem to find a solution $\mathbf{x}$ of the following algebraic problem in $\mathbb{R}^n$:

$$A\mathbf{x} - \lambda\mathbf{x} = \mathbf{b}, \tag{3.20.1}$$

where A is an $n \times n$ matrix. When $\mathbf{b} = 0$, this is an eigenvalue problem for the matrix A. For this equation it is well known that if λ is not an eigenvalue of A, then the equation is solvable for any $\mathbf{b}$. There are no more than n eigenvalues of A. If λ is an eigenvalue of A, then the problem is solvable only for some set of values $\mathbf{b}$ that are orthogonal to all the eigenvectors of the conjugate-transpose matrix A^* that correspond to $\bar{\lambda}$, an eigenvalue of A^*. So to an eigenvalue λ_0 of A there corresponds an eigenvalue $\bar{\lambda}_0$ of A^*; moreover, the dimensions of the subspaces of the corresponding eigenvectors of A and A^* are the same. Furthermore, the situation for the solvability of the dual equation

$$A^*\mathbf{x} - \lambda\mathbf{x} = \mathbf{b}^*$$

is symmetric to the problem involving the operator A.

This was extended by I. Fredholm to the theory of integral equations that are now called Fredholm equations of the second kind:

$$\lambda u(\mathbf{x}) - \int_\Omega K(\mathbf{x},\mathbf{y})u(\mathbf{y})\,d\Omega_{\mathbf{y}} = f(\mathbf{x}).$$

When the operator is compact this equation inherits almost all the qualitative features possessed by equation (3.20.1), except the number of possible eigenvalues: it may be countable, but the only possible point of accumulation is zero. Riesz [Riesz (1918)] and Schauder [Schauder (1930)] extended the Fredholm theory to Banach spaces.

We present a particular case of this theory in a Hilbert space H, which is enough to consider the problem of eigenfrequencies of bounded elastic objects like membranes, plates, shells, or elastic bodies. We recall that the Fredholm integral operator is compact in L^2. Thus we consider the following equation in H:

$$Ax - \lambda x = b,$$

with given $b \in H$. We suppose A to be a compact linear operator in H. Let us introduce the necessary notation. A^* is the adjoint to A, satisfying the equality

$$(Ax, y) = (x, A^*y).$$

Correspondingly we introduce the equation

$$A^*x - \lambda x = b^*.$$

We denote by $N(\lambda)$ the subspace of H spanned by the eigenvectors of A corresponding to a given eigenvalue λ. With the exception of the zero element, every member of this subspace is an eigenvector of A. Indeed any finite linear combination of $x_1, \ldots, x_m \in N(\lambda)$ also belongs to $N(\lambda)$:

$$A\left(\sum_{i=1}^{m} \alpha_i x_i\right) = \sum_{i=1}^{m} \alpha_i A x_i = \sum_{i=1}^{m} \alpha_i \lambda x_i = \lambda\left(\sum_{i=1}^{m} \alpha_i x_i\right).$$

Note that $N(\lambda)$ contains all the eigenvectors corresponding to λ, along with the zero element of H.[6] We denote by $M(\lambda)$ the orthogonal complement of $N(\lambda)$ in H. The corresponding sets for A^* are denoted by $N^*(\lambda)$ and $M^*(\lambda)$. Let us formulate the facts of the Fredholm–Riesz–Schauder theory as

Theorem 3.20.1 *Let A be a compact linear operator in a Hilbert space H. Then*

(1) the spectrum of A consists only of eigenvalues, and thus the remaining points of the complex plane are all regular points of A;
(2) to any nonzero eigenvalue λ of A there corresponds a finite number of linearly independent eigenvectors (i.e., $N(\lambda)$ is finite dimensional);
(3) the only possible point of accumulation of the eigenvalues of A in the complex plane is zero;
(4) if λ is an eigenvalue of A then $\bar{\lambda}$ is an eigenvalue of A^* and vice versa, and the equation

$$Ax - \lambda x = b$$

is solvable if and only if b is orthogonal to the set $N^*(\bar{\lambda})$;
(5) the dimensions of $N(\lambda)$ and $N^*(\bar{\lambda})$ are equal;
(6) A^* is a compact linear operator, and thus
(6a) its spectrum consists only of eigenvalues with zero as the only possible point of accumulation of the eigenvalues;
(6b) to each eigenvalue there corresponds a space of eigenvectors $N^*(\lambda)$ that is finite dimensional;
(6c) the equation

$$A^* x - \lambda x = b^*$$

is solvable if and only if b^* is orthogonal to the subspace $N(\bar{\lambda})$.

The proof will be formulated as a collection of lemmas. We begin by proving statement (2).

[6] An alternative definition of $N(\lambda)$ is as the null space of the operator $A - \lambda I$, i.e., as the set of all $x \in H$ that satisfy the equation $(A - \lambda I)x = 0$.

Lemma 3.20.1 *If λ is any nonzero eigenvalue of A, then $N(\lambda)$ is a closed, finite dimensional subspace of H.*

Proof. To see that $N(\lambda)$ is closed we use the continuity of A. Let x_* be a limit point of $N(\lambda)$. There is a sequence $\{x_n\} \subset N(\lambda)$ such that $x_n \to x_*$ in H. For each n we have $Ax_n = \lambda x_n$, and passage to the limit as $n \to \infty$ gives $Ax_* = \lambda x_*$. Hence $x_* \in N(\lambda)$. We next show that $N(\lambda)$ is finite dimensional. We recall Theorem 3.8.4 which states that any closed and bounded set is compact only in a finite dimensional Hilbert space. So let S be an arbitrary closed and bounded subset of $N(\lambda)$, and choose any sequence $\{x_k\} \subset S$. By compactness of A and the equality $x_k = \lambda^{-1}Ax_k$, we see that $\{x_k\}$ has a Cauchy subsequence. Hence S is precompact. But S is also a closed subset of a complete space H, hence it contains the limits of its Cauchy sequences. We conclude that S is compact, as desired. □

Remark 3.20.1 We do not include into consideration the eigenvalue $\lambda = 0$, because it corresponds to the infinite eigenfrequency of a body. The properties of this eigenvalue differ from the properties of all the rest of the eigenvalues. Take, for example, a one dimensional operator A of the form $Ax = F(x)x_0$ where x_0 is fixed and $F(x)$ is a continuous linear functional. Then by the equation $Ax = \lambda x$, the eigenvalues corresponding to $\lambda = 0$ are those elements x that satisfy $F(x)x_0 = 0$. By the Riesz representation theorem we can express $F(x) = (x, f)$ for some fixed $f \in H$, hence any vector x that is orthogonal to f belongs to $N(0)$. As an even stronger example we may take A to be the zero operator, which is of course compact. In this case the equation $Ax = \lambda x$ becomes $\lambda x = 0$, and with $\lambda = 0$ this holds for any $x \in H$. Here then we have $N(0) = H$. So $\lambda = 0$ was by necessity excluded from statement (2). In statement (3) we see that $\lambda = 0$ is the only possible accumulation point for the set of all eigenvalues.

Statement (3) will be proved as Lemma 3.20.3. In preparation for this we introduce some notation and establish an auxiliary result. Let $\lambda_1, \ldots, \lambda_k$ be eigenvalues of A. We denote by

$$N(\lambda_1)\dot{+}\cdots\dot{+}N(\lambda_k)$$

the space spanned by the union of the eigenvectors that generate the individual eigenspaces $N(\lambda_1), \ldots, N(\lambda_k)$. Our use of the notation for direct sum is justified by the next result which shows, in particular, that eigenspaces corresponding to distinct eigenvalues can intersect only in the zero vector.

Lemma 3.20.2 *Assume $S_i = \{x_1^{(i)}, x_2^{(i)}, \ldots, x_{n_i}^{(i)}\}$ is a linearly independent system of elements in $N(\lambda_i)$ for each $i = 1, \ldots, k$. Then the union $\cup_{i=1}^k S_i$ is linearly independent. If S_i is a basis of $N(\lambda_i)$ for each i, then $\cup_{i=1}^k S_i$ is a basis of $N(\lambda_1) \dot{+} \cdots \dot{+} N(\lambda_k)$.*

Proof. The proof is by induction. We want to show that under the hypothesis of the lemma $\cup_{i=1}^k S_i$ is linearly independent in $N(\lambda_1) \dot{+} \cdots \dot{+} N(\lambda_k)$ for each positive integer k. For $k = 1$ the statement holds trivially. Suppose it holds for $k = n$. Let us take the eigenvalue-eigenvector pairings

$$(\lambda_i, x_p^{(i)}), \quad p = 1, \ldots, n_i, \qquad i = 1, \ldots, n,$$

and renumber everything so that these same pairings are denoted as (λ_j, x_j), $j = 1, \ldots, r$. By assumption then,

$$\sum_{j=1}^{r} \alpha_j x_j = 0 \implies \alpha_j = 0 \text{ for } j = 1, \ldots, r. \tag{3.20.2}$$

We must show that the statement holds for $k = n + 1$. Appending S_{n+1} to $\cup_{i=1}^n S_i$, we now assume that

$$\sum_{j=1}^{r+s} c_j x_j = 0 \tag{3.20.3}$$

and attempt to draw a conclusion regarding the c_j (here s is new notation for the number of elements in S_{n+1}). An application of A to both sides allows us to write

$$\frac{1}{\lambda_{n+1}} \sum_{j=1}^{r+s} c_j \lambda_j x_j = 0$$

and upon subtraction from the previous equation we obtain

$$\sum_{j=1}^{r+s} c_j \left(1 - \frac{\lambda_j}{\lambda_{n+1}}\right) x_j = \sum_{j=1}^{r} c_j \left(1 - \frac{\lambda_j}{\lambda_{n+1}}\right) x_j = 0.$$

We now have $c_j = 0$ for $j = 1, \ldots, r$ by (3.20.2). Substitution into (3.20.3) gives

$$\sum_{j=r+1}^{r+s} c_j x_j = 0;$$

but the eigenvectors participating in this sum are all associated with λ_{n+1} and are linearly independent by assumption. Hence $c_j = 0$ for $j = r + 1, \ldots, r+s$.

The second statement of the lemma follows from the fact that the dimension of the direct sum $N(\lambda_1)\dot{+}\cdots\dot{+}N(\lambda_k)$ is less than or equal to the sum of the dimensions of the constituent eigenspaces $N(\lambda_i)$. Since we do have $n_1 + \cdots + n_k$ linearly independent vectors in the direct sum, we have found a basis. □

Lemma 3.20.3 *The only possible point of accumulation of the eigenvalues of A in the complex plane is $\lambda = 0$.*

Proof. Suppose λ_0 is a limit point of the set of eigenvalues of A, and $|\lambda_0| > 0$. There is a sequence $\{\lambda_n\}$ of distinct eigenvalues of A such that $\lambda_n \to \lambda_0$. For each λ_n take an eigenvector x_n, and denote by H_n the subspace spanned by $\{x_1, \ldots, x_n\}$. Thus $H_n \subseteq H_{n+1}$ for each n. Let $y_1 = x_1/\|x_1\|$. Successively, we can construct another sequence $\{y_n\}$, $n > 1$, as follows. By Lemma 3.20.2 we have $H_n \neq H_{n+1}$, so for each n there exists $y_{n+1} \in H_{n+1}$ such that $\|y_{n+1}\| = 1$ and y_{n+1} is orthogonal to H_n. Indeed, we use the orthogonal decomposition theorem to decompose H_{n+1} into H_n and another nonempty subspace orthogonal to H_n, from which we choose a normalized element. Now consider the sequence $\{y_n/\lambda_n\}$; because it is bounded in H, its image $\{A(y_n/\lambda_n)\}$ contains a Cauchy subsequence. We begin to seek a contradiction to this last statement by writing

$$A\left(\frac{y_{n+m}}{\lambda_{n+m}}\right) - A\left(\frac{y_n}{\lambda_n}\right) = y_{n+m} - \left(y_{n+m} - \frac{1}{\lambda_{n+m}}Ay_{n+m} + \frac{1}{\lambda_n}Ay_n\right) \tag{3.20.4}$$

for $m \geq 1$. On the right the first term y_{n+m} belongs to H_{n+m}; the second (parenthetical) term belongs to H_{n+m-1} because we can write $y_{n+m} = \sum_{k=1}^{n+m} c_k x_k$ and have

$$\begin{aligned} y_{n+m} - \frac{1}{\lambda_{n+m}}Ay_{n+m} &= \sum_{k=1}^{n+m} c_k x_k - \frac{1}{\lambda_{n+m}} A\left(\sum_{k=1}^{n+m} c_k x_k\right) \\ &= \sum_{k=1}^{n+m-1} c_k \left(1 - \frac{\lambda_k}{\lambda_{n+m}}\right) x_k \in H_{n+m-1} \end{aligned}$$

along with the fact that $\lambda_n^{-1}Ay_n \in H_n \subseteq H_{n+m-1}$. Because the two terms on the right side of (3.20.4) are orthogonal the Pythagorean theorem yields

$$\left\| A\left(\frac{y_{n+m}}{\lambda_{n+m}}\right) - A\left(\frac{y_n}{\lambda_n}\right)\right\|^2$$
$$= \|y_{n+m}\|^2 + \left\| y_{n+m} - \frac{1}{\lambda_{n+m}} A y_{n+m} + \frac{1}{\lambda_n} A y_n \right\|^2 \geq 1,$$

for any n and $m \geq 1$, so $\{A(y_n/\lambda_n)\}$ cannot contain a Cauchy sequence.□

Let us proceed to

Lemma 3.20.4 *Let λ be fixed. There are positive constants m_1 and m_2 such that*

$$m_1 \|x\| \leq \|Ax - \lambda x\| \leq m_2 \|x\| \tag{3.20.5}$$

for all $x \in M(\lambda)$.

Proof. We have

$$\|Ax - \lambda x\| \leq \|Ax\| + \|\lambda x\| \leq (\|A\| + |\lambda|)\|x\|,$$

thus establishing the inequality on the right. Proceeding to the inequality on the left, suppose it does *not* hold. Then there is a sequence $\{x_n\} \subset M(\lambda)$ such that $\|x_n\| = 1$ and $\|Ax_n - \lambda x_n\| \to 0$ as $n \to \infty$. Because A is compact, $\{Ax_n\}$ contains a Cauchy subsequence. By the equality

$$\lambda x_n = Ax_n - (Ax_n - \lambda x_n)$$

$\{x_n\}$ also contains a Cauchy subsequence which we again denote as $\{x_n\}$. By completeness of $M(\lambda)$ we have $x_n \to x_0$ for some $x_0 \in M(\lambda)$. Continuity of A gives $Ax_n \to Ax_0$, and from

$$0 = \lim_{n\to\infty} \|Ax_n - \lambda x_n\| = \|Ax_0 - \lambda x_0\|$$

we see that $Ax_0 = \lambda x_0$. This means that $x_0 \in N(\lambda)$. Thus we have $\|x_0\| = 1$, $x_0 \in N(\lambda)$, and $x_0 \in M(\lambda)$; this is impossible since the spaces $N(\lambda)$ and $M(\lambda)$ intersect only in the zero element. □

Lemma 3.20.4 shows that on $M(\lambda)$ we can impose a norm

$$\|x\|_1 = \|Ax - \lambda x\|$$

which is equivalent to the norm of H. The associated inner product is given by

$$(x, y)_1 = (Ax - \lambda x, Ay - \lambda y).$$

Similarly, on $M^*(\bar{\lambda})$ the norm $\|A^*x - \bar{\lambda}x\|$ is equivalent to the norm of H.

Lemma 3.20.5 *The equation*

$$Ax - \lambda x = b \tag{3.20.6}$$

is solvable if and only if b is orthogonal to every vector in $N^*(\bar{\lambda})$; equivalently,

$$R(A - \lambda I) = M^*(\bar{\lambda}). \tag{3.20.7}$$

Similarly, the equation

$$A^*x - \bar{\lambda}x = b^* \tag{3.20.8}$$

is solvable if and only if b^* is orthogonal to every vector in $N(\lambda)$; equivalently,

$$R(A^* - \bar{\lambda}I) = M(\lambda). \tag{3.20.9}$$

Proof. Suppose (3.20.6) is solvable with solution x_0. If $y \in N^*(\bar{\lambda})$ is arbitrary, then

$$(b, y) = (Ax_0 - \lambda x_0, y) = (x_0, A^*y - \bar{\lambda}y) = (x_0, 0) = 0.$$

Conversely, suppose $b \in M^*(\bar{\lambda})$. The functional (x, b) is linear and continuous on H (and so on $M^*(\bar{\lambda})$), hence by the Riesz representation theorem can be represented on $M^*(\bar{\lambda})$ using $(\cdot,\cdot)_1$ as

$$(x, b) = (x, \tilde{b})_1 = (A^*x - \bar{\lambda}x, A^*\tilde{b} - \bar{\lambda}\tilde{b})$$

for some $\tilde{b} \in M^*(\bar{\lambda})$. This equality, being valid for $x \in M^*(\bar{\lambda})$, holds for all $x \in H$ too; indeed bearing $x = x_1 + x_2$, $x_1 \in N^*(\bar{\lambda})$, $x_2 \in M^*(\bar{\lambda})$, we have

$$A^*x - \bar{\lambda}x = A^*x_1 - \bar{\lambda}x_1 + A^*x_2 - \bar{\lambda}x_2 = A^*x_2 - \bar{\lambda}x_2$$

and so, for all $x \in H$,

$$(A^*x - \bar{\lambda}x, A^*\tilde{b} - \bar{\lambda}\tilde{b}) = (A^*x_2 - \bar{\lambda}x_2, A^*\tilde{b} - \bar{\lambda}\tilde{b}) = (x_2, \tilde{b})_1 = (x_2, b) = (x, b)$$

since $(x_1, b) = 0$. Denoting $A^*\tilde{b} - \bar{\lambda}\tilde{b}$ by g we have

$$(A^*x - \bar{\lambda}x, g) = (x, Ag - \lambda g) = (x, b) \text{ for all } x \in H,$$

hence $Ag - \lambda g = b$ and g satisfies (3.20.6). The rest of the lemma is proved analogously. □

By this lemma we have partially addressed part (4) of Theorem 3.20.1.

Lemma 3.20.6 *If N_n is the null space of $(A - \lambda I)^n$, then*

(i) N_n *is a finite dimensional subspace of* H*;*
(ii) $N_n \subseteq N_{n+1}$ *for all* $n = 1, 2, \ldots$*;*
(iii) there exists k *such that* $N_n = N_k$ *for all* $n > k$*.*

Proof.

(i) Writing $(A - \lambda I)^n x = 0$ as

$$(\lambda^n I - n\lambda^{n-1}A + \cdots)x = 0,$$

the sum of the terms beginning with the second is a compact operator $(-B)$ so denoting $\lambda^n = \gamma$ we get an eigenvalue problem $(B - \gamma I)x = 0$ with compact B and so N_n is finite dimensional.

(ii) If $(A - \lambda I)^n x = 0$, then $(A - \lambda I)^{n+1} x = 0$.

(iii) First we show that if $N_{k+1} = N_k$ for some k then $N_{k+m} = N_k$ for $m = 1, 2, 3, \ldots$. Consider the case $m = 2$. By part (ii) we know that $N_k \subseteq N_{k+2}$. Conversely

$$\begin{aligned} x_0 \in N_{k+2} &\implies 0 = (A - \lambda I)^{k+2} x_0 = (A - \lambda I)^{k+1}((A - \lambda I)x_0) \\ &\implies (A - \lambda I)x_0 \in N_{k+1} = N_k \\ &\implies (A - \lambda I)^{k+1} x_0 = 0 \\ &\implies x_0 \in N_{k+1} = N_k, \end{aligned}$$

so $N_{k+2} \subseteq N_k$. Hence $N_{k+2} = N_k$. Now we have $N_{k+1} = N_{k+2}$, and so by the previous argument we get $N_{k+1} = N_{k+3}$, hence $N_{k+3} = N_k$, and so on.

Now suppose there is no k such that $N_k = N_{k+1}$. Then there is a sequence $\{x_n\}$ such that $x_n \in N_n$, $\|x_n\| = 1$, and x_n is orthogonal to N_{n-1}. Since A is compact the sequence $\{Ax_n\}$ must contain a convergent subsequence. But

$$Ax_{n+m} - Ax_n = \lambda x_{n+m} + (Ax_{n+m} - \lambda x_{n+m} - Ax_n)$$

where on the right the first term belongs to N_{n+m} and the second (parenthetical) term belongs to N_{n+m-1}. (To see the latter note that $Ax_n \in N_n$ since

$$(A - \lambda I)^n Ax_n = A(A - \lambda I)^n x_n = 0,$$

and $(A - \lambda I)^{n+m-1}(Ax_{n+m} - \lambda x_{n+m}) = (A - \lambda I)^{n+m}x_{n+m} = 0$.) By orthogonality of these two terms we have

$$\|Ax_{n+m} - Ax_n\|^2 = \|\lambda x_{n+m}\|^2 + \|Ax_{n+m} - \lambda x_{n+m} - Ax_n\|^2 \geq |\lambda|^2.$$

Since $\lambda \neq 0$ we have a contradiction. □

Lemma 3.20.7 *We have $R(A - \lambda I) = H$ if and only if $N(\lambda) = \{0\}$.*

Proof. Let $R(A - \lambda I) = H$ and suppose $N(\lambda) \neq \{0\}$. Take a nonzero $x_0 \in N(\lambda)$. Since $R(A - \lambda I) = H$ we can solve successively the equations in the following infinite system:

$$(A - \lambda I)x_1 = x_0; \quad (A - \lambda I)x_2 = x_1; \quad \cdots \quad (A - \lambda I)x_{n+1} = x_n; \quad \cdots$$

The sequence of solutions $\{x_n\}$ has the property that

$$(A - \lambda I)^n x_n = x_0 \neq 0 \qquad \text{but} \qquad (A - \lambda I)^{n+1}x_n = (A - \lambda I)x_0 = 0.$$

In the terminology of Lemma 3.20.6, these imply that $x_n \notin N_n$ but $x_n \in N_{n+1}$. So there is no finite k such that $N_{k+1} = N_k$, and this contradicts part (iii) of Lemma 3.20.6.

Conversely let $N(\lambda) = \{0\}$. Then $M(\lambda) = H$ hence by equation (3.20.9) we have $R(A^* - \bar{\lambda} I) = H$. By the proof of the converse given above, $N^*(\bar{\lambda}) = \{0\}$ and thus $M^*(\bar{\lambda}) = H$. The proof is completed by reference to (3.20.7). □

We can now establish part (1) of Theorem 3.20.1:

Lemma 3.20.8 *The spectrum of a compact linear operator A consists only of eigenvalues.*

Proof. Suppose λ is not an eigenvalue of A. Then $N(\lambda)$ contains only the zero vector, hence $M(\lambda) = H$ and (3.20.5) applies for all $x \in H$. This means, in conjunction with Theorem 3.11.4, that the operator $(A - \lambda I)^{-1}$ is bounded on the range of $A - \lambda I$, which is H by Lemma 3.20.7. Hence λ is a regular point of the spectrum of A. □

We continue to part (4) of Theorem 3.20.1:

Lemma 3.20.9 *If λ is an eigenvalue of A, then $\bar{\lambda}$ is an eigenvalue of A^*.*

Proof. Suppose λ is an eigenvalue of A but $\bar{\lambda}$ is not an eigenvalue of A^*. Then $N^*(\bar{\lambda}) = \{0\}$ and thus $M^*(\bar{\lambda}) = H$. By equation (3.20.7) we have

$R(A - \lambda I) = H$ hence $N(\lambda) = \{0\}$ by Lemma 3.20.7. This is impossible since an eigenvalue must correspond to at least one eigenvector. □

Finally, part (5) of Theorem 3.20.1 is established as

Lemma 3.20.10 *The spaces $N(\lambda)$ and $N^*(\bar{\lambda})$ have the same dimension.*

Proof. Let the dimensions of $N(\lambda)$ and $N^*(\bar{\lambda})$ be n and m, respectively, and suppose that $n < m$. Choose orthonormal bases $\{x_1, \ldots, x_n\}$ and $\{y_1, \ldots, y_m\}$ of $N(\lambda)$ and $N^*(\bar{\lambda})$, respectively. Let us introduce an auxiliary operator Q by

$$Qx = (A - \lambda I)x + \sum_{k=1}^{n} (x, x_k) y_k \equiv (C - \lambda I)x,$$

where C is a compact linear operator as the sum of the compact operator A and a finite dimensional operator.

First we show that the null space of Q cannot contain nonzero elements. Indeed if $Qx_0 = 0$ then

$$(A - \lambda I)x_0 + \sum_{k=1}^{n} (x_0, x_k) y_k = 0.$$

Because $R(A - \lambda I) = M^*(\bar{\lambda})$ and $M^*(\bar{\lambda})$ is orthogonal to $N^*(\bar{\lambda})$, the terms $(A - \lambda I)x_0 \in M^*(\bar{\lambda})$ and $\sum_{k=1}^{n} (x_0, x_k) y_k \in N^*(\bar{\lambda})$ must separately equal zero; furthermore, since $\{y_k\}$ is a basis we have $(x_0, x_k) = 0$ for $k = 1, \ldots, n$. From $(A - \lambda I)x_0 = 0$ it follows that $x_0 \in N(\lambda)$; because x_0 is orthogonal to all basis elements of $N(\lambda)$, we have $x_0 = 0$.

By Lemma 3.20.7 we have $R(Q) = H$ and thus the equation $Qx = y_{n+1}$ has a solution x_0. But

$$\begin{aligned}
1 &= (y_{n+1}, y_{n+1}) \\
&= (y_{n+1}, Qx_0) \\
&= (y_{n+1}, (A - \lambda I)x_0) + \left(y_{n+1}, \sum_{k=1}^{n} (x_0, x_k) y_k \right) \\
&= ((A^* - \bar{\lambda} I) y_{n+1}, x_0) \\
&= 0,
\end{aligned}$$

a contradiction. Hence $n \geq m$. But A is adjoint to A^* and by the proof above we have $m \geq n$, so $m = n$. □

The proof of Theorem 3.20.1 is now complete.

3.21 Exercises

3.1 Show that a set in a metric space is closed if and only if it contains the limits of all its convergent sequences. That is, S is closed in X iff for any sequence $\{x_n\} \subset S$ such that $x_n \to x$ in X, we have $x \in S$.

3.2 Show that the following sets are closed in any metric space X: (a) any closed ball, (b) the empty set $\emptyset$, (c) X itself, (d) the intersection of any number of closed sets, (e) the union of any finite number of closed sets.

3.3 Suppose a complete metric space X contains a sequence of closed balls $\{B(x_n, r_n)\}_{n=1}^{\infty}$ such that $B(x_{n+1}, r_{n+1}) \subseteq B(x_n, r_n)$ for each n, and such that the radii $r_n \to 0$. Show that there is a unique point $x \in X$ such that $x \in \bigcap_{n=1}^{\infty} B(x_n, r_n)$.

3.4 Verify that if U is a closed linear subspace of a normed space X, then X/U is a normed linear space under the norm $\|\cdot\|_{X/U}$ given by

$$\|x + U\|_{X/U} = \inf_{u \in U} \|x + u\|_X .$$

Prove that if U is a closed subspace of a Banach space X, then X/U is also a Banach space.

3.5 Let M be a closed subspace of a separable normed space X. Show that X/M is separable.

3.6 Let A be a continuous linear operator from X to Y, where X and Y are Banach spaces. Let M be a closed subspace of X that lies within the kernel of A (i.e., if $x \in M$ then $Ax = 0$). Show that A induces an operator from X/M to Y that is also continuous.

3.7 Let A be a compact linear operator acting in a Banach space X, and let M be a closed subspace of X that lies within the kernel of A. Demonstrate that A induces a compact linear operator from X/M to X.

3.8 (a) Show that ℓ^2 is not finite dimensional. (b) The space ℓ^∞ of *uniformly bounded sequences* is the set of all $\mathbf{x}$ having $\|\mathbf{x}\|_\infty < \infty$ where

$$\|\mathbf{x}\|_\infty = \sup_{k \geq 1} |x_k|.$$

Show that we may regard $\|\cdot\|_\infty$ as a limiting case of $\|\cdot\|_p$ as $p \to \infty$. (c) Show that if $p \leq q$, then $\|\mathbf{x}\|_q \leq \|\mathbf{x}\|_p$. Note that this represents an imbedding theorem. (d) Show that $\ell^1 \subseteq \ell^p \subseteq \ell^q \subseteq \ell^\infty$ whenever $q \geq p \geq 1$. (e) Extend this string of inclusions to

$$\ell^1 \subseteq \ell^p \subseteq \ell^q \subseteq c_0 \subseteq c \subseteq \ell^\infty, \qquad 1 \leq p \leq q.$$

(f) Prove that for any $p \in [1, \infty]$, the normed space ℓ^p is a Banach space. (g) Show that the spaces ℓ^p, $1 \leq p < \infty$, are separable. (h) Show that ℓ^∞ is not separable. (i) Show that c_0 is separable.

3.9 The distance function $d(x,y) = |x^3 - y^3|$ is imposed on the set of all real numbers $\mathbb{R}$ to form a metric space. Verify the metric axioms for $d(x,y)$. Show that the resulting space is complete.

3.10 Show that if A is a bounded linear operator then $\|A\|$ is given by the following alternative expressions:

$$\|A\| = \sup_{\|x\|=1} \|Ax\| = \sup_{\|x\|\neq 0} \frac{\|Ax\|}{\|x\|}.$$

Note that we also have

$$\|A\| = \sup_{\|x\|\leq 1} \|Ax\| = \sup_{\|x\|<1} \|Ax\|.$$

3.11 Prove that a system of vectors in a Hilbert space is linearly independent if and only if its Gram determinant does not vanish.

3.12 Show that convergence $\|A_n - A\| \to 0$ in operator norm, that is in $L(X,Y)$ where X is normed and Y is a Banach space, implies uniform convergence $A_n x \to Ax$ on any bounded subset $S \subset D(A)$.

3.13 Let $\{g_n\}$ be an orthonormal sequence in a Hilbert space H, and let $\{c_n\} \in \ell^2$. Show that the series $\sum_{n=0}^{\infty} c_n g_n$ converges in H.

3.14 Derive the differentiation formula

$$\frac{d}{dt}(u(t), v(t)) = \left(\frac{du(t)}{dt}, v(t)\right) + \left(u(t), \frac{dv(t)}{dt}\right).$$

3.15 Show that if $\{x_n\}$ converges weakly to x in a Hilbert space H, then

$$\|x\| \leq \liminf_{n\to\infty} \|x_n\|.$$

3.16 An operator A from a normed space V to a normed space W is said to be *densely defined* if $D(A)$ is dense in V. Assume W is a Banach space, and show that if A is bounded, linear, and densely defined, then A has a unique bounded linear extension to V. Also show that $\|A_e\| = \|A\|$ where A_e is the extension of A.

3.17 Show that in a finite dimensional space weak convergence implies strong convergence.

3.18 Suppose that A and its inverse are both bounded linear operators defined on a normed space X. The *condition number* of A is defined by $\text{cond}(A) = \|A\|\,\|A^{-1}\|$. (a) Show that $\text{cond}(A) \geq 1$. (b) Consider the operator equation $Ax = y$. Given y, let $\hat{x}$ be an approximate solution; denote the "error" by $\varepsilon = x - \hat{x}$ and the "discrepancy" by $r = y - A\hat{x}$. Show that

$$\frac{1}{\text{cond}(A)} \frac{\|r\|}{\|y\|} \leq \frac{\|\varepsilon\|}{\|x\|} \leq \text{cond}(A) \frac{\|r\|}{\|y\|}.$$

3.19 Let T from X to X be a compact operator on an infinite dimensional normed space X. Show that if T has an inverse defined on all of X, then this inverse cannot be bounded.

3.20 (a) Show that every metric space isometry is continuous and one-to-one. (b) Prove that a linear operator $A\colon X \to Y$ between normed spaces is an isometry if and only if $\|Ax\| = \|x\|$ for all $x \in X$. (Notes: (1) We have $\|A\| = 1$ if $X \neq \{0\}$. (2) If A is also an isomorphism between the linear spaces X and Y, then A is called an *isometric isomorphism.*)

3.21 Let $\{g_k\}$ be an orthonormal system in a Hilbert space H. Show that if Parseval's equality

$$\sum_{k=1}^{\infty} |(f, g_k)|^2 = \|f\|^2$$

holds for all $f \in H$, then $\{g_k\}$ is a basis of H.

3.22 Show that the operator d/dx is bounded from $C^{(1)}(-\infty, \infty)$ to $C(-\infty, \infty)$.

3.23 Show that the set of all functions $f(x)$ bounded on $[0, 1]$ and equipped with the norm

$$\|f(x)\| = \sup_{x \in [0,1]} |f(x)|$$

is not separable.

3.24 Show that if X is a normed space and Y is a Banach space then $L(X, Y)$ is a Banach space.

3.25 Assume that X and Y are Banach spaces, $A \in L(X, Y)$ is continuously invertible, and $B \in L(X, Y)$ is such that $\|B\| < \|A^{-1}\|^{-1}$. Then $A + B$ has an inverse $(A + B)^{-1} \in L(Y, X)$ and

$$\|(A + B)^{-1}\| \leq (\|A^{-1}\|^{-1} - \|B\|)^{-1}.$$

3.26 Verify the condition stated for equality to hold in (3.9.1).

3.27 A subset S of a normed space X is said to be *open* if its complement $X \setminus S$ is a closed set. (a) Show that S is open if and only if every point of S is the center of an open ball contained entirely within S. Hence this statement is an equivalent definition of an open set. (b) Show that any open ball is an open set. (c) Show that an operator $f\colon X \to Y$ is continuous if and only if the inverse image of every open set in Y is open in X.

3.28 Give an example of a function that is discontinuous everywhere on its domain of definition.

3.29 (a) Show that if the condition

$$\left(\int_0^1 \int_0^1 |k(s,t)|^2 \, ds \, dt\right)^{1/2} < \infty$$

holds, then the Fredholm integral operator A defined by

$$Au = \int_0^1 k(s,t)u(t)\,dt$$

is a continuous operator on $L^2(0,1)$. (b) Calculate the norm of the *forward shift operator* S on ℓ^2, defined by

$$S\mathbf{x} = S(x_1, x_2, x_3, \dots) = (0, x_1, x_2, \dots).$$

3.30 Consider the operator

$$(Ax)(t) = \int_0^t x^2(s)\,ds$$

acting in $C(0,1)$. Find a closed ball, centered at the origin, on which A is a contraction.

3.31 Consider the subspace S of ℓ^∞ that consists of all sequences $\mathbf{x} = (\xi_i)$ having at most *finite* numbers of nonzero components. Show that S is *not* a Banach space.

3.32 Let A be a bounded linear operator on a Banach space X. Show that if $\|A\| < 1$ then

$$\|(A - I)^{-1}\| \le \frac{1}{1 - \|A\|}.$$

3.33 Show that if X and Y are Banach spaces, then so is the product space $X \times Y$ under the norm

$$\|(x,y)\| = \max\{\|x\|_X, \|y\|_Y\}.$$

3.34 Show that if $x_n \to x$ then $y_n \equiv \frac{1}{n}\sum_{i=1}^n x_i \to x$.

3.35 We have observed that equivalent norms have the same convergence properties. Prove the converse of this statement.

3.36 Show that if $\{x_n\}$ is a Cauchy sequence in a normed space, then the sequence of norms $\{\|x_n\|\}$ converges. (Note that this implies that every Cauchy sequence is bounded.)

3.37 Show that if a metric space X has a dense subspace that is separable, then X is also separable.

3.38 Show that a normed space is complete if and only if every absolutely convergent series converges to an element of the space.

3.39 Show that the operator A acting in ℓ^2 given by

$$A\mathbf{x} = (2^{-1}x_1, 2^{-2}x_2, 2^{-3}x_3, \cdots)$$

is compact.

3.40 Show that the number $\lambda = 0$ belongs to the residual spectrum of the forward-shift operator

$$A\mathbf{x} = A(x_1, x_2, \ldots) = (0, x_1, x_2, \ldots)$$

defined on ℓ^2.

3.41 A sequence of infinite dimensional vectors $\{\mathbf{x}_k\}$ is defined as follows:

$$\mathbf{x}_k = (\underbrace{1, \ldots, 1}_{\substack{\text{first } k \\ \text{positions}}}, 0, 0, \ldots), \qquad k = 1, 2, 3, \ldots.$$

Show that $\{\mathbf{x}_k\}$ is not weakly convergent in ℓ^2.

3.42 Prove that the sequence $\{\sin kx\}$ is weakly convergent to zero in $L^2(0, \pi)$. Then show that it contains no weakly convergent subsequence (and therefore is not weakly compact) in $W^{1,2}(0, \pi)$.

3.43 Use the Hölder inequality to place a bound on the norm of the imbedding operator from $L^p(\Omega)$ into $L^q(\Omega)$, $p \geq q$. Assume Ω is a compact domain in $\mathbb{R}^n$.

3.44 Show that if A is a compact linear operator acting in a Hilbert space H, and $\{x_n\}$ is an orthonormal sequence in H, then $Ax_n \to 0$ as $n \to \infty$.

3.45 Let Ω be a compact set in $\mathbb{R}^n$. Demonstrate that the imbedding

$$C^{(n)}(\Omega) \hookrightarrow C(\Omega)$$

is continuous and compact for $n \geq 1$.

3.46 Suppose a and b are finite. Let P_n be the space consisting of all polynomials on $[a, b]$ having order up to n, supplied with the norm of $C(a, b)$. Describe the space that results when we apply the completion theorem to P_n.

3.47 Show that weak convergence is equivalent to strong convergence in a finite dimensional Hilbert space.

3.48 Use the orthogonal decomposition theorem to show that a closed subspace of a Hilbert space is weakly closed.

3.49 Let S and T be subsets of a metric space. Show that (a) if S is closed and T is open, then $S \setminus T$ is closed, and (b) if S is open and T is closed, then $S \setminus T$ is open.

3.50 Show that if a system is complete in a set S that is dense in a Hilbert space H, then it is complete in H.

3.51 A function f satisfies a Lipschitz condition with constant L if it satisfies the inequality $|f(\mathbf{x}) - f(\mathbf{y})| \leq L|\mathbf{x} - \mathbf{y}|$. Let S be a uniformly bounded collection of functions given on a compact set $\Omega \subset \mathbb{R}^n$ and satisfying a Lipschitz condition on Ω with the same constant L. Show that S is precompact in $C(\Omega)$.

3.52 Let A be a closed linear operator from a normed space X to a normed space Y. Show that A maps compact sets into closed sets.

3.53 Derive inequality (3.10.8).

Chapter 4

Some Applications in Mechanics

In Chapter 1 we studied the tools of the calculus of variations. As a rule, we assumed each corresponding variational problem had a solution. In the same chapter we mentioned the Perron paradox, which demonstrated how careful one should be in using the assumption of existence of some object when studying its properties. Unfortunately a study of the problems of the calculus of variations from the viewpoint of solvability is difficult, even for those problems that seem to be well posed. For example, in nonlinear elasticity for bodies under dead external load, the existence of a minimizer of total potential energy in general is not shown. Fortunately there is a class of variational problems that corresponds to linear boundary value problems for which the problem of existence is solved completely. We shall use mechanical terminology for these problems; in fact, however, some are quite general and can describe objects from fields such as electrodynamics and biology.

4.1 Some Problems of Mechanics from the Viewpoint of the Calculus of Variations; the Virtual Work Principle

We have considered the problem of equilibrium of a membrane as a problem of the calculus of variations. Historically, the membrane was first investigated through the formulation of Poisson's equation

$$-\Delta u(x, y) = f(x, y) \tag{4.1.1}$$

on a 2-D bounded domain Ω. If the edge $\partial\Omega$ of a membrane is fixed (Figure 3.3) in a form described by a given function $a(s)$, then the boundary condition

$$u\big|_{\partial\Omega} = a(s) \tag{4.1.2}$$

can be used to supplement Poisson's equation and formulate a boundary value problem. Using this, we can derive the functional of total potential energy whose minimum points are given by (4.1.1)–(4.1.2). Let us examine one way in which this can be accomplished. First we take a test function that is infinitely differentiable on Ω and zero in some neighborhood of $\partial\Omega$. In what follows we shall consider only simple domains (including most that would be encountered in applications). Let the domain be bounded and possess a piecewise smooth boundary. We shall denote the above set of test functions by $\mathcal{D}$. Let us multiply both sides of (4.1.1) by $\varphi(x, y) \in \mathcal{D}$ and integrate this over Ω:

$$-\int_\Omega \Delta u(x,y)\varphi(x,y)\,dx\,dy = \int_\Omega f(x,y)\varphi(x,y)\,dx\,dy. \tag{4.1.3}$$

Integration by parts on the left gives

$$\int_\Omega \left(\frac{\partial u}{\partial x}\frac{\partial \varphi}{\partial x} + \frac{\partial u}{\partial y}\frac{\partial \varphi}{\partial y}\right) dx\,dy = \int_\Omega f(x,y)\varphi(x,y)\,dx\,dy \tag{4.1.4}$$

since $\varphi(x, y) \equiv 0$ for (x, y) on $\partial\Omega$. If we wish to consider $\varphi(x, y)$ in (4.1.4) as a variation of the solution $u(x, y)$, then it is easily seen that the integral on the left is the first variation of the integral

$$\frac{1}{2}\int_\Omega \left(\left(\frac{\partial u}{\partial x}\right)^2 + \left(\frac{\partial u}{\partial y}\right)^2\right) dx\,dy;$$

we met this integral in Chapter 3 when introducing energy spaces, and called it the internal energy of the membrane due to deformation. The integral on the right in (4.1.4) is linear in φ and thus can be considered as the first variation of the functional

$$\int_\Omega f(x,y)u(x,y)\,dx\,dy,$$

which is the work of external forces on the displacement field $u(x, y)$. Thus we can regard (4.1.4) as a statement of the fact that the first variation of the functional

$$\frac{1}{2}\int_\Omega \left(\left(\frac{\partial u}{\partial x}\right)^2 + \left(\frac{\partial u}{\partial y}\right)^2\right) dx\,dy - \int_\Omega f(x,y)u(x,y)\,dx\,dy$$

is zero. We have met this functional in Chapter 1; it is the expression for the total energy of the membrane: that is, the sum of the internal energy and the potential energy due to the work of external forces. We have shown that an extremal of this functional describes the equilibrium state of the membrane. Lagrange's theorem of classical mechanics states that when the total potential energy of a system of particles takes the minimal value it corresponds to a stable state of equilibrium of the system. Of course, the membrane does not obey the theorems of classical mechanics: it is an object of a different nature. However, Lagrange's theorem is extended to this case.

We have considered a membrane with a clamped edge. We may also consider other boundary conditions known in membrane theory, for example the Neumann condition

$$\left.\frac{\partial u}{\partial n}\right|_{\partial\Omega} = g(s). \tag{4.1.5}$$

On the left we have the derivative in the outward normal direction, and on the right we have a given function. Let us repeat the steps leading to (4.1.4). This time, however, we need not take $\varphi \in \mathcal{D}$; we suppose only that it is sufficiently smooth. Equation (4.1.3) is valid now, but integration by parts on the left, by Green's formulae, brings us an additional term:

$$\int_\Omega \left(\frac{\partial u}{\partial x}\frac{\partial \varphi}{\partial x} + \frac{\partial u}{\partial y}\frac{\partial \varphi}{\partial y}\right) dx\,dy - \int_{\partial\Omega} \frac{\partial u}{\partial n}\varphi(s)\,ds = \int_\Omega f(x,y)\varphi(x,y)\,dx\,dy.$$

By (4.1.5) we have

$$\int_\Omega \left(\frac{\partial u}{\partial x}\frac{\partial \varphi}{\partial x} + \frac{\partial u}{\partial y}\frac{\partial \varphi}{\partial y}\right) dx\,dy = \int_\Omega f(x,y)\varphi(x,y)\,dx\,dy + \int_{\partial\Omega} g(s)\varphi(s)\,ds. \tag{4.1.6}$$

Here we write $\varphi(s)$ to denote the values of $\varphi(x,y)$ on $\partial\Omega$. The last integral on the right side of (4.1.6) looks like the work of the force $g(s)$ acting through a displacement φ on the edge of the membrane, so the meaning of the Neumann condition is that we define a force distribution $g(s)$ acting on the edge.

Note that in this problem statement we neglect inertia; we think of the membrane as a body having zero mass. If external forces that are not self-balanced act on a body free from geometrical restrictions, then mechanics states that the problem of equilibrium cannot be solvable: the body should be moved as a whole and, having zero mass, it should have infinite acceleration. So the self-balance condition is necessary for such problems. For this case we have found that the only kind of free motion as a whole is $u(x) = c$: only for such displacements is the inner energy constant (because of linearity we can put $u(x, y) = 1$). This means that on this displacement the work of external forces must be zero:

$$\int_{\Omega} f(x, y)\, dx\, dy + \int_{\partial\Omega} g(s)\, ds = 0.$$

If we restrict the external forces to those acting on the edge of the membrane so that $f(x, y) = 0$, we have

$$\int_{\partial\Omega} g(s)\, ds = 0,$$

and this is the well-known solvability condition for the Neumann problem. It has a clear mechanical sense: the external forces must be self-balanced.

Note that in classical mechanics the self-balance condition consists of six equations: the three projections each of the resultant force and moment onto the frame axes are all zero. The membrane model is quite approximate, hence does not satisfy all the conditions. This is typical of the approximate models of continuum mechanics. For linear elasticity the self-balance appears exactly as it does in classical mechanics.

We would like to mention that (4.1.6) and similar equations play an important role in what follows and in mechanics as a whole. It can be treated as the formulation of the virtual work principle. The term with minus on the left and the term on the right can be called the work of internal and external forces, respectively. With this interpretation the equation states a fundamental physical law:

On any admissible displacements the sum of the work of internal and external forces of the system in equilibrium is equal to zero.

In this particular case the equation can be obtained as the first variation of the total potential energy functional and so we can begin with formulation of the principle of minimum potential energy. There are body-force systems where the potential of external forces does not exist, and so we cannot use the same principle — however, the virtual work principle remains valid. Continuum mechanics treats the virtual work principle as independent, and relates it to the variational principles of mechanics. Thus the variational part of mechanics contains not only the problems of minimum of some functionals, but also the theory of all the equations that, like (4.1.6), contain some admissible fields of displacements, strains, or stresses. From the viewpoint of the classical calculus of variations, some part of continuum mechanics that is called "the variational problems of mechanics" is not a part of the calculus of variations. In mechanics, they consider as variational anything that involves integro-differential equations containing some virtual variables and from which, using the main lemma of the calculus of variations, it is possible to derive relations such as equilibrium equations or constitutional equations for the material.

Finally, let us note that in deriving (4.1.6) we used a set of smooth admissible variations φ of a solution; we do so even if we try to find a solution with some singularities. If we begin with the principle of minimum potential energy, it is reasonable to consider all the functions for which the terms of the principle make sense; moreover, there is no reason why admissible variations should be smoother than the solution. Later this remark will bring us to the generalized setup of boundary value problems of mechanics.

Similarly, for many problems involving elastic objects (strings, beams, shells, 2-D and 3-D elastic bodies, etc.) we can derive a total potential energy functional whose first variation yields the equilibrium equations for the object. It has the structure $\mathcal{E}-V$ where $\mathcal{E}$ is the inner potential energy of elastic deformation and V is the work of external forces.[1] The condition of minimum of the energy functional gives us the equality of the first variation of the functional to zero on all the admissible variations of corresponding solutions. These integral equations, the equality of the sum of the work of internal and external forces on admissible variations to zero, also express the virtual work principle for corresponding problems, and it is of a more general nature than the principle of minimum potential energy.

We write out corresponding relations (the total potential energy $\mathcal{E} - V$ and the equation of the virtual work principle) for the following objects:

1. *Stretched rod (Figure 3.1):*

$$\frac{1}{2}\int_0^l ES(x)u'^2(x)\,dx - \int_0^l f(x)u(x)\,dx - Fu(l), \tag{4.1.7}$$

$$\int_0^l ES(x)u'(x)v'(x)\,dx = \int_0^l f(x)v(x)\,dx + Fv(l), \tag{4.1.8}$$

[1] In the case of potential forces V is the potential of the forces and, by analogy with elementary physics terminology for gravitational forces, the expression $-V$ can be called the potential energy of the force field.

where $f(x)$ is a force tangential to the rod axis, F is a stretching force at the free end of the rod, and u is the tangential displacement of points of the neutral axis of the rod.

2. *Bent beam (Figure 3.2):*

$$\frac{1}{2}\int_0^l EI(x)w''^2(x)\,dx - \int_0^l f(x)w(x)\,dx - Fw(l), \tag{4.1.9}$$

$$\int_0^l EI(x)w''(x)v''(x)\,dx = \int_0^l f(x)v(x)\,dx + Fv(l), \tag{4.1.10}$$

where w is the transverse displacement of the neutral axis of the beam, $f(x)$ is the transverse distributed force, and F is the transverse force on the end.

3. *Plate (Figure 3.4):*

$$\frac{D}{2}\int_\Omega \left(w_{xx}^2 + w_{yy}^2 + 2\nu w_{xx}w_{yy} + 2(1-\nu)w_{xy}^2\right)\,d\Omega - \int_\Omega Fw\,d\Omega,$$

$$D\int_\Omega \left(w_{xx}v_{xx} + w_{yy}v_{yy} + \nu\left(w_{xx}v_{yy} + w_{yy}v_{xx}\right) + 2(1-\nu)w_{xy}v_{xy}\right)\,d\Omega$$
$$= \int_\Omega Fv\,d\Omega,$$

where D is the plate rigidity, ν is Poisson's ratio, and $w = w(x,y)$ is the deflection at point (x,y) of the domain S occupied by the mid-surface.

4. *3-D linearly elastic body:*

$$\frac{1}{2}\int_V c^{ijkl}e_{kl}(\mathbf{u})e_{ij}(\mathbf{u})\,dV - \int_V \mathbf{F}\cdot\mathbf{u}\,dV - \int_{\partial V_1} \mathbf{f}\cdot\mathbf{u}\,dS, \tag{4.1.11}$$

$$\int_V c^{ijkl}e_{kl}(\mathbf{u})e_{ij}(\mathbf{v})\,dV = \int_V \mathbf{F}\cdot\mathbf{v}\,dV + \int_{\partial V_1} \mathbf{f}\cdot\mathbf{v}\,dS, \tag{4.1.12}$$

where $\mathbf{F}$ are volume external forces and $\mathbf{f}$ are forces on some part of the boundary ∂V_1.

We will use the above formulas to introduce the notion of generalized solution for all these mechanics problems.

4.2 Equilibrium Problem for a Clamped Membrane and its Generalized Solution

As we said, the equilibrium of a membrane with fixed edge can be formulated as the problem of minimum of the functional

$$E_M(u) = \frac{1}{2}\int_\Omega \left(\left(\frac{\partial u}{\partial x}\right)^2 + \left(\frac{\partial u}{\partial y}\right)^2\right) dx\,dy - \int_\Omega f(x,y)u(x,y)\,dx\,dy. \tag{4.2.1}$$

We treat the case for which

$$u\big|_{\partial\Omega} = 0. \tag{4.2.2}$$

Let us consider (4.2.1) in the energy space. We have introduced (§ 3.10) the space $\mathcal{E}_{Mc}$ for a membrane, which is a Hilbert space with an inner product

$$(u,v)_M = \int_\Omega \left(\frac{\partial u}{\partial x}\frac{\partial v}{\partial x} + \frac{\partial u}{\partial y}\frac{\partial v}{\partial y}\right) dx\,dy. \tag{4.2.3}$$

It is clear that the first term in (4.2.1) can be presented as $\frac{1}{2}(u,u)_m = \frac{1}{2}\|u\|_m^2$. The second term

$$\Phi(u) = \int_\Omega f(x,y)u(x,y)\,dx\,dy$$

is a linear functional in u. Let us suppose that $f \in L^p(\Omega)$ for some $p > 1$. By Hölder's inequality we have

$$\begin{aligned}|\Phi(u)| &= \left|\int_\Omega f(x,y)u(x,y)\,dx\,dy\right| \\ &\le \left(\int_\Omega |f(x,y)|^p\,dx\,dy\right)^{1/p} \left(\int_\Omega |u(x,y)|^q\,dx\,dy\right)^{1/q}\end{aligned}$$

with $q = p/(p-1)$. On the energy space by equivalence of the energy norm to the norm of $W^{1,2}(\Omega)$ and Theorem 3.7.3 we have

$$\left(\int_\Omega |u(x,y)|^q\,dx\,dy\right)^{1/q} \le m\,\|u\|_M$$

so

$$|\Phi(u)| \le m\left(\int_\Omega |f(x,y)|^p\,dx\,dy\right)^{1/p} \|u\|_M = m_1\,\|u\|_M\,.$$

Hence $\Phi(u)$ is a linear continuous functional. By the Riesz representation theorem there is a unique $u_0 \in \mathcal{E}_{Mc}$ such that

$$\Phi(u) = (u, u_0)_M. \tag{4.2.4}$$

Thus the energy functional for a membrane with clamped edge can be represented in the energy space as

$$E_M(u) = \frac{1}{2}\|u\|_M^2 - (u, u_0)_M. \tag{4.2.5}$$

Let us consider the problem of minimization of $E_M(u)$ in $\mathcal{E}_{Mc}$.

Theorem 4.2.1 *In the energy space $\mathcal{E}_{Mc}$ the functional $E_M(u)$ attains its minimum at $u = u_0$, and the minimizer is unique.*

Proof. Let us transform the expression for $2E_M(u)$:

$$\begin{aligned} 2E_M(u) &= \|u\|_M^2 - 2(u, u_0)_M \\ &= (u, u)_M - 2(u, u_0)_M + (u_0, u_0)_M - (u_0, u_0)_M \\ &= (u - u_0, u - u_0)_M - (u_0, u_0)_M \\ &= \|u - u_0\|_M^2 - \|u_0\|_M^2, \end{aligned}$$

so

$$\min E_M(u) = -\frac{1}{2}\|u_0\|_M^2.$$

Uniqueness of the minimizer u_0 is evident. □

Let us return to (4.2.4). The equation to find the minimizer of the functional is the same equality to zero of the first variation of the functional $E_M(u)$:

$$\int_\Omega \left(\frac{\partial u_0}{\partial x}\frac{\partial v}{\partial x} + \frac{\partial u_0}{\partial y}\frac{\partial v}{\partial y} \right) dx\,dy = \int_\Omega f(x,y)v(x,y)\,dx\,dy. \tag{4.2.6}$$

It is usually said that (4.2.6) defines a generalized solution $u_0 \in \mathcal{E}_{Mc}$ to the Poisson equation $\Delta u = -f$ with boundary condition (4.2.2) if u_0 satisfies (4.2.6) for any $v \in \mathcal{E}_{Mc}$. This is often called the energy (or weak) solution. Finally, let us note that (4.2.6) expresses the virtual work principle for a membrane with clamped edge.

4.3 Equilibrium of a Free Membrane

For the Neumann problem, the equation of equilibrium (the virtual work principle) is (see (4.1.6))

$$\int_\Omega \left(\frac{\partial u}{\partial x}\frac{\partial \varphi}{\partial x} + \frac{\partial u}{\partial y}\frac{\partial \varphi}{\partial y} \right) dx\,dy = \int_\Omega f(x,y)\varphi(x,y)\,dx\,dy + \int_{\partial\Omega} g(s)\varphi(s)\,ds. \tag{4.3.1}$$

The corresponding total energy functional is evidently

$$E_{M1}(u) = \frac{1}{2}\int_\Omega \left(\left(\frac{\partial u}{\partial x}\right)^2 + \left(\frac{\partial u}{\partial y}\right)^2 \right) dx\,dy \\ - \int_\Omega f(x,y)u(x,y)\,dx\,dy - \int_{\partial\Omega} g(s)u(s)\,ds. \tag{4.3.2}$$

Equation (4.3.1) is then the equality of the first variation of $E_{M1}(u)$ to zero, as follows from general considerations of the calculus of variations. Again, we put the problem of equilibrium of a membrane with given forces $g(s)$ on the edge as a problem of minimum of the energy functional $E_{M1}(u)$ on an energy space. Here we have the option to use a factor space $\mathcal{E}_{Mf}$ (see § 3.10), or its isometric variant where we take the balanced elements satisfying the condition

$$\int_\Omega u(x,y)\,dx\,dy = 0. \tag{4.3.3}$$

On the latter the problem of minimum of the energy is well defined if we require that

$$f(x,y) \in L^{p_1}(\Omega), \qquad g(s) \in L^{p_2}(\partial\Omega), \tag{4.3.4}$$

for some $p_1, p_2 > 1$. But on the factor space the energy functional is not well defined if the forces are not self-balanced with

$$\int_\Omega f(x,y)\,dx\,dy + \int_{\partial\Omega} g(s)\,ds = 0. \tag{4.3.5}$$

If (4.3.5) is not fulfilled, then for different representatives of zero, $u(x,y) = c$, the energy functional $E_{M1}(u)$ takes different values, which is impossible when we seek the minimum of the energy functional. This is a consequence of the fact that in this model we neglect the inertia properties of the membrane. Thus considering the problem of equilibrium on the factor space $E_{M1}(u)$ we get an additional necessary condition (4.3.5) that we have called the condition of self-balance of external forces. This condition does not arise when we adopt the second variant of the energy space, because (4.3.3) is an artificial geometric constraint that was absent from the initial problem statement and has been imposed as an auxiliary restriction. Although we do not need (4.3.5) when we consider the problem in this way, we should nonetheless carry it along since it is required by the initial setup.

Under the restriction (4.3.5) on the forces, we can consider the problem of equilibrium of a free membrane as the problem of minimum of (4.3.2) on the space $\mathcal{E}_{Mf}$ of "usual" functions satisfying (4.3.3). Condition (4.3.4) is sufficient for $E_{M1}(u)$ to be well defined on $\mathcal{E}_{Mf}$. Indeed, we need to demonstrate only that the functional of the work of external forces is well defined in this space. Applying Hölder's inequality we have

$$\begin{aligned}
|\Phi(u)| &= \left| \int_\Omega f(x,y)u(x,y)\,dx\,dy + \int_{\partial\Omega} g(s)u(s)\,ds \right| \\
&\le \left(\int_\Omega |f(x,y)|^{p_1}\,d\Omega \right)^{1/p_1} \left(\int_\Omega |u(x,y)|^{q_1}\,d\Omega \right)^{1/q_1} \\
&\quad + \left(\int_{\partial\Omega} |g(s)|^{p_2}\,ds \right)^{1/p_2} \left(\int_{\partial\Omega} |u(s)|^{q_2}\,d\Omega \right)^{1/q_2} \\
&\le m \left(\|f\|_{L^{p_1}(\Omega)} + \|g\|_{L^{p_2}(\partial\Omega)} \right) \|u\|_M
\end{aligned} \tag{4.3.6}$$

where $\frac{1}{p_1} + \frac{1}{q_1} = 1$, $\frac{1}{p_2} + \frac{1}{q_2} = 1$, and the norm $\|\cdot\|_M$ is defined by (4.2.3). In the last transformation we used Sobolev's imbedding Theorem 3.7.3. Thus $\Phi(u)$ is well defined on $\mathcal{E}_{Mf}$. Linearity of this functional in u is evident, and (4.3.6) guarantees continuity. Hence by the Riesz representation theorem

$$\Phi(u) = (u, u_0)_M$$

where $u_0 \in \mathcal{E}_{Mf}$ is uniquely defined by the external forces f, g. Hence the problem of minimum of $E_{M1}(u)$ can be reformulated as the problem of minimum of

$$E_{M1}(u) = \frac{1}{2} \|u\|_M^2 - (u, u_0)_M. \tag{4.3.7}$$

There is formally no difference between the functionals (4.3.7) and (4.2.5), so we can simply reformulate the results of § 4.2 for this problem as

Theorem 4.3.1 *Let (4.3.4) and (4.3.5) be valid. In the energy space $\mathcal{E}_{Mf}$ the functional $E_{M1}(u)$ attains its minimum at $u = u_0$ and the minimizer is unique.*

The minimizer is a generalized solution of the equilibrium problem for a membrane with free edge. We shall see that all the linear problems of equilibrium we consider reduce to the same problem of minimum of a quadratic functional having the same form

$$E(u) = \frac{1}{2} \|u\|^2 - \Phi(u) \tag{4.3.8}$$

where $\Phi(u)$ is a linear continuous functional. The proof of Theorem 4.2.1 does not depend on the nature of the space in which it is done, so we can immediately formulate

Theorem 4.3.2 *Let $\Phi(u)$ be a linear continuous functional acting in a Hilbert space H. Then the problem of minimum of* (4.3.8) *has a unique solution $u_0 \in H$ defined by the Riesz representation theorem: $\Phi(u) = (u, u_0)$.*

Applications of this theorem appear in the next section.

4.4 Some Other Problems of Equilibrium of Linear Mechanics

All the mechanics problems for which we presented the energy functional and the virtual work principle in § 4.1 ((4.1.7)–(4.1.12)) are of the type of (4.3.8) where the linear functional $\Phi(u)$ is the potential of external forces (or, what is now the same thing, the work of external forces) on the displacement field u. Theorem 4.3.2 asserts the generalized solvability of a corresponding boundary value problem and the uniqueness of its generalized solution if $\Phi(u)$ is continuous. Thus we need to determine when $\Phi(u)$ is continuous. For this we shall use Sobolev's imbedding theorem and the fact that the corresponding energy space is a subspace of a Sobolev space $W^{l,2}(\Omega)$. The corresponding theorems are formulated so similar to Theorem 4.3.1 that we leave them to the reader. We show only the restrictions on external forces to provide continuity of the corresponding potential of external forces as a functional in the corresponding energy space.

Stretched rod

See (4.1.7) and (4.1.8). Here $u(0) = 0$ and

$$\Phi(u) = \int_0^l f(x)u(x)\,dx + Fu(l).$$

In this case $u(x)$ is continuous on $[0,l]$ (we recall that this means that each representative Cauchy sequence for an element of an energy space converges to a continuous function) and so if

$$f(x) \in L(0,l)$$

then

$$\begin{aligned}
|\Phi(u)| &= \left| \int_0^l f(x)u(x)\,dx + Fu(l) \right| \\
&\le \left(\int_0^l |f(x)|\,dx + |F| \right) \max_{x\in[0,l]} |u(x)| \\
&\le m \left(\int_0^l |f(x)|\,dx + |F| \right) \|u\|_R
\end{aligned}$$

where

$$\|u\|_R = \left(\int_0^l ES(x) u'^2(x)\,dx \right)^{1/2}.$$

Bent beam

See (4.1.9) and (4.1.10). Now we can consider different variants of boundary conditions. For clamped edges we formulate

$$w(0) = 0 = w'(0), \qquad\qquad w(l) = 0 = w'(l),$$

and now the energy space for a bent beam with the norm

$$\|w\|_B = \left(\int_0^l EI(x) w''^2(x)\,dx \right)^{1/2}$$

is a subspace of $W^{2,2}(0,l)$ in which functions and their derivatives are continuous on $[0,l]$ and the corresponding operator of imbedding into the space of continuously differentiable functions is continuous. By this we can get a sufficient condition for the potential of external forces $\Phi(w) = \int_0^l f(x)w(x)\,dx + Fw(l)$ of the same type as for a stretched rod:

$$f(x) \in L(0,l).$$

The proof is the same as above. However, in this case it is possible to include in the expression of the potential, and thus into the setup, the point external torques and transverse forces that are common in the strength of materials (they are presented with δ-functions). The proof remains practically the same.

As to other variants of boundary conditions for a bent beam, the difference comes when the beam can move as a rigid whole. Then the situation is quite similar to that for a free membrane. A rigid motion of a free beam (i.e., a function w for which $\|w\|_{BB} = 0$) now has the form $w = a + bx$. Different boundary conditions can restrict the constants a and b (above they are zero). If the beam can move as a rigid body, then there arise additional restrictions which are conditions of self-balance of external forces:

$$\int_0^l f(x)(a+bx)\,dx + F(a+bl) = 0$$

for all admissible a, b. In case the only geometrical boundary constraint is $w(0) = 0$, this reduces to

$$\int_0^l xf(x)\,dx + lF = 0.$$

Plate

As above, there are various possible boundary conditions. When the edge of the plate is clamped,

$$w\big|_{\partial\Omega} = 0 = \frac{\partial w}{\partial n}\Big|_{\partial\Omega}.$$

The norm of the corresponding energy space $\mathcal{E}_{Pc}$, which is

$$\|w\|_P = \left(\int_\Omega \left(w_{xx}^2 + w_{yy}^2 + 2\nu w_{xx}w_{yy} + 2(1-\nu)w_{xy}^2\right)\,d\Omega\right)^{1/2}$$

as was shown in Chapter 3, is equivalent to the norm of $W^{2,2}(\Omega)$ when Ω is compact in $\mathbb{R}^2$. In this case $\mathcal{E}_{Pc}$ is imbedded continuously into $C(\Omega)$. Because of this the potential of external forces is a continuous functional in the energy space when there are not only distributed forces, but lumped forces as well:

$$\Phi(w) = \int_\Omega F(x,y)w(x,y)\,d\Omega + \sum_{k=1}^N F_k w(x_k, y_k).$$

Indeed

$$
\begin{aligned}
|\Phi(w)| &= \left| \int_\Omega F(x,y)w(x,y)\,d\Omega + \sum_{k=1}^{N} F_k w(x_k, y_k) \right| \\
&\le \left(\int_\Omega |F(x,y)|\,dx\,dy + \sum_{k=1}^{N} |F_k| \right) \max_{x,y\in\Omega} w(x,y) \\
&\le m \left(\int_\Omega |F(x,y)|\,dx\,dy + \sum_{k=1}^{N} |F_k| \right) \|w\|_P \\
&= m_1 \|w\|_P \,.
\end{aligned}
$$

So the condition

$$F(x,y) \in L(\Omega) \tag{4.4.1}$$

is sufficient for $\Phi(w)$ to be a linear continuous functional, and thus in this case there exists a unique generalized solution to the problem of equilibrium of the plate with clamped edge.

If the edge of a plate is free of constraints of geometrical type, then there appear motions of the plate as a rigid whole that satisfy

$$\|w\|_P = 0.$$

The corresponding rigid motions are

$$w = ax + by + c$$

where a, b, c are constants. As in the theory of the free membrane, the condition of self-balance of the external forces appears:

$$\Phi(ax+by+c) = \int_\Omega F(x,y)(ax+by+c)\,d\Omega + \sum_{k=1}^{N} F_k(ax_k + by_k + c) = 0. \tag{4.4.2}$$

This holds for all a, b, c, so it represents three equations for the external forces that express equality to zero of the resultant force and resultant moments with respect to the coordinate axes (the reader should write them out and see that this is really so). Condition (4.4.2) must be added to (4.4.1) as a necessary condition for solvability of the problem.

If there are some other geometrical constraints on a plate, then the appearance of the self-balance condition depends on whether the constraints leave some freedom to the plate. For example, if it is fixed at three points that are not on the same straight line, then there are no rigid motions.

But rigid motions arise if only some straight segment in the mid-surface of the plate is fixed, since the plate can rotate about this axis and so some condition of self-balance appears.

Elastic body

We showed that when the boundary of the body is clamped then the energy norm

$$\|\mathbf{u}\|_E = \left(\int_V c^{ijkl} e_{kl}(\mathbf{u}) e_{ij}(\mathbf{u})\, dV \right)^{1/2}$$

is equivalent to the norm of the Sobolev space $\left(W^{1,2}(V)\right)^k$ when V is compact in $\mathbb{R}^k$, $k = 2, 3$. In the 2-D case the imbedding result is exactly as for the membrane, and thus a sufficient condition for generalized solvability is that the Cartesian components of the vector of external forces belong to some $L^p(S)$ with $p > 1$. Mathematical physicists prefer "if and only if" conditions for solvability, and have introduced the so-called negative Sobolev spaces. In terms of these the forces are completely characterized; the only trouble is that in a practical sense this condition gives us no more than if we simply say "the corresponding functional must be continuous in the space", so sufficient conditions are preferable in practice.

For a 3-D elastic body, the imbedding of $W^{1,2}(V)$, when V is compact, is a continuous operator to $L^p(V)$, $1 \le p \le 6$, and to $L^q(S)$, $1 \le q \le 4$, where S is a piecewise smooth surface in Ω. In this case, conditions sufficient for generalized solvability of the problem of equilibrium of a body with clamped boundary are

$$\mathbf{F} \in (L^p(V))^3, \quad p \ge 6/5, \qquad\qquad \mathbf{f} \in (L^q(\partial V))^3, \quad q \ge 4/3.$$

Indeed

$$\begin{aligned}
|\Phi(\mathbf{u})| &= \left| \int_V \mathbf{F} \cdot \mathbf{u}\, dV + \int_{\partial V} \mathbf{f} \cdot \mathbf{u}\, dS \right| \\
&\le \left(\int_V |\mathbf{F}|^{6/5}\, dV \right)^{5/6} \left(\int_V |\mathbf{u}|^6\, dV \right)^{1/6} \\
&\quad + \left(\int_{\partial V} |\mathbf{f}|^{4/3}\, dS \right)^{3/4} \left(\int_{\partial V} |\mathbf{u}|^4\, dS \right)^{1/4} \\
&\le m \left(\left(\int_V |\mathbf{F}|^{6/5}\, dV \right)^{5/6} + \left(\int_{\partial V} |\mathbf{f}|^{4/3}\, dS \right)^{3/4} \right) \|\mathbf{u}\|_E
\end{aligned}$$

where we have used Hölder's inequality and the equivalence of the energy and Sobolev norms.

When we consider the equilibrium of a plate that is free of geometrical constraints, there arise motions of the body as a rigid whole:

$$\mathbf{u} = \mathbf{a} + \mathbf{b} \times \mathbf{r}$$

(we recall that these satisfy $\|\mathbf{u}\|_E = 0$), which imply that for a body free of geometrical constraints the forces must be self-balanced with

$$\int_V \mathbf{F} \cdot (\mathbf{a} + \mathbf{b} \times \mathbf{r})\, dV + \int_{\partial V_1} \mathbf{f} \cdot (\mathbf{a} + \mathbf{b} \times \mathbf{r})\, dS = \mathbf{0}.$$

This equation must be fulfilled for all $\mathbf{a}$ and $\mathbf{b}$, giving six equations which are precisely the conditions of self-balance in classical mechanics: the resultant force and the resultant moments are zero.

In the case of mixed boundary conditions, if the body can move somehow as a rigid whole, then we must retain some subset of the conditions of self-balance of the load. If the body can rotate about a fixed point, for example, then the resultant moment with respect to the fixed point must vanish.

Finally we would like to note the following. The 1-D problems and the plate problem allow us to formulate boundary conditions at a point, and the corresponding boundary value problems in their generalized setup are well posed. But in the problems for the membrane or elastic body, point conditions do not "function": elements are determined only in the sense of L^p, and point conditions make no sense (the setup "neglects" such conditions). So such a setup, with a given value of a function at one point, is not sensible in a generalized (energetic) formulation.

It is sensible to note that when the problem involves elastic support of the type of a Winkler foundation or some interaction of elements with different models like a coupled 3-D elastic body with a plate, the variational statement of the problem includes the sum of internal energies of all the elements of the system. It is necessary to add some geometrical conditions of compatibility of fields of displacements between the bodies involved. The norm of the corresponding energy space must contain all the functionals of internal energies of the bodies (quadratic terms that are non-negative) and sometimes the energy space is quite strange from the point of view of the classical theory of Sobolev spaces. For such "coupled" models, we introduce constraints of geometrical nature on how the coupled elements interact explicitly, but not the conditions for stress terms: the stress conditions on the border are derived in a manner similar to the way in which it is done for the natural boundary conditions in the general theory. This prevents crude errors that are quite common for the setup of similar problems, i.e., when someone tries to write out the equations of force balance for the border elements in cases when the models approximate real stresses in different fashions.

Nonhomogeneous geometrical boundary conditions

We have considered homogeneous boundary conditions of the type $u|_{\partial\Omega} = 0$ because they provide linearity of the corresponding energy space. There are two ways of considering

$$u\big|_{\partial\Omega} = a(s) \tag{4.4.3}$$

where $a(s)$ is a given function. One is to consider the problem of minimizing on a closed cone of all elements satisfying (4.4.3); this is the way it is done in variational inequalities, but we shall consider this later. The other is quite traditional for mathematical physics: we assume the existence of an element with some differential properties that satisfies (4.4.3), and then seek a solution as a sum of this element and another element that satisfies homogeneous boundary conditions. We shall demonstrate this for the membrane problem; for the rest it is done in a similar fashion. First we suppose that there is an element $u^*(x,y) \in W^{1,2}(\Omega)$ (as usual we speak about functions with the understanding that such elements are actually due to the procedure of completion of the set of continuously differentiable functions) and seek the minimum point u of the energy functional

$$E_M(u) = \frac{1}{2}\int_\Omega \left(\left(\frac{\partial u}{\partial x}\right)^2 + \left(\frac{\partial u}{\partial y}\right)^2 \right) dx\,dy - \int_\Omega f(x,y)u(x,y)\,dx\,dy$$

in the form

$$u(x,y) = u^*(x,y) + v(x,y)$$

where $v(x,y) \in \mathcal{E}_{Mc}$. That is, in particular, v satisfies the homogeneous boundary condition $v|_{\partial\Omega} = 0$. Redenoting v by u, we get the following variational problem in $\mathcal{E}_{Mc}$:

$$\frac{1}{2}\int_\Omega \left(\left(\frac{\partial(u+u^*)}{\partial x}\right)^2 + \left(\frac{\partial(u+u^*)}{\partial y}\right)^2\right) dx\,dy \\ - \int_\Omega f(x,y)(u(x,y)+u^*(x,y))\,dx\,dy \to \min.$$

The equality of the first variation to zero, the equation we need to solve to find a generalized solution, is

$$\int_\Omega \left(\frac{\partial u}{\partial x}\frac{\partial \varphi}{\partial x} + \frac{\partial u}{\partial y}\frac{\partial \varphi}{\partial y}\right) dx\,dy = \int_\Omega f(x,y)\varphi(x,y)\,dx\,dy \\ - \int_\Omega \left(\frac{\partial u^*}{\partial x}\frac{\partial \varphi}{\partial x} + \frac{\partial u^*}{\partial y}\frac{\partial \varphi}{\partial y}\right) dx\,dy. \quad (4.4.4)$$

A generalized solution of the equilibrium problem of a membrane with given displacement of the edge is an element $u(x,y) \in \mathcal{E}_{Mc}$ that satisfies (4.4.4) for any $\varphi(x,y) \in \mathcal{E}_{Mc}$. The first term on the right is the same as in the equation for the problem with the homogeneous boundary condition. It is seen that the second term on the right is a linear functional in φ, so we can try to apply Theorem 4.3.2. For this we need to demonstrate that it is a bounded functional. Let us show this. Indeed

$$\left|\int_\Omega \left(\frac{\partial u^*}{\partial x}\frac{\partial \varphi}{\partial x} + \frac{\partial u^*}{\partial y}\frac{\partial \varphi}{\partial y}\right) dx\,dy\right| \le \left\|\frac{\partial u^*}{\partial x}\right\|_{L^2(\Omega)} \left\|\frac{\partial \varphi}{\partial x}\right\|_{L^2(\Omega)} \\ + \left\|\frac{\partial u^*}{\partial y}\right\|_{L^2(\Omega)} \left\|\frac{\partial \varphi}{\partial y}\right\|_{L^2(\Omega)} \\ \le m\,\|u^*\|_{W^{1,2}(\Omega)}\,\|\varphi\|_M\,.$$

Thus by Theorem 4.3.2 there is a unique generalized solution to the problem under consideration. The following question remains. Redenote the above homogeneous part of the solution u_1. Suppose we choose another fixed function u^{**} that takes the same boundary values, and find the homogeneous part of the solution denoted u_2. Do we have uniqueness in the sense that $u_1 + u^* = u_2 + u^{**}$? Denote $u_{21} = u_2 - u_1$ and subtract the equation for u_1 from the equation for u_2 with the same admissible variation

φ. We have

$$\int_\Omega \left(\frac{\partial u_{21}}{\partial x} \frac{\partial \varphi}{\partial x} + \frac{\partial u_{21}}{\partial y} \frac{\partial \varphi}{\partial y} \right) dx\, dy$$
$$= \int_\Omega \left(\frac{\partial (u^{**} - u^*)}{\partial x} \frac{\partial \varphi}{\partial x} + \frac{\partial (u^{**} - u^*)}{\partial y} \frac{\partial \varphi}{\partial y} \right) dx\, dy.$$

But the difference $u^{**} - u^*$ belongs to $\mathcal{E}_{Mc}$ (why?), so since φ is an arbitrary element of $\mathcal{E}_{Mc}$ we have $u_{21} = u^{**} - u^*$. This completes the proof.

A big chapter in the theory of Sobolev spaces is concerned with the so-called trace theorems. These deal with the question of which conditions must be stipulated on the boundary values in order to insure the existence of an element of a Sobolev space taking them as boundary conditions. The corresponding theorems require some smoothness of the boundary of the domain, and are not convenient for practical verification; however, they provide "if and only if" conditions for existence of a continuation of the boundary functions as a function inside the domain, in such a way that the corresponding operator of continuation is continuous. Hence there arise Sobolev spaces $W^{l,p}(\Omega)$ with fractional parameters l.

Finally, we would like to note that the study of generalized solutions is usually the first step in the study of smoothness properties of solutions (see Sobolev [Sobolev (1951)]). The birth of functional analysis was signaled when in this way Hilbert justified the Dirichlet principle (i.e., the same principle of minimum of potential energy) for the solution of Laplace's equation with given boundary data, and showed that there exists an analytical solution of the latter under some restrictions on the given boundary function and the boundary itself. However, there is an important case for which practitioners find precisely the generalized solution. This is the subject of the next section.

4.5 The Ritz and Bubnov–Galerkin Methods

We have seen that all problems in the linear mechanics of solids that we wish to consider have the form

$$E(u) = \frac{1}{2} \|u\|^2 - \Phi(u) \to \min_H$$

where H is a Hilbert (energy) space and $\Phi(u)$ is a linear continuous functional on H. Moreover, with use of the Riesz representation theorem this problem reduces to the minimum problem

$$E(u) = \frac{1}{2} \|u\|^2 - (u, u_0) \to \min_H \tag{4.5.1}$$

with a given $u_0 \in H$. We shall not concretize this for each mechanical problem under consideration, leaving that work to the reader, but shall discuss the problem of finding an approximate solution in abstract form. At first glance the last version of the problem is quite trivial: we know the solution is u_0. However, we should not forget that u_0 is determined only theoretically; the term (u, u_0) stands in place of a functional Φ, and the role of (4.5.1) is to simplify intermediate steps and to help us understand their meaning.

Ritz was the first to think, in practical terms, of the possibility of finding a minimizer, not on the whole space H but on some of its subspaces. In Ritz's time all calculations were done manually, so it was extremely important to find methods that involved as few steps as possible. Thus it was necessary (and still is, despite the capabilities of computers) to find a subspace that has minimal dimension but that can provide good approximation.[2] The finite dimensional subspace was constructed by the choice of basis elements $e_1, e_2, \ldots, e_n$. They should be linearly independent which, as is shown in linear algebra, means that the Gram determinant

$$\begin{vmatrix} (e_1, e_1) & (e_1, e_2) & \cdots & (e_1, e_n) \\ (e_2, e_1) & (e_2, e_2) & \cdots & (e_2, e_n) \\ \vdots & \vdots & \ddots & \vdots \\ (e_n, e_1) & (e_n, e_2) & \cdots & (e_n, e_n) \end{vmatrix} \neq 0. \tag{4.5.2}$$

We also assume the set $e_1, e_2, \ldots, e_n, \ldots$ to be complete in H; that is, any element of H can be approximated within any given accuracy by a finite linear combination of elements from the set. Denote by H_n the space spanned by $e_1, e_2, \ldots, e_n$. We call

$$E(u) = \frac{1}{2} \|u\|^2 - (u, u_0) \to \min_{H_n}$$

[2]The reader notes that the approximate models of mechanics like the theory of shells and plates has the same goal: to reduce the full dimensionality of the problem so they reduce the dimensionality of space coordinates for thin-walled structures from 3-D to 2-D by introducing some hypothesis on the form of deformation or the order of some components strains. The same but more directly, does the Ritz method: it reduces possible forms of deformation of a body to that one which are expected to approximate the real ones more or less accurately.

the Ritz method for the solution of (4.5.1).

Let us denote the minimizer of the problem by $u_n = \sum_{k=1}^{n} c_k e_k$ where the c_k are constants. The equality to zero of the first variation of this functional for all admissible variations $v \in H_n$ is

$$(u_n, v) - (u_0, v) = 0. \tag{4.5.3}$$

Since $e_1, e_2, \dots, e_n$ is a basis of H_n, the last equation is equivalent to the n simultaneous equations

$$\begin{aligned} \left(\sum_{k=1}^{n} c_k e_k, e_1\right) &= (u_0, e_1), \\ \left(\sum_{k=1}^{n} c_k e_k, e_2\right) &= (u_0, e_2), \\ &\vdots \\ \left(\sum_{k=1}^{n} c_k e_k, e_n\right) &= (u_0, e_n), \end{aligned} \tag{4.5.4}$$

called the Ritz system of the nth approximation step. The system can be rewritten as

$$\begin{aligned} (e_1, e_1)c_1 + (e_2, e_1)c_2 + \cdots + (e_n, e_1)c_n &= \Phi(e_1), \\ (e_1, e_2)c_1 + (e_2, e_2)c_2 + \cdots + (e_n, e_2)c_n &= \Phi(e_2), \\ &\vdots \\ (e_1, e_n)c_1 + (e_2, e_n)c_2 + \cdots + (e_n, e_n)c_n &= \Phi(e_n). \end{aligned} \tag{4.5.5}$$

On the right side of (4.5.4) there are given some numbers. It is necessary to find the unknown c_k.

Theorem 4.5.1 *The system of simultaneous equations of the nth approximation has a unique solution $u_n = \sum_{k=1}^{n} c_k e_k$. The sequence $\{u_n\}$ converges strongly to the solution of the problem* (4.5.1).

Proof. It is easy to see that the principal determinant of this system is the transposed Gram determinant so, by the condition (4.5.2), the system (4.5.5) has a unique solution. Let us return to (4.5.3), which we rewrite as

$$(u_n - u_0, v) = 0 \quad \text{for all } v \in H_n.$$

This means $u_n - u_0$ is orthogonal to H_n. Another interpretation is that u_n is an orthogonal projection of u_0 into H_n. Besides, it is easily seen from (4.5.4) that if $e_1, e_2, \ldots, e_n$ is an orthonormal basis of H_n, then (4.5.4) defines the Fourier coefficients $c_k = (u_0, e_k)$ of the solution u_0. Thus, by the Bessel inequality

$$\|u_n\| \leq \|u_0\| .$$

Even if $e_1, e_2, \ldots, e_n$ is not an orthonormal basis of H_n, we always can construct an equivalent orthonormal basis of H_n which consequently defines an orthonormal basis of H. Thus the Fourier expansion of u_0 lies in the space spanned by this basis, and when we find the Ritz approximation u_n it coincides with the first n terms of that Fourier expansion. By the general theory of Fourier expansion, $\{u_n\}$ converges strongly to u_0 in H. □

The only remark needed regarding mechanical problems concerns the problems with free boundary. Such problems may be treated theoretically in factor spaces and in spaces of balanced functions. In numerical calculation by the Ritz method, only the balanced function spaces are appropriate. If we work in the corresponding factor spaces, the solution would contain the same undetermined constants of rigid motions, which means that the corresponding determinant would be zero. Because of rounding errors and other numerical uncertainties, the system of the Ritz method (and any other numerical method) can lose the compatibility present in the initial setup, whereas in the energy space of balanced functions there are no such problems.

4.6 The Hamilton–Ostrogradskij Principle and the Generalized Setup of Dynamical Problems of Classical Mechanics

One of the main variational principles of classical dynamics, the Hamilton–Ostrogradskij principle, is not minimal. It asserts that the real motion of a system of material points, described by generalized coordinates

$$\mathbf{q}(t) = (q_1(t), q_2(t), \ldots, q_n(t))$$

and under the influence potential forces, occurs in such manner that among all the motions from the initial position $\mathbf{q}_0$ taken at time instant t_0 to the final position $\mathbf{q}_1$ taken at time instant t_1, the real motion yields an extremal for the *action* functional

$$\int_{t_0}^{t_1} L(\mathbf{q}, \dot{\mathbf{q}}, t)\, dt. \tag{4.6.1}$$

Here an overdot denotes differentiation with respect to time t. The function L is called the kinetic potential and is given by

$$L = K - E \tag{4.6.2}$$

where K and E are the kinetic and potential energies, respectively, of the system. The first variation of this functional is

$$\delta \int_{t_0}^{t_1} L(\mathbf{q}, \dot{\mathbf{q}}, t)\, dt = \int_{t_0}^{t_1} \sum_{i=1}^{n} \left(\frac{\partial L(\mathbf{q}, \dot{\mathbf{q}}, t)}{\partial q_i} \delta q_i + \frac{\partial L(\mathbf{q}, \dot{\mathbf{q}}, t)}{\partial \dot{q}_i} \delta \dot{q}_i \right) dt \tag{4.6.3}$$

where all variations δq_i of the generalized coordinates are considered as independent functions (cf., Chapter 1), and $\delta q_i(t_0) = 0 = \delta q_i(t_1)$ for $i = 1, 2, \ldots, n$. From the equality of the first variation to zero we obtain Lagrange's equations of motion

$$\frac{d}{dt} \frac{\partial L}{\partial \dot{q}_i} - \frac{\partial L}{\partial q_i} = 0 \tag{4.6.4}$$

which form the basis of Lagrangian mechanics and of physics as a whole. In general the action does not attain a minimum or maximum. Normally for Lagrange's equations (if not in Hamiltonian form) a Cauchy problem is formulated in which equations (4.6.4) are supplemented with initial data

$$\mathbf{q}(t_0) = \mathbf{q}_0, \qquad \dot{\mathbf{q}}(t_0) = \mathbf{q}_{01}. \tag{4.6.5}$$

If we consider (4.6.3) as a generalized setup for some problem for (4.6.4), we see that (4.6.3) with the boundary conditions $\mathbf{q}(t_0) = \mathbf{q}_0$, $\mathbf{q}(t_1) = \mathbf{q}_1$, $\delta\mathbf{q}(t_0) = 0 = \delta\mathbf{q}(t_1)$ is formulated for a boundary value problem. How do we reformulate (4.6.3) and requirements for $q_i(t)$ to get a generalized setup for the Cauchy problem (4.6.4), (4.6.5)? We would like to do this because the same operation will be done when we go from equilibrium problems to the dynamic problems of the mechanics of solids. Let us take a special class of variations $\delta\mathbf{q}(t)$ that are continuously differentiable and have $\delta\mathbf{q}(t_1) = \mathbf{0}$. Denote this class D_1. Take $\delta\mathbf{q}(t) \in D_1$, multiply (4.6.4) by $\delta q_i(t)$, sum over i, and integrate over $[t_0, t_1]$:

$$\int_{t_0}^{t_1} \sum_{i=1}^{n} \left(\frac{d}{dt} \frac{\partial L}{\partial \dot{q}_i} - \frac{\partial L}{\partial q_i} \right) \delta q_i\, dt = 0.$$

Integration by parts (the operation inverse to the standard one done in the calculus of variations) gives

$$\int_{t_0}^{t_1} \sum_{i=1}^{n} \left(\frac{\partial L}{\partial \dot{q}_i} \delta \dot{q}_i + \frac{\partial L}{\partial q_i} \delta q_i \right) dt - \sum_{i=1}^{n} \left(\frac{\partial L}{\partial \dot{q}_i} \delta q_i \right) \Bigg|_{t=t_0} = 0. \qquad (4.6.6)$$

In the second sum, the terms given at t_0, there stand values (4.6.5) so they do not contain q_i; in the integrand there participate only $q_i(t)$ and $\dot{q}_i(t)$ whereas (4.6.4) contain second derivatives of $q_i(t)$. Thus the requirements for $q_i(t)$ in (4.6.6) are less than in (4.6.4), and it is sensible to formulate a generalized setup of the Cauchy problem using (4.6.6) because now in (4.6.6) we need not appoint values for $\mathbf{q}$ and $\dot{\mathbf{q}}$ at instant t_1 in advance. It is clear that from (4.6.6), using the standard procedure of the calculus of variations we can obtain (4.6.4) if require (4.6.6) to hold for any $\delta \mathbf{q}(t) \in D_1$.

Next we need to define a space in which we seek a solution. Usually this would be a space where in the norm there is integration over time, and this means we cannot stipulate on a generalized solution the point condition $\dot{\mathbf{q}}(t_0) = \mathbf{q}_1$, it comes into the definition through the second sum term of (4.6.6). The first initial condition $\mathbf{q}(t_0) = \mathbf{q}_0$ could be stipulated as a separate one. We do not formulate exact statements now because, first of all, the form of the norm depends on the form of L and the statements would depend on this. More important is that the generalized setup is not introduced in classical mechanics. We have engaged in these considerations only to prepare ourselves for the more complex problems of continuum mechanics, for which all of the pertinent details will be repeated.

4.7 Generalized Setup of Dynamic Problems for a Membrane

In continuum mechanics the Hamilton–Ostrogradskij principle can be formulated in the form (4.6.1), (4.6.2) as well:

$$\delta \int_{t_0}^{t_1} L\, dt = 0, \qquad L = K - E,$$

where for each of the objects we have considered in equilibrium — beam, membrane, plate, elastic body — E is the energy functional we used (the difference between the elastic energy of an object and the potential of external forces acting on the object); here the state of the body at t_0 and t_1 must coincide with the real states of the body. The kinetic energy K is given in the common manner

$$K = \frac{1}{2} \int_S \rho \dot{\mathbf{u}}^2 \, dS$$

where S is the domain taken by the object in a coordinate frame and ρ is the specific density of the material. For example, in the case of a 3-D elastic body the equation of the Hamilton–Ostrogradskij principle looks like

$$\delta \int_{t_0}^{t_1} \left\{ \frac{1}{2} \int_V \rho \dot{\mathbf{u}}^2 \, dV - \left[\frac{1}{2} \int_V c^{ijkl} e_{kl}(\mathbf{u}) e_{ij}(\mathbf{u}) \, dV - \left(\int_V \mathbf{F} \cdot \mathbf{u} \, dV + \int_{\partial V} \mathbf{f} \cdot \mathbf{u} \, dS \right) \right] \right\} dt = 0$$

for any admissible variation of displacement vector $\delta \mathbf{u}$. Here $\mathbf{u}$ must satisfy the geometrical boundary conditions of the problem, $\delta \mathbf{u} = \delta \mathbf{u}(\mathbf{x}, t)$ the homogeneous geometrical boundary conditions and, besides,

$$\delta \mathbf{u}(\mathbf{x}, t_0) = \mathbf{0} = \delta \mathbf{u}(\mathbf{x}, t_1).$$

So this formulation corresponds to a boundary value problem as if the values of $\mathbf{u}(\mathbf{x}, t)$ are given at $t = t_0$ and $t = t_1$.

Now we would like to derive a generalized setup of the Cauchy problem for the dynamic problems under consideration. It is clear that the corresponding energy spaces should include the terms with integrals for the kinetic energy and, besides, if we would like to use the tools of Hilbert spaces, integration over time should be incorporated into the norm. The form of the integrand of the part of the action for E remains the same, so we need only consider what happens to the kinetic energy term. We begin with the universal equation that is the virtual work principle in statics. To simplify the calculations we consider a membrane; the remaining problems can be treated in a similar fashion. We combine the virtual work principle with d'Alembert's principle, which asserts that the system of external forces can be balanced by the inertia forces. For a membrane the work of external forces complemented by the forces of inertia on a virtual displacement $v(\mathbf{x}, t)$ is

$$\int_\Omega (f(\mathbf{x}, t) - \rho \ddot{u}(\mathbf{x}, t)) \, v(\mathbf{x}, t) \, d\Omega, \qquad d\Omega = dx \, dy.$$

Thus, for a membrane with clamped edge, the virtual work principle gives

$$\int_\Omega \left(\frac{\partial u}{\partial x} \frac{\partial v}{\partial x} + \frac{\partial u}{\partial y} \frac{\partial v}{\partial y} \right) d\Omega = \int_\Omega (f(\mathbf{x}, t) - \rho \ddot{u}(\mathbf{x}, t)) \, v(\mathbf{x}, t) \, d\Omega. \tag{4.7.1}$$

Of course we could begin with the differential equations of motion and obtain the same result step by step, but we take a shorter route. We suppose all functions are smooth enough to provide for the necessary transformations, and that the virtual displacement v satisfies

$$v(\mathbf{x}, T) = 0.$$

Let us integrate (4.7.1) over time and integrate by parts in the last term:

$$\int_0^T \int_\Omega \left(\frac{\partial u}{\partial x}\frac{\partial v}{\partial x} + \frac{\partial u}{\partial y}\frac{\partial v}{\partial y} \right) d\Omega\, dt = \int_0^T \int_\Omega f(\mathbf{x}, t) v(\mathbf{x}, t)\, d\Omega\, dt$$
$$+ \int_0^T \int_\Omega \rho \dot{u}(\mathbf{x}, t)\dot{v}(\mathbf{x}, t)\, d\Omega\, dt + \int_\Omega \rho u_1^*(\mathbf{x}) v(\mathbf{x}, 0)\, d\Omega. \quad (4.7.2)$$

Here $u_1^*(\mathbf{x})$ is an initial condition for $u(\mathbf{x}, t)$:

$$u(\mathbf{x}, t)\big|_{t=t_0} = u_0^*(\mathbf{x}), \qquad \dot{u}(\mathbf{x}, t)\big|_{t=t_0} = u_1^*(\mathbf{x}).$$

We shall use (4.7.2) for the generalized setup of the dynamic problem for a membrane. To do this we need to introduce proper functional spaces.

An energy space for a clamped membrane (dynamic case)

Without loss of generality we can set $t_0 = 0$ and denote $t_1 = T$. It is clear that the expression for an inner product in this space should include some terms from (4.7.2). Let it be given by

$$(u, v)_{[a,b]} = \int_a^b \int_\Omega \rho \dot{u}(\mathbf{x}, t)\dot{v}(\mathbf{x}, t)\, d\Omega\, dt + \int_a^b \int_\Omega \left(\frac{\partial u}{\partial x}\frac{\partial v}{\partial x} + \frac{\partial u}{\partial y}\frac{\partial v}{\partial y} \right) d\Omega\, dt. \quad (4.7.3)$$

The energy space denoted as $\mathcal{E}_{Mc}(a, b)$ is the completion of the set of twice continuously differentiable functions that satisfy the boundary condition

$$u|_{\partial\Omega} = 0, \quad (4.7.4)$$

with respect to the norm $\|u\| = (u, u)_{[a,b]}$. Denote $Q_{a,b} = \Omega \times [a, b]$.

Lemma 4.7.1 *$\mathcal{E}_{Mc}(a, b)$ is a closed subspace of $W^{1,2}(Q_{a,b})$. The norm of $\mathcal{E}_{Mc}(a, b)$ is equivalent to the norm of $W^{1,2}(Q_{a,b})$.*

Proof. It suffices to prove the last statement of the lemma for twice differentiable functions satisfying (4.7.4). The inequality

$$(u, u)_{[a,b]} \le M \|u\|^2_{W^{1,2}(Q_{a,b})}$$

is evident. Let us show that the inverse inequality with a positive constant m is valid as well. From the Friedrichs inequality it follows that

$$\int_a^b \|u\|^2_{W^{1,2}(\Omega)}\,dt \le m \int_a^b \int_\Omega \left(\left(\frac{\partial u}{\partial x}\right)^2 + \left(\frac{\partial u}{\partial y}\right)^2 \right) d\Omega\,dt.$$

Adding to both sides the term

$$\int_a^b \int_\Omega \rho \dot{u}^2(\mathbf{x},t)\,d\Omega\,dt$$

after easy transformations, we get the needed inequality. □

By Sobolev's imbedding theorem, from Lemma 4.7.1 it follows that $\mathcal{E}_{Mc}(a,b)$ is continuously imbedded into $L^6(Q(a,b))$ and at any fixed $t \in [a,b]$ into $L^4(\Omega)$, so we can pose an initial condition for u to satisfy in the sense of $L^4(\Omega)$. However we now demonstrate a general result that shows the meaning in which we can state the initial condition.

Let H be a separable Hilbert space. Consider the set of functions of the parameter $t \in [a,b]$ that take values in H. In what follows $H = L^2(\Omega)$. The theory of such functions is quite similar to the usual theory of functions in one variable. In particular, we can introduce the space $C(H;a,b)$ of all functions continuous on $[a,b]$ and taking values in H. Its properties are the same as those of $C(a,b)$: if H is separable it is a separable Banach space with the norm of an element $x(t)$ given by

$$\|x\|_{C(H;a,b)} = \max_{t\in[a,b]} \|x(t)\|_H \,.$$

For functions with values in H we can introduce the notion of derivative as

$$x'(t) = \lim_{\Delta t \to 0} \frac{x(t+\Delta t) - x(t)}{\Delta t},$$

as well as derivatives of higher order. The definite Riemann integral

$$\int_c^d x(t)\,dt$$

is the limit of Riemann sums that must not depend on the manner in which $[c,d]$ is partitioned. Analogous to the spaces $C^{(k)}(a,b)$, for functions with values in H we can introduce spaces $C^{(k)}(H;a,b)$ (we leave this to the reader). Finally we can introduce an analogue to $L^2(a,b)$, denoted by

$L^2(H;a,b)$. This is a Hilbert space with an inner product

$$(x,y)_{L^2(H;a,b)} = \int_a^b (x(t),y(t))_H \, dt, \tag{4.7.5}$$

and is the completion of $C(H;a,b)$ in the norm induced by (4.7.5). Note that $L^2(L^2(\Omega);a,b)$ is $L^2(Q_{a,b})$. Quite similarly, we can introduce a Sobolev space $W^{1,2}(H;a,b)$ as the completion of $C^{(1)}(H;a,b)$ with respect to the norm induced by

$$(x,y)_{W^{1,2}(H;a,b)} = \int_a^b \{(x(t),y(t))_H + (x'(t),y'(t))_H\} \, dt.$$

Lemma 4.7.2 *$W^{1,2}(H;a,b)$ is continuously imbedded into $C(H;a,b)$.*

The proof mimics that of the similar result for $W^{1,2}(a,b)$, so we leave it to the reader. Lemma 4.7.2 states that we can formulate the initial condition for $u(\mathbf{x},t)$ at a fixed t in the sense of $L^2(\Omega)$ since the element of $\mathcal{E}_{Mc}(a,b)$, by the form of the norm, belongs to $W^{1,2}(L^2(\Omega);a,b)$ as well. However, for the formulation of the initial boundary value problem we need a stronger result. This is a particular imbedding theorem in a Sobolev space that is useful for hyperbolic boundary value problems.

Lemma 4.7.3 *If $\{u_n\}$ converges weakly to u_0 in $\mathcal{E}_{Mc}(a,b)$, then it also converges to u_0 uniformly with respect to t in the norm of $C(L^2(\Omega);a,b)$.*

Proof. By equivalence on $\mathcal{E}_{Mc}(a,b)$ of the norm of $\mathcal{E}_{Mc}(a,b)$ to the norm of $W^{1,2}(Q_{a,b})$, and Sobolev's imbedding theorem, we state that

$$\|u_n\|_{[a,b]} \le m \tag{4.7.6}$$

and that

$$\|u_n - u_0\|_{L^2(Q_{a,b})} \to 0 \quad \text{as } n \to \infty. \tag{4.7.7}$$

So u_n converges to u_0 strongly in $L^2(Q_{a,b})$. Now we need a special bound for an element of $W^{1,2}(L^2(\Omega);a,b)$, into which $W^{1,2}(Q_{a,b})$ imbeds continuously. We derive the estimate for elements that are smooth in time t, and then extend to all the elements. Let $c \in [a,b)$ and $\Delta > 0$ be such that $c + \Delta \in [a,b)$. Let $t,s \in [c,c+\Delta]$. We begin with a simple identity

$$v(\mathbf{x},t) = v(\mathbf{x},s) + \int_s^t \frac{\partial v(\mathbf{x},\theta)}{\partial \theta} \, d\theta,$$

from which

$$\int_\Omega v^2(\mathbf{x},t)\,d\Omega = \int_\Omega \left(v(\mathbf{x},s) + \int_s^t \frac{\partial v(\mathbf{x},\theta)}{\partial\theta}\,d\theta \right)^2 d\Omega$$
$$\leq 2\int_\Omega v^2(\mathbf{x},s)\,d\Omega + 2\int_\Omega \left(\int_s^t \frac{\partial v(\mathbf{x},\theta)}{\partial\theta}\,d\theta \right)^2 d\Omega.$$

Let us integrate this with respect to s over $[c, c+\Delta]$. Dividing through by Δ we get

$$\int_\Omega v^2(\mathbf{x},t)\,d\Omega \leq \frac{2}{\Delta}\int_c^{c+\Delta}\int_\Omega v^2(\mathbf{x},s)\,d\Omega\,ds$$
$$+ \frac{2}{\Delta}\int_c^{c+\Delta}\int_\Omega \left(\int_s^t 1\cdot \frac{\partial v(\mathbf{x},\theta)}{\partial\theta}\,d\theta \right)^2 d\Omega\,ds.$$

Applying Hölder's inequality to the last term on the right we have

$$\int_\Omega v^2(\mathbf{x},t)\,d\Omega \leq \frac{2}{\Delta}\int_{Q_{c,c+\Delta}} v^2(\mathbf{x},s)\,d\Omega\,ds$$
$$+ \frac{2}{\Delta}\int_c^{c+\Delta}\int_\Omega \left(\int_s^t 1^2\,d\theta \int_s^t \left(\frac{\partial v(\mathbf{x},\theta)}{\partial\theta} \right)^2 d\theta \right) d\Omega\,ds.$$

Finally, direct integration in the last integral and simple estimates bring us to the desired inequality

$$\int_\Omega v^2(\mathbf{x},t)\,d\Omega \leq \frac{2}{\Delta}\int_{Q_{c,c+\Delta}} v^2(\mathbf{x},\theta)\,d\Omega\,d\theta + \Delta\int_{Q_{c,c+\Delta}} \left(\frac{\partial v(\mathbf{x},\theta)}{\partial\theta} \right)^2 d\Omega\,d\theta, \tag{4.7.8}$$

which is the basis for the proof of Lemma 4.7.3. By the completion procedure (4.7.8) extends to any element of $\mathcal{E}_{a,b}$. We write it out for $u_n - u_0$:

$$\int_\Omega (u_n(\mathbf{x},t) - u_0(\mathbf{x},t))^2\,d\Omega \leq \frac{2}{\Delta}\int_{Q_{c,c+\Delta}} (u_n(\mathbf{x},\theta) - u_0(\mathbf{x},\theta))^2\,d\Omega\,d\theta$$
$$+ \Delta\int_{Q_{c,c+\Delta}} \left(\frac{\partial\,(u_n(\mathbf{x},\theta) - u_0(\mathbf{x},\theta))}{\partial\theta} \right)^2 d\Omega\,d\theta. \tag{4.7.9}$$

Let $\varepsilon > 0$ be an arbitrarily small positive number. To prove the lemma it is enough to find a number N such that the right-hand side of (4.7.9) is less than ε for any $t \in [c, c+\Delta]$. Let us put $\Delta = \varepsilon/2m$ where m is the constant from (4.7.6). Then the last integral is less than $\varepsilon/2$. By (4.7.7) we can find

N such that

$$\frac{2}{\Delta}\int_{Q_{c,c+\Delta}} (u_n(\mathbf{x},t) - u_0(\mathbf{x},t))^2 \, d\Omega \, dt \le \frac{\varepsilon}{2}$$

independent of $t \in [c, c+\Delta]$. Since this is independent of $c \in [a,b]$ we establish the result for all $t \in [a,b]$. $\square$

Generalized setup

Without loss of generality we consider the initial problem on $[0,T]$ for fixed but arbitrary T. In this case we use the energy space $\mathcal{E}_{Mc}(0,T)$. In addition, we need to define a closed subspace which is the completion of the subset of twice continuously differentiable functions satisfying (4.7.4) that vanish at $t = T$. We denote this by D_0^T.

Definition 4.7.1 $u(\mathbf{x},t) \in \mathcal{E}_{Mc}(0,T)$ is called a generalized solution of the dynamic problem of a clamped membrane if it satisfies the equation

$$\int_0^T \int_\Omega \left(\frac{\partial u}{\partial x}\frac{\partial v}{\partial x} + \frac{\partial u}{\partial y}\frac{\partial v}{\partial y} \right) d\Omega \, dt = \int_0^T \int_\Omega f(\mathbf{x},t) v(\mathbf{x},t) \, d\Omega \, dt$$
$$+ \int_0^T \int_\Omega \rho \dot{u}(\mathbf{x},t) \dot{v}(\mathbf{x},t) \, d\Omega \, dt + \int_\Omega \rho u_1^*(\mathbf{x}) v(\mathbf{x},0) \, d\Omega \quad (4.7.10)$$

with any $v(\mathbf{x},t) \in D_0^T$ and the first initial condition

$$u(\mathbf{x},t)\big|_{t=0} = u_0^*(\mathbf{x}) \quad (4.7.11)$$

in the sense of $L^2(\Omega)$, that is, $\int_\Omega (u(\mathbf{x},0) - u_0^*(\mathbf{x}))^2 \, d\Omega = 0$.

Let us suppose that

(i) $u_0^*(\mathbf{x}) \in W^{1,2}(\Omega)$ and satisfies (4.7.4),
(ii) $u_1^*(\mathbf{x}) \in L^2(\Omega)$, and
(iii) $f(\mathbf{x},t) \in L^2(Q_{0,T})$.

It is easy to demonstrate that under these restrictions all the terms of (4.7.10) are sensible. Our goal now is to prove the following.

Theorem 4.7.1 *Under restrictions (i)–(iii) there exists (in the sense of Definition 4.7.1) a generalized solution to the dynamic problem for a clamped membrane, and it is unique.*

The proof splits into several lemmas. First we construct an approximate method of solution of the problem under consideration, a variant of the

Bubnov–Galerkin method called the Faedo–Galerkin method. Then we justify its convergence. Finally, we give an independent proof of uniqueness.

The Faedo–Galerkin method.

Suppose there is a complete system of elements of $\mathcal{E}_{Mc}$, any finite set of which is a linearly independent system. In applications these are normally the smooth functions except in the finite element method where they are piecewise smooth. Take the first n elements of the system. We always can "orthonormalize" the latter system with respect to the inner product of $L^2(\Omega)$:

$$\rho \int_\Omega \varphi_i(\mathbf{x})\varphi_j(\mathbf{x})\,d\Omega = \delta_{ij} = \begin{cases} 1, & i = j, \\ 0, & i \neq j. \end{cases} \tag{4.7.12}$$

This is done only to simplify calculations (and to get the final equations in normal form); it is not necessary in principle. We seek the nth approximation of the solution in the form

$$u_n(\mathbf{x}, t) = \sum_{k=1}^{n} c_k(t)\varphi_k(\mathbf{x}) \tag{4.7.13}$$

where the $c_k(t)$ are time functions that satisfy the following system of the Faedo–Galerkin equations, which are implied by (4.7.1) in which we put u_n instead of u and consequently φ_i instead of v:

$$\int_\Omega \left(\frac{\partial u_n}{\partial x}\frac{\partial \varphi_i}{\partial x} + \frac{\partial u_n}{\partial y}\frac{\partial \varphi_i}{\partial y} \right) d\Omega = \int_\Omega \left(f(\mathbf{x}, t) - \rho \ddot{u}_n(\mathbf{x}, t)\right) \varphi_i(\mathbf{x})\,d\Omega \tag{4.7.14}$$

for $i = 1, \ldots, n$. These can be written as

$$\rho \int_\Omega \ddot{u}_n(\mathbf{x}, t)\varphi_i(\mathbf{x})\,d\Omega = -(u_n, \varphi_i)_M + \int_\Omega f(\mathbf{x}, t)\varphi_i(\mathbf{x})\,d\Omega, \quad i = 1, \ldots, n.$$

Finally, using (4.7.13) and (4.7.12), let us rewrite this as

$$\ddot{c}_i(t) = -\sum_{k=1}^{n} c_k(t)(\varphi_k, \varphi_i)_M + \int_\Omega f(\mathbf{x}, t)\varphi_i(\mathbf{x})\,d\Omega, \quad i = 1, \ldots, n. \tag{4.7.15}$$

This is a system of simultaneous ordinary differential equations for which we must formulate initial conditions. The condition $\dot{u}(\mathbf{x}, t)|_{t=0} = u_1(\mathbf{x})$ and (4.7.12) imply

$$\dot{c}_i(0) = \rho^{-1/2} \int_\Omega u_1^*(\mathbf{x})\varphi_i(\mathbf{x})\,d\Omega, \quad i = 1, \ldots, n. \tag{4.7.16}$$

From (4.7.11) we derive the following conditions for $c_i(0)$. Let us solve the problem

$$\left\| u_0^* - \sum_{k=1}^{n} a_k \varphi_k \right\|_M^2 \to \min_{a_1,\ldots,a_n} . \tag{4.7.17}$$

We know this is solvable; moreover, its solution $d_1, \ldots, d_n$ gives us $\sum_{k=1}^n d_k \varphi_k$, the orthogonal projection in $\mathcal{E}_{Mc}$ of u_0 onto the subspace spanned by $\varphi_1, \ldots, \varphi_n$. Thus the second set of initial conditions is

$$c_i(0) = d_i, \quad i = 1, \ldots, n. \tag{4.7.18}$$

So the setup of the nth approximation of the Faedo–Galerkin method consists of (4.7.15) supplemented with (4.7.16) and (4.7.18). We begin by establishing the properties of this Cauchy problem.

Unique solvability of the Cauchy problem for the nth approximation of the Faedo–Galerkin method.
We would like to understand what we can say about the solution of the Cauchy problem (4.7.15), (4.7.16), (4.7.18). The simultaneous equations (4.7.15) are linear in the unknown $c_i(t)$. The load terms $\int_\Omega f(\mathbf{x},t)\varphi_i(\mathbf{x})\,d\Omega$ belong to $L^2(0,T)$; indeed, by Schwarz's inequality

$$\begin{aligned}\int_0^T \left(\int_\Omega f(\mathbf{x},t)\varphi_i(\mathbf{x})\,d\Omega\right)^2 dt &\le \int_0^T \left(\int_\Omega f^2(\mathbf{x},t)\,d\Omega\right)\left(\int_\Omega \varphi_i^2(\mathbf{x})\,d\Omega\right) dt \\ &= \|\varphi_i\|_{L^2(\Omega)}^2 \, \|f\|_{L^2(Q_{0,T})}^2 .\end{aligned}$$

From general ODE theory the Cauchy problem (4.7.15), (4.7.16), (4.7.18) has a unique solution on $[0,T]$ with arbitrary T such that

$$c_i''(t) \in L^2(0,T)$$

and $c_i(t), c_i'(t)$ are continuous on $[0,T]$. This can be shown by the traditional way of proving such results, in which a Cauchy problem is transformed into a system of integral equations (by double integration of the equations in time taking into account the initial conditions). For the integral equations the existence of a unique continuous solution can be shown with use of Banach's contraction principle, and then differentiation in time yields the remaining properties. Now we obtain the estimate of the solution that we need to prove the above theorem. The estimate for the solution $c_i(t)$,

$i = 1, \dots, n$, is

$$\max_{t\in[0,T]} \left(\sum_{k=1}^{n} (c_k'(t))^2 + \left\| \sum_{k=1}^{n} c_k(t)\varphi_k \right\|_M^2 \right) \le m.$$

Indeed, let us multiply the ith equation in (4.7.15) by $c_i'(t)$ and sum over i:

$$\sum_{i=1}^{n} \ddot{c}_i(t)\dot{c}_i(t) = -\sum_{i=1}^{n}\sum_{k=1}^{n} (c_k(t)\varphi_k, \dot{c}_i(t)\varphi_i)_M + \sum_{i=1}^{n} \int_\Omega f(\mathbf{x},t)\dot{c}_i(t)\varphi_i(\mathbf{x})\, d\Omega. \tag{4.7.19}$$

The term on the left is

$$\begin{aligned}\sum_{i=1}^{n} \ddot{c}_i(t)\dot{c}_i(t) &= \frac{1}{2}\frac{d}{dt}\sum_{i=1}^{n} \dot{c}_i(t)\dot{c}_i(t) = \frac{1}{2}\rho\frac{d}{dt}\sum_{i,j=1}^{n} \dot{c}_i(t)\dot{c}_j(t)\int_\Omega \varphi_i(\mathbf{x})\varphi_j(\mathbf{x})\, d\Omega \\ &= \frac{d}{dt}\left(\frac{1}{2}\rho\int_\Omega \dot{u}_n(\mathbf{x},t)\dot{u}_n(\mathbf{x},t)\, d\Omega\right).\end{aligned}$$

Similarly

$$\sum_{i=1}^{n}\sum_{k=1}^{n} (c_k(t)\varphi_k, \dot{c}_i(t)\varphi_i)_M = \frac{1}{2}\frac{d}{dt}(u_n(\mathbf{x},t), u_n(\mathbf{x},t))_M$$

and

$$\sum_{i=1}^{n} \int_\Omega f(\mathbf{x},t)\dot{c}_i(t)\varphi_i(\mathbf{x})\, d\Omega = \int_\Omega f(\mathbf{x},t)\dot{u}_n(\mathbf{x},t)\, d\Omega.$$

So (4.7.19) can be presented as

$$\begin{aligned}\frac{d}{dt}\left(\frac{1}{2}\rho\int_\Omega \dot{u}(\mathbf{x},t)\dot{u}_n(\mathbf{x},t)\, d\Omega\right) + \frac{1}{2}\frac{d}{dt}(u_n(\mathbf{x},t), u_n(\mathbf{x},t))_M \\ = \int_\Omega f(\mathbf{x},t)\dot{u}_n(\mathbf{x},t)\, d\Omega,\end{aligned}$$

or rewritten as

$$\frac{1}{2}\frac{d}{dt}\left(\rho \|\dot{u}_n(\mathbf{x},t)\|_{L^2(\Omega)}^2 + \|u_n(\mathbf{x},t)\|_M^2\right) = \int_\Omega f(\mathbf{x},t)\dot{u}_n(\mathbf{x},t)\, d\Omega.$$

Integrating over time t (renaming t by s) we have

$$\frac{1}{2}\int_0^t \frac{d}{ds}\left(\rho\left\|\dot{u}_n(\mathbf{x},s)\right\|^2_{L^2(\Omega)} + \left\|u_n(\mathbf{x},s)\right\|^2_M\right) ds = \int_0^t \int_\Omega f(\mathbf{x},s)\dot{u}_n(\mathbf{x},s)\,d\Omega\,ds$$

or

$$\frac{1}{2}\left(\rho\left\|\dot{u}_n(\mathbf{x},t)\right\|^2_{L^2(\Omega)} + \left\|u_n(\mathbf{x},t)\right\|^2_M\right) = \frac{1}{2}\left(\rho\left\|\dot{u}_n(\mathbf{x},0)\right\|^2_{L^2(\Omega)} + \left\|u_n(\mathbf{x},0)\right\|^2_M\right) + \int_0^t \int_\Omega f(\mathbf{x},s)\dot{u}_n(\mathbf{x},s)\,d\Omega\,ds.$$

Taking into account the way in which we derived the initial conditions for u_n, we have

$$\left\|\dot{u}_n(\mathbf{x},0)\right\|_{L^2(\Omega)} \le \left\|u_1^*(\mathbf{x})\right\|_{L^2(\Omega)}, \qquad \left\|u_n(\mathbf{x},0)\right\|_M \le \left\|u_0^*(\mathbf{x})\right\|_M.$$

We can then state that

$$\frac{1}{2}\left(\rho\left\|\dot{u}_n(\mathbf{x},t)\right\|^2_{L^2(\Omega)} + \left\|u_n(\mathbf{x},t)\right\|^2_M\right) \le \frac{1}{2}\left(\rho\left\|u_1^*(\mathbf{x})\right\|^2_{L^2(\Omega)} + \left\|u_0^*(\mathbf{x})\right\|^2_M\right) + \int_0^t \int_\Omega f(\mathbf{x},s)\dot{u}_n(\mathbf{x},s)\,d\Omega\,ds.$$

Using the elementary inequality $|ab| \le a^2/2\varepsilon + \varepsilon b^2/2$ we get

$$\left|\int_0^t \int_\Omega f(\mathbf{x},s)\dot{u}_n(\mathbf{x},s)\,d\Omega\,ds\right| \le \frac{1}{2\varepsilon}\int_0^t \int_\Omega f^2(\mathbf{x},s)\,d\Omega\,ds + \frac{\varepsilon}{2}\int_0^t \int_\Omega \dot{u}_n^2(\mathbf{x},s)\,d\Omega\,ds \le \frac{1}{2\varepsilon}\int_0^T \int_\Omega f^2(\mathbf{x},s)\,d\Omega\,ds + \frac{\varepsilon T}{2}\max_{s\in[0,T]}\int_\Omega \dot{u}_n^2(\mathbf{x},s)\,d\Omega$$

so

$$\frac{1}{2}\left(\rho \left\|\dot{u}_n(\mathbf{x},t)\right\|^2_{L^2(\Omega)} + \left\|u_n(\mathbf{x},t)\right\|^2_M\right) \le \frac{1}{2}\left(\rho \left\|u_1^*(\mathbf{x})\right\|^2_{L^2(\Omega)} + \left\|u_0^*(\mathbf{x})\right\|^2_M\right) + \frac{1}{2\varepsilon}\int_0^T\int_\Omega f^2(\mathbf{x},s)\,d\Omega\,ds + \frac{\varepsilon T}{2}\max_{s\in[0,T]}\int_\Omega \dot{u}_n^2(\mathbf{x},s)\,d\Omega.$$

Putting $\varepsilon = \rho/(2T)$ and taking the maximum of the left-hand side of the last inequality we get

$$\max_{t\in[0,T]}\frac{1}{2}\left(\rho \left\|\dot{u}_n(\mathbf{x},t)\right\|^2_{L^2(\Omega)} + \left\|u_n(\mathbf{x},t)\right\|^2_M\right) \le \frac{1}{2}\left(\rho \left\|u_1^*(\mathbf{x})\right\|^2_{L^2(\Omega)} + \left\|u_0^*(\mathbf{x})\right\|^2_M\right) + \frac{T}{\rho}\int_0^T\int_\Omega f^2(\mathbf{x},s)\,d\Omega\,ds + \frac{\rho}{4}\max_{t\in[0,T]}\int_\Omega \dot{u}_n^2(\mathbf{x},t)\,d\Omega$$

so

$$\max_{t\in[0,T]}\frac{1}{4}\left(\rho \left\|\dot{u}_n(\mathbf{x},t)\right\|^2_{L^2(\Omega)} + \left\|u_n(\mathbf{x},t)\right\|^2_M\right) \le \frac{1}{2}\left(\rho \left\|u_1^*(\mathbf{x})\right\|^2_{L^2(\Omega)} + \left\|u_0^*(\mathbf{x})\right\|^2_M\right) + \frac{T}{\rho}\int_0^T\int_\Omega f^2(\mathbf{x},s)\,d\Omega\,ds.$$

This is the needed estimate, which can be written out as

$$\max_{t\in[0,T]}\left(\rho \left\|\dot{u}_n(\mathbf{x},t)\right\|^2_{L^2(\Omega)} + \left\|u_n(\mathbf{x},t)\right\|^2_M\right) \le m$$

where the constant m does not depend on the number n. In particular, from this follows the rougher estimate

$$\int_0^T\left(\rho \left\|u_n(\mathbf{x},t)\right\|^2_{L^2(\Omega)} + \left\|u_n(\mathbf{x},t)\right\|^2_M\right)dt \le m_1$$

which can be written in terms of (4.7.3) as

$$(u_n, u_n)_{[0,T]} \le m_1. \tag{4.7.20}$$

Convergence of the Faedo–Galerkin method

Now we show that there is a subsequence of $\{u_n(\mathbf{x},t)\}$ that converges to a generalized solution of the problem under consideration. Because of (4.7.20) $\{u_n\}$ contains a subsequence that converges weakly to an element $u_0(\mathbf{x},t)$. We shall show that $u_0(\mathbf{x},t)$ is a generalized solution. By Lemma 4.7.2 we can consider it as a function continuous in t on $[0,T]$ with values in $L^2(\Omega)$. Let us renumber this subsequence, denoting it by $\{u_n\}$ (in fact, by the uniqueness theorem proved later, the whole sequence converges weakly so there is no need for renumbering; however, at this moment we are not assured of uniqueness). So, now we know that $u_n(\mathbf{x},t)$ tends to $u_0(\mathbf{x},t)$ weakly in $\mathcal{E}_{Mc}(0,T)$. First we show that u_0 satisfies (4.7.11). Indeed, by the method of constructing the Faedo–Galerkin approximations u_n, we see that $\{u_n(\mathbf{x},0)\}$ converges to the initial value $u_0^*(\mathbf{x})$ strongly in $W^{1,2}(\Omega)$ and thus in $L^2(\Omega)$. On the other hand, by Lemma 4.7.3 $\{u_n(\mathbf{x},t)\}$ converges to $u_0(\mathbf{x},t)$ in the norm of $C(L^2(\Omega);0,T)$. Thus (4.7.11) holds for $u_0(\mathbf{x},t)$. Let us verify that (4.7.10) for $u = u_0(\mathbf{x},t)$ holds for any $v(\mathbf{x},t) \in D_0^T$. First we reduce the set of admissible v to a subset of D_0^T defined as follows. Let

$$v_k(t,\mathbf{x}) = \sum_{k=1}^{n} d_k(t)\varphi_k(\mathbf{x}), \qquad k \le n$$

where the $d_k(t)$ are continuously differentiable and $d_k(T) = 0$. Denote the set of all such finite sums by D_{0f}^T. This set is dense in D_0^T and thus, to complete the proof of Theorem 4.7.1, it is enough to demonstrate the validity of (4.7.10) for $u = u_0(\mathbf{x},t)$ when $v \in D_{0f}^T$. Let us return to (4.7.14) for u_n:

$$\int_\Omega \left(\frac{\partial u_n}{\partial x}\frac{\partial \varphi_i}{\partial x} + \frac{\partial u_n}{\partial y}\frac{\partial \varphi_i}{\partial y} \right) d\Omega = \int_\Omega \left(f(\mathbf{x},t) - \rho \ddot{u}_n(\mathbf{x},t)\right)\varphi_i(\mathbf{x})\, d\Omega,$$
$$i = 1,\dots,n.$$

Multiplying the ith equation by $d_i(t)$ and summing from $i = 1$ to k we get

$$\int_\Omega \left(\frac{\partial u_n}{\partial x}\frac{\partial v_k}{\partial x} + \frac{\partial u_n}{\partial y}\frac{\partial v_k}{\partial y} \right) d\Omega = \int_\Omega \left(f(\mathbf{x},t) - \rho \ddot{u}_n(\mathbf{x},t)\right) v_k(\mathbf{x},t)\, d\Omega$$

for $k \leq n$. Let us integrate this with respect to t:

$$\int_0^T \int_\Omega \left(\frac{\partial u_n}{\partial x} \frac{\partial v_k}{\partial x} + \frac{\partial u_n}{\partial y} \frac{\partial v_k}{\partial y} \right) d\Omega \, dt = \int_0^T \int_\Omega \left(f(\mathbf{x}, t) - \rho \ddot{u}_n(\mathbf{x}, t) \right) v_k(\mathbf{x}, t) \, d\Omega \, dt.$$

Integrating by parts in the last term we get

$$\int_0^T \int_\Omega \left(\frac{\partial u_n}{\partial x} \frac{\partial v_k}{\partial x} + \frac{\partial u_n}{\partial y} \frac{\partial v_k}{\partial y} \right) d\Omega \, dt = \int_0^T \int_\Omega f(\mathbf{x}, t) v_k(\mathbf{x}, t) \, d\Omega \, dt + \int_0^T \int_\Omega \rho \dot{u}_n(\mathbf{x}, t) \dot{v}_k(\mathbf{x}, t) \, d\Omega \, dt + \int_\Omega \rho \dot{u}_n(\mathbf{x}, 0) v_k(\mathbf{x}, 0) \, d\Omega.$$

Let us now fix $v_k(\mathbf{x}, t)$ and let $n \to \infty$. Because of the properties of u_n we have

$$\int_0^T \int_\Omega \left(\frac{\partial u_0}{\partial x} \frac{\partial v_k}{\partial x} + \frac{\partial u_0}{\partial y} \frac{\partial v_k}{\partial y} \right) d\Omega \, dt = \int_0^T \int_\Omega f(\mathbf{x}, t) v_k(\mathbf{x}, t) \, d\Omega \, dt + \int_0^T \int_\Omega \rho \dot{u}_0(\mathbf{x}, t) \dot{v}_k(\mathbf{x}, t) \, d\Omega \, dt + \int_\Omega \rho u_1^*(\mathbf{x}) \dot{v}_k(\mathbf{x}, 0) \, d\Omega,$$

as is required by Definition 4.7.1.

Uniqueness of the generalized solution

Theorem 4.7.2 *A generalized solution of the dynamic problem for a membrane with clamped edge is unique.*

Proof. Let us suppose there exist two generalized solutions to the problem under consideration, denoted as u' and u''. Subtracting term by term the equations (4.7.10) for these solutions and introducing $u = u'' - u'$, we get

$$\int_0^T \int_\Omega \rho \dot{u}(\mathbf{x}, t) \dot{v}(\mathbf{x}, t) \, d\Omega \, dt - \int_0^T \int_\Omega \left(\frac{\partial u}{\partial x} \frac{\partial v}{\partial x} + \frac{\partial u}{\partial y} \frac{\partial v}{\partial y} \right) d\Omega \, dt = 0 \tag{4.7.21}$$

for any $v \in D_0^T$. Also,

$$u(\mathbf{x}, t)\big|_{t=0} = 0$$

holds in the sense of $L^2(\Omega)$. Let us introduce an auxiliary function

$$w(\mathbf{x},t) = \begin{cases} \int_\tau^t u(\mathbf{x},\vartheta)\,d\vartheta, & t \in [0,\tau], \\ 0, & t > \tau. \end{cases}$$

First we note that on $[0,\tau]$

$$\frac{\partial w(\mathbf{x},t)}{\partial t} = u(\mathbf{x},t).$$

This and other similar relations between w, u are established by simple differentiation of the representative functions of corresponding Cauchy sequences; then, the limit passage justifies that they hold for the elements themselves. It is seen that $w(\mathbf{x},t)$ belongs to D_0^T. Moreover, it has generalized derivatives $\partial^2 w/\partial t\partial x = \partial u/\partial x$, $\partial^2 w/\partial t\partial y = \partial u/\partial y$ in $L^2(Q_{0,\tau})$. Next, $\partial^2 w/\partial t^2 = \partial u/\partial t \in L^2(Q_{0,\tau})$. Finally, as follows from Lemma 4.7.2, w and its first derivatives belong to $C(L^2(\Omega);0,\tau)$ (the reader should verify this). Let us put $v = w$ in (4.7.21). This equality can be written as

$$\int_0^T \int_\Omega \rho \frac{\partial u(\mathbf{x},t)}{\partial t} u(\mathbf{x},t)\,d\Omega\,dt - \int_0^T \int_\Omega \left(\frac{\partial^2 w}{\partial x \partial t}\frac{\partial w}{\partial x} + \frac{\partial^2 w}{\partial y \partial t}\frac{\partial w}{\partial y} \right) d\Omega\,dt = 0,$$

and rewritten as

$$\frac{1}{2}\int_0^T \int_\Omega \frac{\partial}{\partial t}\left\{ \rho u^2(\mathbf{x},t) - \left(\frac{\partial w}{\partial x}\right)^2 - \left(\frac{\partial w}{\partial y}\right)^2 \right\} d\Omega\,dt = 0.$$

Integrating over t we get

$$\int_\Omega \left\{ \rho u^2(\mathbf{x},t) - \left(\frac{\partial w}{\partial x}\right)^2 - \left(\frac{\partial w}{\partial y}\right)^2 \right\} d\Omega \Bigg|_{t=0}^{t=\tau} = 0.$$

Using the initial condition for u and the definition of w we have

$$\int_\Omega \rho u^2(\mathbf{x},\tau)\,d\Omega + \int_\Omega \left\{ \left(\frac{\partial w(\mathbf{x},t)}{\partial x}\right)^2 + \left(\frac{\partial w(\mathbf{x},t)}{\partial y}\right)^2 \right\} d\Omega \Bigg|_{t=0} = 0.$$

Here all integrands are positive so

$$\int_\Omega \rho u^2(\mathbf{x},\tau)\,d\Omega = 0.$$

Since τ is an arbitrary point of $[0,T]$ we have $u = 0$. □

Let us recall that because of uniqueness it can be shown (by way of contradiction) that the whole Faedo–Galerkin sequence of approximations $\{u_n\}$ converges weakly to the generalized solution of the problem under consideration in the energy space.

4.8 Other Dynamic Problems of Linear Mechanics

Let us briefly consider the changes necessary in order to treat various other dynamical problems of mechanics.

We begin with a mixed problem for the membrane. If a portion of the edge is free from clamping and loading, how must our approach change? Only in the definition of the energy space. The removal of restrictions on the free part of the boundary simply requires us to use a wider energy space; then everything carries through as before, and the same theorems are formally established.

When on some part Γ_1 of the edge a load $f(s,t)$ is given, then the equation for generalized solution appears as follows:

$$\begin{aligned}\int_0^T \int_\Omega \left(\frac{\partial u}{\partial x}\frac{\partial v}{\partial x} + \frac{\partial u}{\partial y}\frac{\partial v}{\partial y} \right) d\Omega\, dt = &\int_0^T \int_\Omega f(\mathbf{x},t) v(\mathbf{x},t)\, d\Omega\, dt \\ &+ \int_0^T \int_{\Gamma_1} \varphi(s,t) v(s,t)\, ds\, dt \\ &+ \int_0^T \int_\Omega \rho \dot{u}(\mathbf{x},t)\dot{v}(\mathbf{x},t)\, d\Omega\, dt \\ &+ \int_\Omega \rho u_1^*(\mathbf{x}) v(\mathbf{x},0)\, d\Omega. \qquad (4.8.1)\end{aligned}$$

For solvability we also need

$$\varphi(s,t) \in W^{1,2}(L^2(\Gamma_1); 0, T).$$

Under this restriction it is possible to demonstrate an *a priori* estimate of the generalized solution, and thus to prove existence of a generalized solution. The formulation and proof of uniqueness remain practically unchanged (except for the definition and notation for the energy space).

We shall not consider in detail all the other problems of dynamics for the objects we studied in statics. The introduction of the main equation of motion always repeats all the steps we performed for the membrane. The corresponding energy space formulation, in which the inner product is denoted by $(\cdot,\cdot)_\mathcal{E}$, yields

$$\int_0^T (u(t),v(t))_\mathcal{E}\,dt = \int_0^T\int_\Omega f(\mathbf{x},t)v(\mathbf{x},t)\,d\Omega\,dt + \int_0^T\int_{\Gamma_1} \varphi(s,t)v(s,t)\,ds\,dt + \int_0^T\int_\Omega \rho\dot{u}(\mathbf{x},t)\dot{v}(\mathbf{x},t)\,d\Omega\,dt + \int_\Omega \rho u_1^*(\mathbf{x})v(\mathbf{x},0)\,d\Omega$$

which parallels (4.8.1) for the membrane. All the reasoning leading to the main theorems remains the same; again, the differences lie only in the definitions of the appropriate energy spaces. We leave it to the reader to formulate and prove the existence and uniqueness of generalized solutions for initial-boundary value problems in the theory of plates and for 3-D and 2-D elastic bodies.

4.9 The Fourier Method

One of the main methods for solving dynamical problems is that developed by Fourier. The method facilitates the description of transient processes. Normally the class of loads considered analytically is not wide, and it is possible to find a partial solution that "removes" the effect of the load; it then remains to find how the behavior of a non-loaded object changes from some arbitrary initial state. For solving the latter problem, Fourier proposed a method of separation of variables. As an example let us consider the dynamic problem for a string, described by

$$\frac{\partial^2 u}{\partial t^2} = \frac{\partial^2 u}{\partial x^2}, \qquad x \in [0,\pi] \tag{4.9.1}$$

with initial and boundary conditions

$$u(0,t) = 0 = u(\pi,t), \qquad u(x,0) = u_0(x), \qquad \frac{\partial u(x,0)}{\partial t} = u_1(x). \tag{4.9.2}$$

We seek a particular solution to (4.9.1) in the form $u(x,t) = T(t)v(x)$. From (4.9.1) we have

$$\frac{T''(t)}{T(t)} = \frac{X''(x)}{X(x)} = -\lambda^2.$$

The value λ can only be constant since each fraction of the equality depends on only one of the independent variables x or t. We need to find non-trivial solutions of this form. The equation

$$X''(x) + \lambda^2 X(x) = 0 \tag{4.9.3}$$

with the necessary boundary conditions

$$X(0) = 0 = X(\pi) \tag{4.9.4}$$

has nontrivial solutions only when $\lambda = k$, k being a positive or negative integer; that is, $X_k(x) = c \sin kx$. There are no other non-trivial solutions to (4.9.3)–(4.9.4), which is typical of eigenvalue problems for distributed systems. Using this, we then find an adjoint solution for the equation

$$T''(t) + k^2 T(t) = 0,$$

whose general solution is

$$T_k(t) = c_{k0} \cos kt + c_{k1} \sin kt.$$

Hence Fourier obtained a general solution to the string (or wave) equation as

$$\sum_{k=1}^{\infty} (c_{k0} \cos kt + c_{k1} \sin kt) \sin kx. \tag{4.9.5}$$

Finally, we can look for coefficients that satisfy (4.9.2). So a central role in Fourier theory is played by the eigenvalue problem, the problem of finding nontrivial solutions to a boundary value problem with a parameter, (4.9.3)–(4.9.4). A similar problem arises in all linear mechanical problems, and in a similar fashion. In fact, we could begin at once to seek a class of particular solutions of the form $e^{i\mu t} v(x)$ where $v(0) = 0 = v(\pi)$. Now we have the same eigenvalue problem for $v(x)$:

$$v''(x) + \mu^2 v(x) = 0.$$

Moreover, when we seek a general solution as a sum of particular real solutions, we come to the same expression (4.9.5). The same can be said for any of the linear mechanical problems considered earlier. Thus in every case we come to a particular eigenvalue boundary value problem, then to the problem of finding the coefficients of the corresponding Fourier series of the type (4.9.5), and finally to the problem of convergence. This will be considered in detail in the next few sections.

4.10 An Eigenfrequency Boundary Value Problem Arising in Linear Mechanics

For each problem considered earlier, the dynamic equations with use of the D'Alembert principle have the form

$$(u, \eta)_{\mathcal{E}} = -\int_{\Omega} \rho \frac{\partial^2 u}{\partial t^2} \eta \, d\Omega \tag{4.10.1}$$

where $(\cdot, \cdot)_{\mathcal{E}}$ is a scalar product in the corresponding energy space and η is an admissible virtual displacement. To formulate the eigenfrequency problem accompanying this equation, we put $u = e^{i\mu t} v(\mathbf{x})$ in (4.10.1) and obtain

$$(v, \eta)_{\mathcal{E}} = \rho\mu^2 \int_{\Omega} v\eta \, d\Omega. \tag{4.10.2}$$

Let us put $\rho = 1$ (this can be done by appropriate choice of dimensional units, for example; it is done only to simplify the calculations). Since we now consider complex-valued u, we let η be complex as well. The second multiplier in a complex inner product is complex conjugated, so (4.10.2) takes the form

$$(v, \eta)_{\mathcal{E}} = \mu^2 \int_{\Omega} v\overline{\eta} \, d\Omega. \tag{4.10.3}$$

Equation (4.10.3) defines the general form of the eigenfrequency problems for the elastic objects considered in this chapter.

Definition 4.10.1 If (4.10.3) has a nonzero solution v at some μ, then v is called an *eigensolution* (*eigenvector*) and μ is called the corresponding *eigenfrequency*. The value $\lambda = 1/\mu$ is called the eigenvalue of the object.

Remark 4.10.1 We could arrive at the same eigenvalue problem by considering heat transfer described by

$$\frac{\partial T}{\partial t} = \Delta T$$

with zero temperature T on the boundary of the domain. If we seek a solution in the form $T(\mathbf{x}, t) = e^{-\mu t} v(\mathbf{x})$ in generalized statement, we get the equation that coincides with (4.10.3) governing eigen-oscillations of a membrane taking the same domain in the plane; the only discrepancy is the form of the parameter in the equation: it is μ for heat transfer whereas it is μ^2 for the membrane. Next, introducing $\lambda = 1/\mu$ in the heat problem we get a parameter that is usually called the eigenvalue. However we will keep our terminology since it makes more mechanical sense. Next there is a discrepancy between our terminology and that which is common in textbooks of mathematical physics: we call eigenfrequencies the quantities that are called eigenvalues in mathematical physics; the reason is that in mathematical physics they normally consider the equation in $L^2(\Omega)$ so $A = \Delta$ is considered as an unbounded operator in $L^2(\Omega)$ and the terminology is borrowed from standard spectral theory. But in our approach this differential operator corresponds to the identity operator in an energy space.

We have arrived at the problem (4.10.3) formulated in a complex energy space. The next lemma allows us to return to real spaces.

Lemma 4.10.1 *All possible eigenfrequencies of the problem* (4.10.3) *are real.*

Proof. The result follows from the fact that $(v, v)_\mathcal{E}$ and $\int_\Omega v\bar{v}\,d\Omega$ are positive numbers for any v, hence so is $\mu^2 = (v, v)_\mathcal{E} / \int_\Omega v\bar{v}\,d\Omega$. □

Since (4.10.3) is linear in v, now we can consider separately its real and imaginary parts, and so consider it only in a real energy space. Thus the equation we shall study is formulated in a real energy space, and so the eigenfrequency problem is as follows:

Eigenvalue Problem. Find a nonzero u belonging to a real energy space $\mathcal{E}$ that satisfies the equation

$$(u, v)_\mathcal{E} = \mu^2 \int_\Omega uv\,d\Omega \tag{4.10.4}$$

for any $v \in \mathcal{E}$.

We require that $\mathcal{E}$ is a Hilbert space and that there is a constant $m > 0$ such that

$$\|u\|_\mathcal{E} \geq m\,\|u\|_{W^{1,2}(\Omega)} \tag{4.10.5}$$

for any $u \in \mathcal{E}$. All the energy spaces we introduced had this property; in the case of a 3-D elastic body, u is a vector function, and in the integral on the right of (4.10.4) uv must mean a dot product of the displacement vectors $\mathbf{u}$ and $\mathbf{v}$.

Let us transform (4.10.4) into an operator form using the Riesz representation theorem. At any fixed $u \in \mathcal{E}$, the integral $\int_\Omega uv\,d\Omega$ is a functional linear in v. Schwarz's inequality, Sobolev's imbedding theorem, and (4.10.5) give us

$$\begin{aligned}\left|\int_\Omega uv\,d\Omega\right| &\le \|u\|_{L^2(\Omega)}\,\|v\|_{L^2(\Omega)}\\ &\le m_1\,\|u\|_{W^{1,2}(\Omega)}\,\|v\|_{W^{1,2}(\Omega)}\\ &\le m_2\,\|u\|_{\mathcal{E}}\,\|v\|_{\mathcal{E}}\,,\end{aligned} \tag{4.10.6}$$

which means this functional is continuous for $v \in \mathcal{E}$. Thus it can be represented as an inner product in $\mathcal{E}$:

$$\int_\Omega uv\,d\Omega = (w, v)_{\mathcal{E}}, \tag{4.10.7}$$

where $w \in \mathcal{E}$ is uniquely defined by u. (The second position of v in the inner product is unimportant since it is symmetric in its arguments.) Since to any $u \in \mathcal{E}$ there corresponds $w \in \mathcal{E}$, we have defined an operator acting in $\mathcal{E}$. Denoting this by A we have

$$w = Au.$$

With this notation (4.10.4) takes the form

$$(u, v)_{\mathcal{E}} = \mu^2(Au, v)_{\mathcal{E}}.$$

Since $v \in \mathcal{E}$ is arbitrary we get

$$u = \mu^2 Au.$$

Although A has been introduced theoretically, we should be able to establish some of its properties through the defining equality

$$(Au, v)_{\mathcal{E}} = \int_\Omega uv\,d\Omega. \tag{4.10.8}$$

Let us begin.

Lemma 4.10.2 *The operator A is linear and continuous on $\mathcal{E}$.*

Proof. For linearity it is enough to establish the equality

$$A(\alpha_1 u_1 + \alpha_2 u_2) = \alpha_1 Au_1 + \alpha_2 Au_2 \tag{4.10.9}$$

for any real numbers α_i and elements $u_i \in \mathcal{E}$. By (4.10.8) we have

$$\begin{aligned}(A(\alpha_1 u_1 + \alpha_2 u_2), v)_{\mathcal{E}} &= \int_\Omega (\alpha_1 u_1 + \alpha_2 u_2) v \, d\Omega \\ &= \alpha_1 \int_\Omega u_1 v \, d\Omega + \alpha_2 \int_\Omega u_2 v \, d\Omega.\end{aligned}$$

On the other hand

$$(Au_i, v)_{\mathcal{E}} = \int_\Omega u_i v \, d\Omega, \qquad i = 1, 2,$$

and thus

$$(A(\alpha_1 u_1 + \alpha_2 u_2), v)_{\mathcal{E}} = \alpha_1 (Au_1, v)_{\mathcal{E}} + \alpha_2 (Au_2, v)_{\mathcal{E}} .$$

From this (4.10.9) follows by the arbitrariness of v. To prove continuity of A let us use (4.10.6), from which

$$|(Au, v)_{\mathcal{E}}| = \left| \int_\Omega uv \, d\Omega \right| \leq m_2 \|u\|_{\mathcal{E}} \|v\|_{\mathcal{E}} .$$

Setting $v = Au$, we get for an arbitrary u

$$|(Au, Au)_{\mathcal{E}}| \leq m_2 \|u\|_{\mathcal{E}} \|Au\|_{\mathcal{E}} .$$

It follows that

$$\|Au\|_{\mathcal{E}} \leq m_2 \|u\|_{\mathcal{E}} ,$$

and this completes the proof. □

Definition 4.10.2 An operator B is called *strictly positive* in a Hilbert space H if

$$(Bx, x) \geq 0$$

for any $x \in H$, and from the equality $(Bx, x) = 0$ it follows that $x = 0$.

Lemma 4.10.3 *The operator A is strictly positive in $\mathcal{E}$.*

Proof. Clearly

$$(Au, u)_{\mathcal{E}} = \int_\Omega u^2 \, d\Omega \geq 0.$$

If $(Au, u)_{\mathcal{E}} = 0$, then $u = 0$ in $L^2(\Omega)$ and thus in $\mathcal{E}$. □

Lemma 4.10.4 *The operator A is self-adjoint.*

Proof. From the symmetry in the arguments u, v in the definition (4.10.8) and continuity of A, the proof follows immediately. Indeed,

$$(Au, v)_{\mathcal{E}} = \int_\Omega uv \, d\Omega = \int_\Omega vu \, d\Omega = (Av, u)_{\mathcal{E}} = (u, Av)_{\mathcal{E}}.$$

□

The last property we wish to establish is

Lemma 4.10.5 *The operator A is compact.*

Proof. It is enough to demonstrate that for any weakly Cauchy sequence $\{u_n\}$ the corresponding $\{Au_n\}$ is a strongly Cauchy sequence. Let $\{u_n\}$ be a weakly Cauchy sequence in $\mathcal{E}$. By (4.10.5) it is a weakly Cauchy sequence in $W^{1,2}(\Omega)$ and thus, by Sobolev's imbedding theorem, it is a strongly Cauchy sequence in $L^2(\Omega)$. Let us use an inequality following from (4.10.6):

$$\left| \int_\Omega uv \, d\Omega \right| \le m_3 \|u\|_{L^2(\Omega)} \|v\|_{\mathcal{E}},$$

so that

$$|(A(u_n - u_m), v)_{\mathcal{E}}| = \left| \int_\Omega (u_n - u_m) v \, d\Omega \right| \le m_3 \|u_n - u_m\|_{L^2(\Omega)} \|v\|_{\mathcal{E}}.$$

Putting $v = A(u_n - u_m)$ we get

$$|(A(u_n - u_m), A(u_n - u_m))_{\mathcal{E}}| \le m_3 \|u_n - u_m\|_{L^2(\Omega)} \|A(u_n - u_m)\|_{\mathcal{E}}$$

so that

$$\|A(u_n - u_m)\|_{\mathcal{E}} \le m_3 \|u_n - u_m\|_{L^2(\Omega)} \to 0 \quad \text{as } n, m \to \infty.$$

This completes the proof. □

4.11 The Spectral Theorem

The results of this section are general despite their formulation in energy spaces. They apply in any separable Hilbert space $\mathcal{E}$, whether or not the space relates to any mechanical problem. We suppose A is a self-adjoint, strictly positive, compact operator acting in a real Hilbert space $\mathcal{E}$. The inner product in $\mathcal{E}$ is denoted $(u, v)_{\mathcal{E}}$. Because A is self-adjoint and strictly positive, the bilinear functional $(Au, v)_{\mathcal{E}}$ has all the properties of an inner product. Let us denote this inner product by

$$(u, v)_A = (Au, v)_{\mathcal{E}}$$

and its corresponding norm by $\|u\|_A = (u, u)_A^{1/2}$.

Since $\mathcal{E}$ is incomplete with respect to the new norm we can apply the completion theorem. The completion of $\mathcal{E}$ with respect to the norm $\|u\|_A$ is denoted by $\mathcal{E}_A$ and is called the energy space of the operator A. But, unlike the earlier energy spaces, this energy space for the problems under consideration does not relate to the system energy. Looking at the form of the inner product in $\mathcal{E}_A$ for A from the previous section, we see that it is an inner product in $L^2(\Omega)$. Moreover, from the general theory of the L^p spaces it is known that infinitely differentiable functions whose support is compact in Ω (so they are zero on the boundary of Ω) are dense in $L^2(\Omega)$. This means the resulting space $\mathcal{E}_A$ for the problems of the previous section is $L^2(\Omega)$ (more precisely, we can put their elements into one-to-one correspondence in such a way that all the distances between elements are preserved).

In what follows we need

Definition 4.11.1 A functional F is called *weakly continuous* at a point u if for any sequence $\{u_n\}$ weakly convergent to u we have $F(u_n) \to F(u)$ as $n \to \infty$. A functional is weakly continuous on a domain M if it is weakly continuous at each point $u \in M$.

By definition a linear weakly continuous functional is continuous, and vice versa.

Lemma 4.11.1 *A functional $F(u)$ weakly continuous on the unit ball $\|u\|_{\mathcal{E}} \leq 1$ of a Hilbert space $\mathcal{E}$ takes its minimal and maximal values on this ball.*

Proof. This is similar to a classical theorem of calculus on the extremes of a continuous function given on a compact set. We prove the statement for maxima of F. The result for minima will then follow by consideration of $-F$. Let $\{u_n\}$ be a sequence in the unit ball, denoted by B, such that

$$F(u_n) \to \sup_{\|u\|_{\mathcal{E}} \leq 1} F(u) \quad \text{as } n \to \infty.$$

Since $\{u_n\}$ lies in B it contains a weakly convergent subsequence $\{u_{n_k}\}$. Since B is weakly closed in $\mathcal{E}$ this subsequence has a weak limit u^* belonging to B. The value $F(u^*)$ is finite and since F is weakly continuous we have

$$F(u_{n_k}) \to F(u^*) = \sup_{\|u\|_{\mathcal{E}} \leq 1} F(u),$$

so u^* is the needed point. □

Lemma 4.11.2 *Let A be a compact linear operator in a Hilbert space $\mathcal{E}$. Then $F(u) = (Au, u)_\mathcal{E}$ is a weakly continuous functional in $\mathcal{E}$.*

Proof. Let $\{u_n\}$ be weakly convergent to u. Consider

$$\begin{aligned}|(Au_n, u_n)_\mathcal{E} - (Au, u)_\mathcal{E}| &= |(Au_n, u_n)_\mathcal{E} - (Au, u_n)_\mathcal{E} + (Au, u_n)_\mathcal{E} - (Au, u)_\mathcal{E}| \\ &\le |(Au_n, u_n)_\mathcal{E} - (Au, u_n)_\mathcal{E}| + |(Au, u_n)_\mathcal{E} - (Au, u)_\mathcal{E}| \\ &\le \|A(u_n - u)\|_\mathcal{E} \|u_n\|_\mathcal{E} + |(Au, u_n - u)_\mathcal{E}| \\ &\to 0 \quad \text{as } n \to \infty.\end{aligned}$$

For the first addendum this happened since $\|u_n\|_\mathcal{E}$ is bounded and $A(u_n - u) \to 0$ strongly in $\mathcal{E}$. The second addendum tends to zero since it is a linear continuous functional in $u_n - u$. □

For a strictly positive operator all the eigenvalues are nonnegative (why?) and so we will denote them as λ^2: $Ax = \lambda^2 x$. This is done to preserve the terminology of mechanics, where the corresponding value $\mu = 1/\lambda$ is called an eigenfrequency of the object. Now let us formulate the main result of this section.

Theorem 4.11.1 *Let A be a self-adjoint, strictly positive, compact operator acting in a real separable Hilbert space. Then*

(i) A has a countable set of eigenfrequencies with no finite limit point;
(ii) to each eigenfrequency of A there corresponds a finite dimensional set of eigenvectors $\{\varphi_k\}$; we can choose eigenvectors constituting an orthonormal basis;
(iii) the union of all orthonormal bases $\{\varphi_k\}$ corresponding to the eigenfrequencies of A is orthonormal in $\mathcal{E}$;
(iv) the same union $\{\varphi_k\}$ is an orthogonal basis in $\mathcal{E}_A$;
(v) for any $u \in \mathcal{E}$ there holds

$$Au = \sum_{k=1}^{\infty} \lambda_k^2 (u, \varphi_k)_\mathcal{E} \varphi_k, \qquad A\varphi_k = \lambda_k^2 \varphi_k. \tag{4.11.1}$$

We subdivide the proof into Lemmas 4.11.1 through 4.11.7. Statements (i) and (ii) are consequences of the Fredholm–Riesz–Schauder theory of compact operators. Statement (iii) is a consequence of the fact that the operator A is self-adjoint. So we know some properties of the eigenvalues of A, but it remains unknown whether the set of eigenvectors is nonempty. First we demonstrate the existence of one such eigenvector.

Lemma 4.11.3 *For a self-adjoint strictly positive compact linear operator A acting in $\mathcal{E}$*

$$\lambda_1^2 = \sup_{\|u\|_{\mathcal{E}} \leq 1} (Au, u)_{\mathcal{E}}$$

is an eigenvalue of A. It is also the largest eigenvalue of A, and the lowest eigenfrequency of A is $\mu_1 = 1/\lambda_1$.

Proof. If λ^2 is an eigenvalue then $Au = \lambda^2 u$ and it follows that $(Au, u)_{\mathcal{E}} = \lambda^2 \|u\|_{\mathcal{E}}^2$. So for $\|u\|_{\mathcal{E}} \leq 1$ we have $(Au, u)_{\mathcal{E}} \leq \lambda_1^2$, and thus all the eigenvalues are non-negative and less than or equal to $\lambda_1^2 > 0$. Let us demonstrate that λ_1^2 is an eigenvalue of A. By Lemmas 4.11.1 and 4.11.2 we know that $\sup(Au, u)_{\mathcal{E}}$ is attained on some point φ_1 of the ball $\|u\|_{\mathcal{E}} \leq 1$. Since the form $(Au, u)_{\mathcal{E}}$ is homogeneous in u, we know that φ_1 belongs to the unit sphere $\|u\|_{\mathcal{E}} = 1$:

$$\lambda_1^2 = (A\varphi_1, \varphi_1)_{\mathcal{E}}, \qquad \|\varphi_1\|_{\mathcal{E}} = 1.$$

We show that φ_1 is an eigenvector of A. It is clear that λ_1^2 can be defined as the maximum of the form $(Au, u)_{\mathcal{E}}$ on the unit sphere $\|u\|_{\mathcal{E}} = 1$. Because of homogeneity the same can be said about the functional

$$G(u) = \frac{(Au, u)_{\mathcal{E}}}{\|u\|_{\mathcal{E}}^2} = (Av, v)_{\mathcal{E}}, \qquad v = \frac{u}{\|u\|_{\mathcal{E}}}, \qquad \|v\|_{\mathcal{E}} = 1.$$

Thus $G(u)$ takes the same set of values as $(Au, u)_{\mathcal{E}}$ on the unit sphere $\|u\|_{\mathcal{E}} = 1$ and, moreover, it attains its maximal value equal to λ_1^2 at the same point φ_1. Consider $G(\varphi_1 + \alpha w)$ for a fixed $w \in \mathcal{E}$. This is a function continuously differentiable in α in some neighborhood of $\alpha = 0$, and attaining its maximum at $\alpha = 0$. Thus

$$\left. \frac{dG(\varphi_1 + \alpha w)}{d\alpha} \right|_{\alpha=0} = 0.$$

Calculating this we get

$$(A\varphi_1, w)_{\mathcal{E}} - \frac{(A\varphi_1, \varphi_1)_{\mathcal{E}}}{\|\varphi_1\|_{\mathcal{E}}} (\varphi_1, w)_{\mathcal{E}} = 0;$$

that is,

$$\left(A\varphi_1 - \lambda_1^2 \varphi_1, w\right)_{\mathcal{E}} = 0$$

for arbitrary $w \in \mathcal{E}$. This means φ_1 is an eigenvector and λ_1^2 is an eigenvalue of A. □

Now we are going to describe a procedure for finding other eigenvectors and eigenvalues of A, using the established property that the set of all eigenvectors of A has an orthonormal basis. We know how to find the first eigenvector. For the rest we shall use the procedure whose ith step is as follows. Let $\varphi_1, \ldots, \varphi_n$ be mutually orthogonal eigenvectors determined by the procedure. Denote by $\mathcal{E}_{n\perp}$ the orthogonal complement in $\mathcal{E}$ of the subspace of $\mathcal{E}$ spanned by $\varphi_1, \ldots, \varphi_n$. Consider the operator A given on $\mathcal{E}_{n\perp}$. We can repeat the reasoning of Lemma 4.11.3 and find an eigenvalue denoted by λ_{n+1}^2 and an eigenvector φ_{n+1} of the restriction of A to $\mathcal{E}_{n\perp}$. So

$$\left(A\varphi_{n+1} - \lambda_{n+1}^2\varphi_{n+1}, w\right)_{\mathcal{E}} = 0 \qquad (4.11.2)$$

holds for any $w \in \mathcal{E}_{n\perp}$. Now we show that this holds for any $w \in \mathcal{E}$. By the orthogonal decomposition theorem, it is enough to prove that (4.11.2) holds when w is any of the previous eigenvectors $\varphi_1, \ldots, \varphi_n$. Since for any $i < n+1$

$$(\varphi_{n+1}, \varphi_i)_{\mathcal{E}} = 0 \quad \text{and} \quad (A\varphi_{n+1}, \varphi_i)_{\mathcal{E}} = (\varphi_{n+1}, A\varphi_i)_{\mathcal{E}} = \lambda_i^2\,(\varphi_{n+1}, \varphi_i)_{\mathcal{E}} = 0,$$

it follows that (4.11.2) holds for any $w \in \mathcal{E}$. Hence we really did obtain the next eigenpair.

Lemma 4.11.4 *For an infinite dimensional space $\mathcal{E}$, the eigenvalues of A are countable. The corresponding eigenfrequencies $\mu_i = 1/\lambda_i$, $\lambda_i > 0$, are such that $\mu_i \le \mu_{i+1} \to +\infty$ as $i \to \infty$.*

Proof. The above procedure can terminate only when we get some subspace $\mathcal{E}_{n\perp}$ on the unit ball of which $\sup(Au, u)_{\mathcal{E}} = 0$. But then $\mathcal{E}_{n\perp}$ contains only the zero element since A is strictly positive. So $\mathcal{E}$ is finite dimensional, a contradiction. The rest of the lemma follows from the method of constructing the eigenvalues. □

Lemma 4.11.5 *The set of all the constructed eigenvectors $\varphi_1, \ldots, \varphi_n, \ldots$ is an orthonormal basis of $\mathcal{E}$.*

Proof. Take any $u \in \mathcal{E}$ and consider the remainder of the Fourier series

$$u_n = u - \sum_{k=1}^{n} (u, \varphi_k)_{\mathcal{E}} \varphi_k.$$

We see that $(u_n, \varphi_k)_{\mathcal{E}} = 0$ for $k \le n$, and thus $u_n \in \mathcal{E}_{n\perp}$. From Fourier expansion theory we know that $\{\sum_{k=1}^{n}(u, \varphi_k)_{\mathcal{E}}\varphi_k\}$ is convergent, hence so is $\{u_n\}$. Suppose, contrary to the statement of the lemma, that the strong limit of $\{u_n\}$ is $u_0 \neq 0$. By the procedure for finding eigenvalues and the fact that u_n is in $\mathcal{E}_{n\perp}$, we have

$$\frac{(Au_n, u_n)_{\mathcal{E}}}{\|u_n\|_{\mathcal{E}}^2} \leq \lambda_{n+1}^2.$$

Passage to the limit in n implies

$$\frac{(Au_0, u_0)_{\mathcal{E}}}{\|u_0\|_{\mathcal{E}}^2} \leq 0;$$

hence $u_0 = 0$, and this completes the proof. □

Lemma 4.11.6 *For any $u \in \mathcal{E}$ there holds* (4.11.1), *i.e.,*

$$Au = \sum_{k=1}^{\infty} \lambda_k^2 (u, \varphi_k)_{\mathcal{E}} \varphi_k, \qquad A\varphi_k = \lambda_k^2 \varphi_k.$$

Proof. The Fourier series $u = \sum_{k=1}^{\infty}(u, \varphi_k)_{\mathcal{E}}\varphi_k$ is strongly convergent. Applying a compact (and hence continuous) operator A we get

$$Au = \sum_{k=1}^{\infty}(u, \varphi_k)_{\mathcal{E}} A\varphi_k = \sum_{k=1}^{\infty} \lambda_k^2 (u, \varphi_k)_{\mathcal{E}} \varphi_k,$$

as required. □

The last non-proven statement of the theorem follows from

Lemma 4.11.7 *The set $\psi_k = \varphi_k/\lambda_k$, $\lambda_k > 0$, $k = 1, 2, 3, \ldots$, is an orthonormal basis of $\mathcal{E}_A$.*

Proof. Mutual orthogonality of the ψ_k in $\mathcal{E}_A$ follows from the equality chain

$$(\psi_i, \psi_j)_A = (A\psi_i, \psi_j)_{\mathcal{E}} = \left(\frac{1}{\lambda_i} A\varphi_i, \frac{\varphi_j}{\lambda_j}\right)_{\mathcal{E}} = \frac{\lambda_i^2}{\lambda_i \lambda_j}(\varphi_i, \varphi_j)_{\mathcal{E}}.$$

Hence the set is orthonormal as well. For the proof it is enough to demonstrate that for any $u \in \mathcal{E}$ Parseval's equality in $\mathcal{E}_A$ holds. This is shown by the chain of transformations

$$\begin{aligned}(u,u)_A &= (Au,u)_\mathcal{E} = \left(\sum_{k=1}^{\infty}(u,\varphi_k)_\mathcal{E}A\varphi_k, u\right)_\mathcal{E} = \sum_{k=1}^{\infty}(u,\varphi_k)_\mathcal{E}(A\varphi_k,u)_\mathcal{E}\\ &= \sum_{k=1}^{\infty}\left(u,\frac{A\varphi_k}{\lambda_k^2}\right)_\mathcal{E}(A\varphi_k,u)_\mathcal{E} = \sum_{k=1}^{\infty}(u,A\psi_k)_\mathcal{E}(A\psi_k,u)_\mathcal{E}\\ &= \sum_{k=1}^{\infty}(u,\psi_k)_A^2.\end{aligned}$$

□

4.12 The Fourier Method, Continued

We have obtained general results on the structure of the spectrum and the properties of the eigenvalue problem for a strictly positive, self-adjoint, compact linear operator A. This eigenvalue problem includes all the eigenvalue problems of linear mechanics that we have considered.

In § 4.9 we began to study the Fourier method for dynamical linear problems. We tried to find a general solution of a general linear initial-boundary value problem for a body free of external load. However, the fact that the eigenvectors of A, satisfying

$$\lambda_k^2(\varphi_k, v)_\mathcal{E} = \int_\Omega \varphi_k(\mathbf{x})v(\mathbf{x})\,d\Omega,$$

constitute an orthogonal basis in $\mathcal{E}$ and $\mathcal{E}_A$ simultaneously, allows us to consider the problem for a loaded body as well. Here the Fourier method appears to relate to the Faedo–Galerkin method for a special basis, namely for the eigenvectors of the operator A which is now well defined by the relation (4.10.7). Let us recall that for the basis

$$(\varphi_i,\varphi_j)_\mathcal{E} = \delta_{ij} = \begin{cases}1, & i=j,\\ 0, & i\neq j,\end{cases} \qquad \int_\Omega \varphi_i(\mathbf{x})\varphi_j(\mathbf{x})\,d\Omega = \lambda_i^2\delta_{ij}. \tag{4.12.1}$$

Let us review the general notations of this section. In $\mathcal{E}(0,T)$ an inner product is defined as

$$(u,v)_{[0,T]} = \int_0^T (u,v)_\mathcal{E}\,dt + \int_0^T\int_\Omega \dot{u}(\mathbf{x},t)\dot{v}(\mathbf{x},t)\,d\Omega\,dt$$

(changing the dimensions we put $\rho = 1$) and D_0^T denotes the subspace that is the completion of that subset of the base of $\mathcal{E}(0,T)$ which are equal to zero at $t = T$. A generalized solution $u \in \mathcal{E}(0,T)$ is defined by

$$\int_0^T (u,v)_{\mathcal{E}}\,dt = \int_0^T \int_\Omega f(\mathbf{x},t)v(\mathbf{x},t)\,d\Omega\,dt + \int_0^T \int_\Omega \dot{u}(\mathbf{x},t)\dot{v}(\mathbf{x},t)\,d\Omega\,dt + \int_\Omega u_1^*(\mathbf{x})v(\mathbf{x},0)\,d\Omega \qquad (4.12.2)$$

for any $v \in D_0^T$. Note that the initial condition for the first time derivative, that is u_1^*, is taken into account in (4.12.2); we do not require it to hold separately. Another initial condition

$$u(\mathbf{x},t)\big|_{t=0} = u_0^*(\mathbf{x})$$

must be satisfied in the sense of $L^2(\Omega)$; see Definition 4.7.1. The boundary conditions are hidden inside the definition of $\mathcal{E}$. We recall that we require $u_0^*(\mathbf{x}) \in \mathcal{E}$, $u_1^*(\mathbf{x}) \in \mathcal{E}_A$, $f(\mathbf{x},t) \in L^2(\Omega \times [0,T])$. Now we return to the Faedo–Galerkin method with the basis elements φ_k, $k = 1, 2, \ldots$, that are eigenvectors of A with the properties we studied earlier. Let us seek the nth Faedo–Galerkin approximation $u_n = \sum_{k=1}^n c_k(t)\varphi_k$ to the generalized solution given by the equations

$$\ddot{c}_i(t)\int_\Omega \varphi_i^2(\mathbf{x})\,d\Omega = -(\varphi_i,\varphi_i)_{\mathcal{E}}c_i(t) + \int_\Omega f(\mathbf{x},t)\varphi_i(\mathbf{x})\,d\Omega, \qquad i = 1,\ldots,n \qquad (4.12.3)$$

or, because of (4.12.1),

$$\ddot{c}_i(t) + \mu_i^2 c_i(t) = f_i(t), \quad \mu_i = 1/\lambda_i, \qquad i = 1,\ldots,n \qquad (4.12.4)$$

where

$$f_i(t) = \mu_i^2 \int_\Omega f(\mathbf{x},t)\varphi_i(\mathbf{x})\,d\Omega$$

and eigenfrequencies $\mu_i = 1/\lambda_i \to \infty$. We see that equations (4.12.4) are mutually independent. Let us derive the initial conditions for these equations. Denoting $c_i(0) = d_{0i}$, $\dot{c}_i(0) = d_{1i}$, and remembering that d_{0i} are defined by

$$\left\| u_0^* - \sum_{k=1}^n d_{0k}\varphi_k \right\|_{\mathcal{E}}^2 \to \min$$

we get

$$d_{0i}(\varphi_i,\varphi_i)_{\mathcal{E}} = (u_0^*,\varphi_i)_{\mathcal{E}}$$

so

$$c_i(0) = d_{0i} = (u_0^*, \varphi_i)_\mathcal{E} = (u_0^*, \mu_i A\varphi_i)_\mathcal{E} = \mu_i \int_\Omega u_0^*(\mathbf{x})\varphi_i(\mathbf{x})\,d\Omega. \quad (4.12.5)$$

Similarly, minimizing

$$\left\| u_1^* - \sum_{k=1}^{n} d_{1k}\varphi_k \right\|_A^2 \to \min$$

we obtain

$$d_{1i}(\varphi_i, \varphi_i)_A = (\varphi_i, u_1^*)_A$$

or

$$\dot{c}_i(0) = d_{1i} = \mu_i^2(\varphi_i, u_1^*)_A = \mu_i^2 \int_\Omega u_1^*(\mathbf{x})\varphi_i(\mathbf{x})\,d\Omega \quad (4.12.6)$$

so we see that the initial conditions are split as well. Because of the mutual orthogonality and basis properties of $\{\varphi_i\}$ in $\mathcal{E}$ and $\mathcal{E}_A$ we can rewrite the corresponding Parseval equalities

$$\sum_{i=1}^{\infty} d_{0i}^2 = \|u_0^*\|_\mathcal{E}^2 \quad (4.12.7)$$

and

$$\sum_{i=1}^{\infty} d_{1i}^2(\varphi_i, \varphi_i)_A = \sum_{i=1}^{\infty} d_{1i}^2\lambda_i^2 = \|u_1^*\|_A^2. \quad (4.12.8)$$

The solution of the problem (4.12.4), (4.12.5), (4.12.6) is

$$c_i(t) = d_{0i}\cos(\mu_i t) + d_{1i}\sin(\mu_i t) + \frac{1}{\mu_i}\int_0^t f_i(\tau)\sin\mu_i(t-\tau)\,d\tau.$$

It is easily seen that $c_i(t)$ is continuously differentiable on $[0, T]$. Note that unlike the case of general complete system of basis elements the coefficients of the Faedo–Galerkin method do not depend on the number of the step. Let us see the behavior of corresponding partial sums of formal series

$$u(\mathbf{x}, t) = \sum_{i=1}^{\infty} \Bigg(d_{0i}\cos(\mu_i t) + d_{1i}\sin(\mu_i t) + \frac{1}{\mu_i}\int_0^t f_i(\tau)\sin\mu_i(t-\tau)\,d\tau \Bigg)\varphi_i(\mathbf{x}). \quad (4.12.9)$$

Let us note that the part

$$u(\mathbf{x},t) = \sum_{i=1}^{\infty} \left(d_{0i}\cos(\mu_i t) + d_{1i}\sin(\mu_i t)\right)\varphi_i(\mathbf{x})$$

is a formal solution for the dynamic problem for a load-free elastic body by the Fourier method. From the above we know these partial sums weakly converge to a generalized solution of the dynamic problem. So in a certain way $u(\mathbf{x},t)$ given formally by (4.12.9) is this solution. We will establish the properties of the series (4.12.9) and thus those of the generalized solution.

Let us consider the convergence of series (4.12.9). For this multiply the identity (4.12.3) term by term by $\dot{c}_i(t)$ and then sum up the equalities in i:

$$\sum_{i=1}^{n} \ddot{c}_i(t)\dot{c}_i(t)\int_\Omega \varphi_i^2(\mathbf{x})\,d\Omega + \sum_{i=1}^{n} c_i(t)\dot{c}_i(t)(\varphi_i,\varphi_i)_\varepsilon$$
$$= \sum_{i=1}^{n}\int_\Omega f(\mathbf{x},t)\dot{c}_i(t)\varphi_i(\mathbf{x})\,d\Omega$$

or

$$\frac{1}{2}\frac{d}{dt}\left(\sum_{i=1}^{n} \dot{c}_i^2(t)\int_\Omega \varphi_i^2(\mathbf{x})\,d\Omega + \sum_{i=1}^{n} c_i^2(t)(\varphi_i,\varphi_i)_\varepsilon\right)$$
$$= \int_\Omega f(\mathbf{x},t)\left(\sum_{i=1}^{n}\dot{c}_i(t)\varphi_i(\mathbf{x})\right)d\Omega.$$

Note that we are repeating the way in which the estimate of the Faedo–Galerkin approximation was obtained. So redenoting t by τ and integrating the last equality in τ over $[0,t]$ we get

$$\frac{1}{2}\left(\sum_{i=1}^{n}\dot{c}_i^2(t)\int_\Omega \varphi_i^2(\mathbf{x})\,d\Omega + \sum_{i=1}^{n} c_i^2(t)(\varphi_i,\varphi_i)_\mathcal{E}\right)$$
$$= \frac{1}{2}\left(\sum_{i=1}^{n}\dot{c}_i^2(0)\int_\Omega \varphi_i^2(\mathbf{x})\,d\Omega + \sum_{i=1}^{n} c_i^2(0)(\varphi_i,\varphi_i)_E\right)$$
$$+ \int_0^t \int_\Omega f(\mathbf{x},\tau)\left(\sum_{i=1}^{n}\dot{c}_i(\tau)\varphi_i(\mathbf{x})\right)d\Omega\,d\tau$$
$$\le \frac{1}{2}\sum_{i=1}^{n}\left(d_{1i}^2\lambda_i^2 + d_{0i}^2\right) + T\int_0^t\int_\Omega f^2(\mathbf{x},\tau)\,d\Omega\,d\tau$$
$$+ \frac{1}{4T}\int_0^t\int_\Omega\left(\sum_{i=1}^{n}\dot{c}_i(\tau)\varphi_i(\mathbf{x})\right)^2 d\Omega\,d\tau$$
$$= \frac{1}{2}\sum_{i=1}^{n}\left(d_{1i}^2\lambda_i^2 + d_{0i}^2\right) + T\int_0^t\int_\Omega f^2(\mathbf{x},\tau)\,d\Omega\,d\tau$$
$$+ \frac{1}{4T}\int_0^t\sum_{i=1}^{n}\dot{c}_i^2(\tau)\left(\int_\Omega \varphi_i^2(\mathbf{x})\,d\Omega\right)^2 d\tau.$$

Here we used the elementary inequality $|ab| \le a^2/(4T) + Tb^2$ and mutual orthogonality of the φ_i in $\mathcal{E}_A = L^2(\Omega)$. Taking maximum values on $[0,T]$ in the last inequalities we get

$$\frac{1}{2}\max_{t\in[0,T]}\left(\sum_{i=1}^{n}\dot{c}_i^2(t)\int_\Omega \varphi_i^2(\mathbf{x})\,d\Omega + \sum_{i=1}^{n} c_i^2(t)(\varphi_i,\varphi_i)_\mathcal{E}\right)$$
$$\le \frac{1}{2}\sum_{i=1}^{n}\left(d_{1i}^2\lambda_i^2 + d_{0i}^2\right) + T\int_0^T\int_\Omega f^2(\mathbf{x},\tau)\,d\Omega\,d\tau$$
$$+ \frac{1}{4T}T\max_{\tau\in[0,T]}\sum_{i=1}^{n}\dot{c}_i^2(\tau)\left(\int_\Omega \varphi_i^2(\mathbf{x})\,d\Omega\right)^2$$

so

$$\frac{1}{2}\max_{t\in[0,T]}\left(\frac{1}{2}\sum_{i=1}^{n}\dot{c}_i^2(t)\int_\Omega \varphi_i^2(\mathbf{x})\,d\Omega + \sum_{i=1}^{n} c_i^2(t)(\varphi_i,\varphi_i)_\mathcal{E}\right)$$
$$\le \frac{1}{2}\sum_{i=1}^{n}\left(d_{1i}^2\lambda_i^2 + d_{0i}^2\right) + T\int_0^T\int_\Omega f^2(\mathbf{x},\tau)\,d\Omega\,d\tau. \tag{4.12.10}$$

The right-hand side of (4.12.10), because of (4.12.7) and (4.12.8), is bounded by a some constant M independent of n. Because of the properties of orthogonality of the basis elements and the form of the norm of a partial sum for series (4.12.9) that is $u_n(\mathbf{x}, t) = \sum_{i=1}^{n} c_i(t)\varphi_i(\mathbf{x})$ we have that the sequence $\{u_n\}$ converges in $C(\mathcal{E}; 0, T)$ and $\{du_n/dt\}$ converges in $C(\mathcal{E}_A; 0, T) = C(L^2(\Omega); 0, T)$. Thus the series (4.12.9), which is also a generalized solution to the problem under consideration, belongs to $C(\mathcal{E}; 0, T)$, whereas its time derivative $\partial u/\partial t$ belongs to $C(\mathcal{E}_A; 0, T)$. Simultaneously we justified convergence of the Fourier method for a free-load dynamical problem for an elastic body. We note that assuming existence of time derivatives of the force term f, in the same manner we can demonstrate that the solution has additional time derivatives. Moreover, for the free-load case we can demonstrate that the time derivative of any order of the solution is in $C(\mathcal{E}_A; 0, T)$.

4.13 Equilibrium of a von Kármán Plate

So far we have considered only linear problems of mechanics. Of course, such problems represent only the simplest approximations of the actual objects and processes of nature: although weakly nonlinear processes can often be analyzed with sufficient accuracy through the use of linear models, many important physical effects are inherently nonlinear. It is fortunate that the speed of machine computation has increased to the point where more realistic simulations of such effects have become possible. But the availability of numerical methods has also underscored the importance of analytical considerations. To work effectively we must know whether a solution exists and to which class of functions it belongs. We should also understand the differences between various methods of numerical solution and be prepared to place rigorous bounds on the error.

An important nonlinear problem, and one that can be regarded as a touchstone for many numerical methods, is the problem of equilibrium of a thin elastic plate under transverse load q. The plate is described by two nonlinear equations,

$$D\Delta^2 w = [f, w] + q, \tag{4.13.1}$$

$$\Delta^2 f = -[w, w], \tag{4.13.2}$$

given over a 2-D region Ω that represents the mid-surface of the plate. Here $w = w(x, y)$ is the transverse displacement of a point (x, y) of the mid-surface, $f = f(x, y)$ is the Airy stress function, D is the rigidity coefficient of the plate, and the notation $[u, v]$ is defined by

$$[u, v] = u_{xx}v_{xx} + u_{yy}v_{yy} - 2u_{xy}v_{xy}$$

where the subscripts x and y denote the partial derivatives $\partial/\partial x$ and $\partial/\partial y$, respectively. With suitably chosen dimensionless variables we can get $D = 1$. We shall consider the problem with the boundary conditions

$$w\big|_{\partial\Omega} = 0 = \left.\frac{\partial w}{\partial n}\right|_{\partial\Omega} \tag{4.13.3}$$

and

$$f\big|_{\partial\Omega} = 0 = \left.\frac{\partial f}{\partial n}\right|_{\partial\Omega}. \tag{4.13.4}$$

Conditions (4.13.3) mean that the edge of the plate is fixed against transverse displacement and rotation, and (4.13.4) means that the lateral boundary is not subjected to tangential load. In mechanics, condition (4.13.4) is derived for a simply connected domain. As usual we consider the domain Ω to be compact and to have a piecewise smooth boundary so that Sobolev's imbedding theorem for $W^{2,2}(\Omega)$ is applicable. If we neglect the term $[f, w]$ in (4.13.1), we get the linear equation of equilibrium of a plate under transverse load as was considered in Chapter 3. We would like to apply the tools of generalized setup of mechanical problems. Let us begin with the pair of integro-differential equations

$$a(w, \zeta) = B(f, w, \zeta) + \int_\Omega q\zeta \, d\Omega, \tag{4.13.5}$$

$$a(f, \eta) = -B(w, w, \eta), \tag{4.13.6}$$

where

$$a(u, v) = \int_\Omega (u_{xx}(v_{xx} + \mu v_{yy}) + 2(1-\mu)u_{xy}v_{xy} + u_{yy}(v_{yy} + \mu v_{xx})) \, d\Omega,$$

μ is Poisson's ratio for the material, $0 < \mu < 1/2$, and

$$B(u, v, \varphi) = \int_\Omega ((u_{xy}v_y - u_{yy}v_x)\varphi_x + (u_{xy}v_x - u_{xx}v_y)\varphi_y) \, d\Omega.$$

From a variational perspective, (4.13.5) and (4.13.6) would appear to constitute the first variation of some functional; we could regard ζ and η as arbitrary admissible smooth variations of w and f. Because such a viewpoint would bring us back to (4.13.1) and (4.13.2), we could try (4.13.5) and

(4.13.6) as equations appropriate for the generalized setup. There may be other forms of the bilinear functional $a(u,v)$ that yield the same equations (4.13.1) and (4.13.2) as a consequence of the variational technique; however, for other types of boundary conditions that differ from (4.13.3) this would lead us to incorrect natural boundary conditions. If we wish to consider boundary conditions for f including tangential load, then we need to take a different form of the left-hand side in (4.13.6) (see, for example, Vorovich [Vorovich (1999)]). But for conditions (4.13.4) we can forget about the physical meaning of the Airy function and use the same form of $a(u,v)$ in the generalized equation. Thus we are going to use (4.13.5) and (4.13.6) for the generalized setup of the equilibrium problem for von Kármán's plate. Our experience with the linear equilibrium problem for a plate suggests that we exploit the form $a(u,v)$ as an inner product in "energy" spaces for w and f. This means, by the results for a linear plate, that the solution will be sought in the subspace of $W^{2,2}(\Omega)$ consisting of the functions satisfying the boundary conditions (4.13.3). We need to see whether all the terms of (4.13.5) and (4.13.6) are sensible when all the functions included therein reside in the energy spaces (note that we now consider dimensionless versions of the equations). Of course, we need to suppose q satisfies at least the same conditions as for the generalized setup of the corresponding linear plate problem. For definiteness, let $q \in L(\Omega)$. We will check that the other terms in the equations are sensible. It is necessary to consider only the trilinear form $B(u,v,w)$. Apply Hölder's inequality for three functions to a typical term:

$$\left|\int_\Omega u_{xx} v_y w_x \, d\Omega\right| \le \left(\int_\Omega u_{xx}^2 \, d\Omega\right)^{1/2} \left(\int_\Omega v_y^4 \, d\Omega\right)^{1/4} \left(\int_\Omega w_x^4 \, d\Omega\right)^{1/4}$$
$$\le m \|u\|_P \|v\|_P \|w\|_P , \tag{4.13.7}$$

where we have used the fact that in $\mathcal{E}_{Pc}$ the norm

$$\|w\|_P = (a(w,w))^{1/2}$$

is equivalent to the norm of $W^{2,2}(\Omega)$ and elements of $W^{2,2}(\Omega)$ have the first derivatives belonging to $L^p(\Omega)$ with any finite $p > 1$, in particular for $p = 4$, which is necessary in Hölder's inequality. So all the terms of the equations have sense in the energy space. Thus, we can state the following definition:

Definition 4.13.1 A generalized solution to the equilibrium problem is a pair w, f that belongs to $\mathcal{E}_{Pc} \times \mathcal{E}_{Pc}$ and satisfies (4.13.5)–(4.13.6) for any

ζ, η from $\mathcal{E}_{Pc}$.

Equation (4.13.6) is linear in f. Using this we will eliminate f from the explicit statement of the problem. The right-hand side of (4.13.6) is linear in η; estimates of the type (4.13.7) give us

$$|B(w,w,\eta)| \le m \, \|w\|_P^2 \, \|\eta\|_P \,. \tag{4.13.8}$$

This means $B(w,w,\eta)$ is continuous in η so we can apply the Riesz representation theorem and state that for any fixed $w \in \mathcal{E}_{Pc}$

$$-B(w,w,\eta) = (C,\eta)_P = a(C,\eta). \tag{4.13.9}$$

This $C \in \mathcal{E}_{Pc}$, uniquely defined by w, is considered as the value of an operator in $\mathcal{E}_{Pc}$ at w: $C = C(w)$. Then (4.13.6) is rewritten as

$$a(f,\eta) = a(C(w),\eta)$$

and thus $f = C(w)$. We will make further use of this.

Let us call a *nonlinear* operator in a Hilbert space *completely continuous* if it takes any weakly Cauchy sequence into a strongly Cauchy sequence.

Lemma 4.13.1 *The operator $C(w)$ is completely continuous in $\mathcal{E}_{Pc}$.*

The proof is based on the elementary property of the trilinear form $B(u,v,w)$, as given in the following lemma.

Lemma 4.13.2 *For $u,v,w \in \mathcal{E}_{Pc}$, there hold the following properties of symmetry:*

$$\begin{aligned} B(u,v,w) &= B(w,u,v) = B(v,w,u) = B(v,u,w) \\ &= B(w,v,u) = B(u,w,v). \end{aligned} \tag{4.13.10}$$

Proof. We introduced the energy spaces as completions of the sets of functions satisfying appropriate boundary conditions and having all the continuous derivatives (in this case up to second order) that are included in the form of the energy of the body. However, the set of infinitely differentiable functions is dense in subspaces of $C^{(k)}(\Omega)$, and this means we can use it as a base to get a corresponding energy space (in other words, among representative Cauchy sequences of an element of an energy space there are those which consist of infinitely differentiable functions only). The validity of relations (4.13.10) is verified by direct integration by parts for functions u, v, w having all the third continuous derivatives (they cancel mutually after transformations). Taking then representative Cauchy sequences for elements u, v, w of $\mathcal{E}_{Pc}$ that have infinitely differentiable members we get the needed property by the limit passage in the equalities (4.13.10) written for the members. Equation (4.13.8) justifies the limit passage. □

Proof. (For Lemma 4.13.1). By (4.13.10) and definition (4.13.9), for any $\eta \in \mathcal{E}_{Pc}$ we have

$$(C(w), \eta)_P = -B(w, w, \eta) = -B(\eta, w, w). \tag{4.13.11}$$

Let $\{w_n\}$ be a weakly Cauchy sequence in $\mathcal{E}_{Pc}$ and thus $\|w_n\|_P < c_0$ with c_0 independent of n. We must show that $\{C(w_n)\}$ is a strongly Cauchy sequence. From (4.13.11) it follows that

$$|(C(w_{n+m}) - C(w_n), \eta)_P| = |B(\eta, w_{n+m}, w_{n+m}) - B(\eta, w_n, w_n)|. \tag{4.13.12}$$

Consider a typical pair of corresponding members of the right-hand side of this:

$$\begin{aligned}
&\left| \int_\Omega \eta_{xx}(w_{n+m_y} w_{n+m_x} - w_{n_y} w_{n_x})\, d\Omega \right| \\
&= \left| \int_\Omega \eta_{xx}(w_{n+m_y} w_{n+m_x} - w_{n+m_y} w_{n_x} + w_{n+m_y} w_{n_x} - w_{n_y} w_{n_x})\, d\Omega \right| \\
&\le \left| \int_\Omega \eta_{xx} w_{n+m_y}(w_{n+m_x} - w_{n_x})\, d\Omega \right| + \left| \int_\Omega \eta_{xx} w_{n_x}(w_{n+m_y} - w_{n_y})\, d\Omega \right|.
\end{aligned}$$

Applying Hölder's inequality to each term on the right as in (4.13.7), we have

$$\begin{aligned}
&\left| \int_\Omega \eta_{xx}(w_{n+m_y} w_{n+m_x} - w_{n_y} w_{n_x})\, d\Omega \right| \\
&\le \left(\int_\Omega \eta_{xx}^2\, d\Omega \right)^{1/2} \left(\int_\Omega w_{n+m_y}^4\, d\Omega \right)^{1/4} \left(\int_\Omega (w_{n+m_x} - w_{n_x})^4\, d\Omega \right)^{1/4} \\
&\quad + \left(\int_\Omega \eta_{xx}^2\, d\Omega \right)^{1/2} \left(\int_\Omega w_{n_x}^4\, d\Omega \right)^{1/4} \left(\int_\Omega (w_{n+m_y} - w_{n_y})^4\, d\Omega \right)^{1/4} \\
&\le M \|\eta\|_P\, c_0 \left(\|w_{n+m_x} - w_{n_x}\|_{L^4(\Omega)} + \left\|w_{n+m_y} - w_{n_y}\right\|_{L^4(\Omega)} \right)
\end{aligned}$$

with a constant M defined by the imbedding theorem for $\mathcal{E}_{Pc}$. Doing this for each of corresponding pair in the right-hand side of (4.13.12) we get

$$\begin{aligned}
&|(C(w_{n+m}) - C(w_n), \eta)_P| \\
&\leq M_1 \|\eta\|_P \left(\|w_{n+m_x} - w_{nx}\|_{L^4(\Omega)} + \left\|w_{n+m_y} - w_{ny}\right\|_{L^4(\Omega)} \right)
\end{aligned}$$

Putting $\eta = C(w_{n+m}) - C(w_n)$ we get

$$\begin{aligned}
&|(C(w_{n+m}) - C(w_n), C(w_{n+m}) - C(w_n))_P| \\
&\leq M_1 \|C(w_{n+m}) - C(w_n)\|_P \\
&\qquad \cdot \left(\|w_{n+m_x} - w_{nx}\| \,|_{L^4(\Omega)} + \left\|w_{n+m_y} - w_{ny}\right\|_{L^4(\Omega)} \right)
\end{aligned}$$

or

$$\begin{aligned}
&\|C(w_{n+m}) - C(w_n)\|_P \\
&\leq M_1 \left(\|w_{n+m_x} - w_{nx}\|_{L^4(\Omega)} + \left\|w_{n+m_y} - w_{ny}\right\|_{L^4(\Omega)} \right). \qquad (4.13.13)
\end{aligned}$$

But by Sobolev's imbedding theorem for $W^{2,2}(\Omega)$ that is applicable to its subspace $\mathcal{E}_{Pc}$, we have that for a sequence $\{w_n\}$ weakly convergent in $\mathcal{E}_{Pc}$,

$$\|w_{n+m_x} - w_{nx}\|_{L^4(\Omega)} + \left\|w_{n+m_y} - w_{ny}\right\|_{L^4(\Omega)} \to 0 \quad \text{as } n \to \infty.$$

This implies the needed statement of the lemma:

$$\|C(w_{n+m}) - C(w_n)\|_P \to 0 \quad \text{as } n \to \infty.$$

□

Now we return to the generalized setup of the problem and eliminate $f = C(w)$ from the statement. Then (4.13.5)–(4.13.6) reduce to the single equation

$$(w, \zeta)_P = B(C(w), w, \zeta) + \int_\Omega q\zeta \, d\Omega. \qquad (4.13.14)$$

Let us present (4.13.14) in operator form. Consider the right-hand side of (4.13.14), $B(C(w), w, \zeta) + \int_\Omega q\zeta \, d\Omega$, as a functional in ζ at a fixed w. It is linear in ζ. Next we get

$$\begin{aligned}
\left| B(C(w), w, \zeta) + \int_\Omega q\zeta \, d\Omega \right| &\leq m_1 \|C(w)\|_P \|w\|_P \|\zeta\|_P + \max_\Omega |\zeta| \int_\Omega |q| \, d\Omega \\
&\leq m_2 \|\zeta\|_P
\end{aligned}$$

where we have used a consequence of inequality (4.13.7), the inequality

$$\|C(w)\|_P \le M_1 \|w\|_P^2$$

that can be obtained in the same fashion as (4.13.13) with use of Sobolev's imbedding theorem in $W^{2,2}(\Omega)$. This means $B(C(w), w, \zeta) + \int_\Omega q\zeta \, d\Omega$ is a continuous linear functional in $\zeta \in \mathcal{E}_{Pc}$. Applying the Riesz representation theorem we get

$$B(C(w), w, \zeta) + \int_\Omega q\zeta \, d\Omega = (G, \zeta)_P$$

where $G \in \mathcal{E}_{Pc}$ is uniquely defined by $w \in \mathcal{E}_{Pc}$. Thus G can be considered as the result of an operator $G = G(w)$ acting in $\mathcal{E}_{Pc}$. Then (4.13.14) becomes

$$(w, \zeta)_P = (G(w), \zeta)_P$$

and so, because of the arbitrariness of $\zeta \in \mathcal{E}_{Pc}$, we get an operator equation

$$w = G(w) \tag{4.13.15}$$

where G is a nonlinear operator in $\mathcal{E}_{Pc}$. Let us establish the properties of G.

Lemma 4.13.3 *The operator G is completely continuous in $\mathcal{E}_{Pc}$; that is, it takes any weakly Cauchy sequence into a strongly Cauchy sequence.*

The proof practically repeats all the steps of the proof of Lemma 4.13.1 (in fact it is easier since C is a completely continuous operator) so we leave it to the reader.

To use the tools of the calculus of variations we should present (4.13.15) as the equality of the first variation of some functional to zero. As we will see, the functional we mean is

$$F(w) = \frac{1}{2}a(w, w) + \frac{1}{4}a(C(w), C(w)) - \int_\Omega qw \, d\Omega. \tag{4.13.16}$$

Let us introduce

Definition 4.13.2 Suppose a functional Φ at point x in a real Hilbert space H can be represented as

$$\Phi(x + y) - \Phi(x) = (K(x), y)_H + o(\|y\|_H) \tag{4.13.17}$$

for any y, $\|y\|_H \le \varepsilon$ with some small $\varepsilon > 0$. The correspondence from x to $K(x)$ is called the *gradient* of Φ and is denoted as $\operatorname{grad} \Phi(x) = K(x)$.

The reader sees that this is a way of representation of the first variation of a functional in a real Hilbert space which was the central point of the first chapter. In many cases, the main term of the representation can be found by formal differentiation in a parameter t:

$$(K(x), y)_H = \frac{d}{dt}\Phi(x+ty)\bigg|_{t=0}. \tag{4.13.18}$$

For example, the gradient of the functional $\frac{1}{2}\|x\|_H^2$ is the identity operator. Indeed,

$$\frac{d}{dt}\left(\frac{1}{2}(x+ty, x+ty)\right)_H\bigg|_{t=0} = (x,y)_H.$$

The reader can check this by direct calculation according to Definition 4.13.2. As in Chapter 1, we have

Lemma 4.13.4 *Suppose a functional $\Phi(x)$ has at any point x of a real Hilbert space H a continuous gradient $K(x)$. If $\Phi(x)$ attains a minimum at x_0, then $K(x_0) = 0$.*

Proof. For any fixed y and small t, by (4.13.17) we have

$$0 \le \Phi(x_0 + ty) - \Phi(x_0) = t(K(x_0), y)_H + o(|t|).$$

From this inequality we conclude, as is standard reasoning for Chapter 1, that $(K(x_0), y)_H = 0$. Hence $K(x_0) = 0$ by the arbitrariness of y. □

Note that we derived a version of the Euler equation for an abstract functional. The points x at which $K(x) = 0$ are called *critical points* of $\Phi(x)$. Thus points of minimum of a smooth functional Φ are its critical points. Let us apply this to our equation.

Theorem 4.13.1 *Let $q \in L(\Omega)$. There exists a generalized solution $w_0, f_0 \in \mathcal{E}_{Pc}$ to the equilibrium problem for von Kármán's plate with boundary conditions* (4.13.3), (4.13.4). *The element w_0 is a point of minimum of the functional $F(w)$ defined by* (4.13.16).

We present the proof as several lemmas.

Lemma 4.13.5 *At any $w \in \mathcal{E}_{Pc}$ we have*

$$\operatorname{grad} F(w) = w - G(w).$$

Proof. Let us consider $F(w + t\zeta)$ at any fixed $w, \zeta \in \mathcal{E}_{Pc}$. In t this is a simple polynomial so we can define $\operatorname{grad} F$ by (4.13.18). Consider

$$\begin{aligned}\frac{d}{dt}F(w+t\zeta)\bigg|_{t=0} &= \frac{d}{dt}\bigg(\frac{1}{2}a(w+t\zeta, w+t\zeta)\\ &\quad + \frac{1}{4}a(C(w+t\zeta), C(w+t\zeta)) - \int_\Omega q(w+t\zeta)\,d\Omega\bigg)\bigg|_{t=0}\\ &= a(w,w) + \frac{1}{2}a\left(\frac{dC(w+t\zeta)}{dt}, C(w)\right)\bigg|_{t=0} - \int_\Omega q\zeta\,d\Omega.\end{aligned} \tag{4.13.19}$$

From (4.13.11) with use of symmetry of its right-hand side in w we have

$$a\left(\frac{dC(w+t\zeta)}{dt}, \eta\right)\bigg|_{t=0} = -\frac{d}{dt}B(\eta, w+t\zeta, w+t\zeta)\bigg|_{t=0} = -2B(\eta, w, \zeta).$$

So

$$a\left(\frac{dC(w+t\zeta)}{dt}, C(w)\right)\bigg|_{t=0} = -2B(C(w), w, \zeta).$$

Combining this with (4.13.19) we get

$$\begin{aligned}\frac{d}{dt}F(w+t\zeta)\bigg|_{t=0} &= a(w,\zeta) - B(C(w), w, \zeta) - \int_\Omega q\zeta\,d\Omega\\ &= (w - G(w), \zeta)_P,\end{aligned}$$

which completes the proof. □

From this and the above we get

Lemma 4.13.6 *Any critical point w in $\mathcal{E}_{Pc}$ of functional F given by* (4.13.16) *implies the pair $w, f = C(w)$ is a generalized solution of the problem under consideration.*

Now we are going to show that there is a point at which $F(w)$ attains its minimum. First we note that this minimum point is in a ball centered at the origin whose radius is defined only by the load q. This follows from the inequality chain

$$\begin{aligned}2F(w) &\geq a(w,w) - 2\left|\int_\Omega qw\,d\Omega\right|\\ &\geq \|w\|_P^2 - 2\max_\Omega |w| \int_\Omega |q|\,d\Omega\\ &\geq \|w\|_P^2 - M_0\,\|w\|_P\,,\end{aligned} \tag{4.13.20}$$

where the constant M_0 is defined by the norm of q in $L(\Omega)$ and the norm of the imbedding operator from $\mathcal{E}_{Pc}$ to $C(\Omega)$. Since $F(0) = 0$ and outside of the sphere $\|w\|_P = M_0 + 1$ we have $F(w) \geq M_0 + 1$ and thus

Lemma 4.13.7 *If there is a minimum point of the functional F, then it belongs to the ball $\|w\|_P < M_0 + 1$. Moreover, the functional F is growing in $\mathcal{E}_{Pc}$; that is, $F(w) \to \infty$ when $\|w\|_P \to \infty$.*

The fact that F is a growing functional follows immediately from inequality (4.13.20). Now we need to prove that F attains its limit point.

Lemma 4.13.8 *The functional $\Phi(w) = \frac{1}{4}a(C(w), C(w)) - \int_\Omega qw\,d\Omega$ is weakly continuous in $\mathcal{E}_{Pc}$, thus the functional $F(w)$ is represented as*

$$F(w) = \frac{1}{2}\|w\|_P^2 + \Phi(w)$$

with a weakly continuous functional Φ.

Proof. Evident since $\int_\Omega qw\,d\Omega$ is a continuous linear functional and C is a completely continuous operator. □

The proof of Theorem 4.13.1 is completed by the following result due to Tsitlanadze:

Theorem 4.13.2 *Let $f(x)$ be a growing functional in a Hilbert space H that has the form*

$$f(x) = \|x\|_H^2 + \varphi(x)$$

where $\varphi(x)$ is a weakly continuous functional in H. Then

(i) there is a point x_0 at which $f(x)$ attains its absolute minimum, $f(x_0) \leq f(x)$ for any $x \in H$;
(ii) any sequence $\{x_n\}$ minimizing f, that is $\lim_{n\to\infty} f(x_n) = f(x_0)$, contains a subsequence that strongly converges to x_0.

Proof. On any ball $\varphi(x)$ is bounded and thus $f(x)$ is bounded as well. Because $f(x)$ is growing we state that a possible minimum point is inside a closed ball B of a radius R. Let a be the infimum of values of $f(x)$. Then

$$\inf_{x\in H} f(x) = \inf_{\|x\|_H \leq R} f(x) = a.$$

Take a minimizing sequence $\{x_n\}$ of f. We can consider it is inside B and thus contains a weakly convergent subsequence that we redenote by $\{x_n\}$ again. Without loss of generality, we can consider the sequence of norms of x_n to converge to b, such that $b \leq R$. Since a closed ball centered at the origin is weakly closed we get that $\{x_n\}$ converges weakly to an element $x_0 \in B$. It is enough to show now that $\{x_n\}$ converges strongly to x_0. We know that if for a weak Cauchy sequence the sequence of norms of the elements converges to the norm of the weak limit element then it converges strongly. Thus we need to demonstrate only that $\|x_0\|_H = b$. Let us show this. It is clear that

$$\|x_0\|_H \leq b.$$

Indeed, because of weak convergence of $\{x_n\}$ to x_0 we have

$$\|x_0\|_H^2 = \lim_{n\to\infty}(x_n, x_0)_H \leq \|x_0\|_H \lim_{n\to\infty} \|x_n\|_H = b\,\|x_0\|_H\,.$$

Next, because of weak continuity of φ we have $\lim_{n\to\infty}\varphi(x_n) = \varphi(x_0)$ and thus

$$a = \lim_{n\to\infty} f(x_n) = \lim_{n\to\infty}\left(\|x_n\|_H^2 + \varphi(x_n)\right) = b^2 + \varphi(x_0).$$

But

$$f(x_0) = \|x_0\|_H^2 + \varphi(x_0) \geq a$$

so $\|x_0\|_H^2 \geq b^2$ which means that $\|x_0\|_H = b$. All statements of the theorem are proven. □

By this theorem the proof of Theorem 4.13.1 is also completed. Note that Theorem 4.13.2 prepared everything to formulate the theorem on convergence of the Ritz approximations to a generalized solution of the problem under consideration. We leave this to the reader.

4.14 A Unilateral Problem

Now let us consider a simple problem of deformation of a membrane constrained by a surface underneath it. During deformation, the membrane cannot penetrate through this surface. This problem belongs to the class of *unilateral problems*, and can be formulated as a problem involving a so-called variational inequality. By this approach we obtain problems with free boundaries; i.e., the boundary of the domain over which some equations are applicable is determined during solution, not in advance. Our previous use of the term "free" indicated a lack of geometrical constraints on the displacements, whereas for this problem there is an obstacle and the border of contact between this obstacle and the membrane is undetermined (free). Now we begin.

Consider a membrane under load f occupying a compact domain Ω with clamped edge. Let us suppose that "under" the membrane there is an obstacle described by a function $\varphi = \varphi(x, y)$ such that the points of the membrane cannot go through this surface:

$$u(x, y) \geq \varphi(x, y) \tag{4.14.1}$$

for all $(x, y) \in \Omega$. We will suppose that the clamped edge of the membrane is described by

$$u\big|_{\partial\Omega} = a(s). \tag{4.14.2}$$

Of course we should suppose some compatibility between the boundary condition and the obstacle; that is, on the boundary we should have

$$\varphi\big|_{\partial\Omega} \leq a(s). \tag{4.14.3}$$

We wish to find a solution to this problem. First of all it is clear that now it can appear a domain in which the membrane "lays" on the obstacle φ. This set is called the *coincidence set* since on this set the membrane takes the form of the obstacle. It is clear mechanically that on the coincidence set the membrane equation should not be applied (in fact the equation holds but it contains an unknown force reaction of the obstacle) whereas outside the coincidence set it should be applied. Mechanical considerations normally work quite well; however, in this case we do not yet know how to define the coincidence set, its border, or the conditions for a solution on the latter.

Classical setup of the problem

Let us try to use the tools of calculus of variations to determine these. We would like to obtain a classical statement of the problem, hence we suppose that all the functions we will use are sufficiently smooth. Since the mechanics of this problem guarantees the applicability of the principle of minimum of total energy, a solution is a minimizer of the energy functional

$$F(u) = \frac{1}{2}\int_\Omega \left(\left(\frac{\partial u}{\partial x}\right)^2 + \left(\frac{\partial u}{\partial y}\right)^2 \right) d\Omega - \int_\Omega f u \, d\Omega$$

over the set of functions satisfying (4.14.1) and (4.14.2). Let us suppose there is a solution of this problem belonging to $C^{(2)}(\Omega)$, so we will find equations for a minimizer over the subset of $C^{(2)}(\Omega)$ consisting of functions satisfying (4.14.1) and (4.14.2). We denote this subset by C_φ. Note that we must assume $\varphi \in C^{(2)}(\Omega)$ as well. Later we will "forget" this requirement. Thus we need to find equations governing a minimizer $u \in C_\varphi$ of $F(u)$ over C_φ. It is clear that the set C_φ is convex in $C^{(2)}(\Omega)$, which means that if u_1 and u_2 belong to C_φ then for any $t \in [0,1]$ we have $(1-t)u_1 + tu_2 \in C_\varphi$. Let us take an arbitrary $v \in C_\varphi$. Because of convexity of C_φ we see that $u + t(v-u) = (1-t)u + tv$ belongs to C_φ for any $t \in [0,1]$ as well. So by the principle of minimum of total energy we have

$$F(u + t(v-u)) \geq F(u)$$

for any $v \in C_\varphi$ and $t \in [0,1]$. Remembering the notation

$$(u,v)_M = \int_\Omega \left(\frac{\partial u}{\partial x}\frac{\partial v}{\partial x} + \frac{\partial u}{\partial y}\frac{\partial v}{\partial y} \right) d\Omega \tag{4.14.4}$$

we have

$$\frac{1}{2}(u + t(v-u), u + t(v-u))_M - \frac{1}{2}(u,u)_M - t\int_\Omega f(v-u)\, d\Omega \geq 0$$

or

$$t\left[(u, v-u)_M - \int_\Omega f(v-u)\, d\Omega\right] + \frac{1}{2}t^2(v-u, v-u)_M \geq 0 \tag{4.14.5}$$

for any $t \in [0,1]$. This implies that for a fixed v the coefficient at t must be nonnegative:

$$(u, v-u)_M - \int_\Omega f(v-u)\, d\Omega \geq 0. \tag{4.14.6}$$

Indeed if we suppose this coefficient is negative then choosing sufficiently small t we get that (4.14.5) is not fulfilled since t^2 tends to zero faster than t when $t \to 0$. Hence a minimizer u must satisfy (4.14.6) for any $v \in C_\varphi$. Such inequalities are called *variational inequalities*. Denote $\eta = v - u$. It is clear that on the boundary

$$\eta\big|_{\partial\Omega} = 0. \tag{4.14.7}$$

Then (4.14.6) takes the form

$$(u, \eta)_M - \int_\Omega f\eta\, d\Omega \geq 0. \tag{4.14.8}$$

We see that on the left side of (4.14.8) there is the first variation of functional F with virtual displacement η. In the calculus of variations, from (4.14.8) we stated that the first variation is equal to zero for any η. This was done because η was sufficiently arbitrary; this time, however, we have $\eta \geq 0$ on the coincidence set for u, so we cannot use the trick involving a sign change on η in order to obtain an equality in (4.14.8). Let us derive the differential equations from (4.14.8). Traditional integration by parts with regard for (4.14.7) yields

$$\int_{\Omega} (-\Delta u - f)\, \eta \, d\Omega \geq 0. \tag{4.14.9}$$

If we restrict the support of η to the coincidence set of u denoted by Ω_{φ}, all we get from this is

$$-\Delta u - f \geq 0$$

inside Ω_{φ}. This means that on Ω_{φ} there is a reaction of the supporting obstacle applied to the membrane. Recall that on the coincidence set we have $u = \varphi$. We consider u to be of the class of $C^{(2)}(\Omega)$, and thus on the boundary of Ω_{φ} that we denote by Γ_{φ} we have that all the first derivatives of u and φ are equal:

$$\nabla(u - \varphi)\big|_{\Gamma_{\varphi}} = 0.$$

This is the equation we can use to determine the position of Γ_{φ}. Let consider what happens outside of the coincidence set Ω_{φ}. Here the only restriction for η is some smallness of its negative values. For sufficiently small η with compact support lying in $\Omega \backslash \Omega_{\varphi}$ we have equality to zero in (4.14.9). Thus the usual tools of the calculus of variations imply that in $\Omega \backslash \Omega_{\varphi}$ there holds the Poisson equation

$$\Delta u = -f$$

as was expected above from mechanical considerations. Let us summarize the setup of the problem:

$$\begin{aligned}
&\Delta u = -f \text{ on } \Omega \backslash \Omega_{\varphi},\\
&\Delta u + f \leq 0,\ u = \varphi \text{ on } \Omega_{\varphi},\\
&\nabla(u - \varphi) = 0 \text{ on } \Gamma_{\varphi},\\
&u = a \text{ on } \partial\Omega.
\end{aligned}$$

We can write out the equations of equilibrium on Ω as

$$(\Delta u + f)(u - \varphi) = 0 \text{ in } \Omega.$$

Generalized setup

It is difficult to prove the existence of a classical solution to the above problem. When the coincidence set is of complex shape or the load is non-smooth, the energy approach to the solution is quite appropriate. For the setup of the problem we shall use an energy space where the elements are sets of equivalent Cauchy sequences, so we need to explain the meaning of inequality (4.14.1). We begin with the inequality $u(x, y) \geq 0$. We say that $u(x, y) \geq 0$, $u \in W^{1,2}(\Omega)$, if there is a representative Cauchy sequence of $u(x, y)$ such that each of its terms $u_n(x, y) \geq 0$. We say that $u(x, y) \geq \varphi(x, y)$ if $u(x, y) - \varphi(x, y) \geq 0$. If $\varphi(x, y) \in W^{1,2}(\Omega)$, then the set of functions $u(x, y) \geq \varphi(x, y)$ is closed in $W^{1,2}(\Omega)$ and in any closed subspace of this space. Let us assume that the obstacle function $\varphi(x, y) \in W^{1,2}(\Omega)$ and satisfies (4.14.3). Now we need to find a minimizer $u = u(x, y) \in W^{1,2}(\Omega)$ of

$$F(u) = \frac{1}{2}\int_\Omega \left(\left(\frac{\partial u}{\partial x}\right)^2 + \left(\frac{\partial u}{\partial y}\right)^2 \right) d\Omega - \int_\Omega f u \, d\Omega$$

over a subset W_φ of elements of $W^{1,2}(\Omega)$ satisfying

$$u\big|_{\partial\Omega} = a(s)$$

and

$$u(x, y) \geq \varphi(x, y).$$

This minimizer is called a generalized solution of the unilateral problem for the clamped membrane. We suppose $\varphi \in W^{1,2}(\Omega)$ and $f \in L^p(\Omega)$ for some $p > 1$. In this case the problem of minimization of $F(u)$ over W_φ is well defined. In the same manner as above we get that a minimizer $u \in W_\varphi$ satisfies the variational inequality (4.14.6) for all $v \in W_\varphi$. We would like to reduce the problem to the case we have studied. Let us assume there is an element $g = g(x, y) \in W^{1,2}(\Omega)$ that satisfies the same boundary condition as a solution,

$$g(x, y)\big|_{\partial\Omega} = a(s), \tag{4.14.10}$$

and introduce another unknown function w by the equality

$$u = w + g.$$

From the properties of u it follows that

$$w(x, y)\big|_{\partial\Omega} = 0.$$

We see that $w \in W^{1,2}(\Omega)$ and thus $w \in \mathcal{E}_{Mc}$. To formulate the setup of the problem in terms of w, it is clear that w should satisfy

$$w(x, y) \geq \varphi(x, y) - g(x, y). \tag{4.14.11}$$

Let us denote the subset of $\mathcal{E}_{Mc}$ consisting of elements satisfying (4.14.11) by $W_{\varphi-g}$. The functional $F(u)$ reduces to the functional

$$F_1(w) = \frac{1}{2}\int_\Omega \left(\left(\frac{\partial(w+g)}{\partial x} \right)^2 + \left(\frac{\partial(w+g)}{\partial y} \right)^2 \right) d\Omega - \int_\Omega f(w+g)\, d\Omega.$$

Since f and g are fixed, the problem of minimization of $F(u)$ becomes the problem of minimization of functional

$$\Phi(w) = \frac{1}{2}\int_\Omega \left(\left(\frac{\partial(w+g)}{\partial x} \right)^2 + \left(\frac{\partial(w+g)}{\partial y} \right)^2 \right) d\Omega - \int_\Omega fw\, d\Omega$$

over the set $W_{\varphi-g}$.

Let us formulate the problem explicitly: given $\varphi, g \in W^{1,2}(\Omega)$ such that (4.14.10) and (4.14.3) are valid, find a minimizer of $\Phi(w)$ over $W_{\varphi-g}$.

Using the notation (4.14.4) we can rewrite the expression for $\Phi(w)$ as

$$\Phi(w) = \frac{1}{2}(w + g, w + g)_M - \int_\Omega fw\, d\Omega.$$

Let w^* be a minimizer of $\Phi(w)$ over $W_{\varphi-g}$. We repeat reasoning with which we derived (4.14.6), namely, let us fix an arbitrary $w \in W_{\varphi-g}$. Then

$$\Phi(w^* + t(w - w^*)) \geq \Phi(w^*)$$

for any $t \in [0, 1]$. For such t it follows that

$$\frac{1}{2}(w^* + t(w - w^*) + g, w^* + t(w - w^*) + g)_M$$
$$-\frac{1}{2}(w^* + g, w^* + g)_M - t\int_\Omega f(w - w^*)\, d\Omega \geq 0$$

or

$$t\left\{(w^*, w - w^*)_M + (g, w - w^*)_M - \int_\Omega f(w - w^*)\,d\Omega\right\} \\ + \frac{1}{2}t^2(w - w^*, w - w^*)_M \geq 0.$$

Since this holds for any $t \in [0, 1]$ we conclude that the coefficient of t must be non-negative:

$$(w^*, w - w^*)_M \geq \int_\Omega f(w - w^*)\,d\Omega - (g, w - w^*)_M$$

for all $w \in W_{\varphi - g}$. This is a necessary condition for w^* to be a minimizer of $\Phi(w)$ over $W_{\varphi-g}$.

Theorem 4.14.1 *There exists a generalized solution to the unilateral problem for the membrane with clamped edge, it is the only minimizer w^* of $\Phi(w)$ over $W_{\varphi-g}$.*

Proof. Let us show uniqueness of the minimizer w^*. Suppose to the contrary that there are two minimizers w_1^* and w_2^*. Then

$$(w_i^*, w - w_i^*)_M \geq \int_\Omega f(w - w_i^*)\,d\Omega - (g, w - w_i^*)_M.$$

We put $w = w_2^*$ in the inequality for w_1^* and $w = w_1^*$ in the inequality for w_2^*; adding the results we get

$$(w_1^* - w_2^*, w_2^* - w_1^*)_M \geq 0,$$

which is possible only when $w_1^* = w_2^*$ since $w_i^* \in \mathcal{E}_{Mc}$.

Now let us show existence of a minimizer of $\Phi(w)$. It is clear that $\Phi(w)$ is bounded from below on $W_{\varphi-g}$ (why?). Let $d = \inf \Phi(w)$ over $W_{\varphi-g}$, and let $\{w_n\}$ be a minimizing sequence for $\Phi(w)$ in $W_{\varphi-g}$:

$$\Phi(w_n) \to d \quad \text{as } n \to \infty.$$

Now we show that $\{w_n\}$ is a Cauchy sequence. Indeed, consider

$$(w_n - w_m, w_n - w_m)_M = 2(w_n, w_n)_M + 2(w_m, w_m)_M \\ - 4\left(\frac{1}{2}(w_n + w_m), \frac{1}{2}(w_n + w_m)\right)_M. \tag{4.14.12}$$

An elementary transformation shows that

$$2(w_n, w_n)_M + 2(w_m, w_m)_M - 4\left(\frac{1}{2}(w_n + w_m), \frac{1}{2}(w_n + w_m)\right)_M$$
$$= 4\Phi(w_n) + 4\Phi(w_m) - 8\Phi\left(\frac{1}{2}(w_n + w_m)\right). \tag{4.14.13}$$

Next $\Phi(w_n) = d + \varepsilon_n$ where $\varepsilon_n \to 0$ as $n \to \infty$. Because $W_{\varphi-g}$ is convex the element $\frac{1}{2}(w_n + w_m)$ belongs to $W_{\varphi-g}$, so $\Phi\left(\frac{1}{2}(w_n + w_m)\right) \geq d$, hence (4.14.12)–(4.14.13) imply

$$(w_n - w_m, w_n - w_m)_M \leq 2(d + \varepsilon_n) + 2(d + \varepsilon_m) - 4d$$
$$= 2(\varepsilon_n + \varepsilon_m) \to 0 \quad \text{as } n, m \to \infty.$$

This completes the proof. □

We have proved solvability of a unilateral problem for a clamped membrane. Since all the problems we considered for plates, rods, and elastic bodies have the same structure, and since in the reasoning for the membrane we used only the structure of the energy functional, we can immediately reformulate unilateral problems for all the objects we just mentioned (of course, for a 3-D body we can stipulate the unilateral condition only on the boundary). We leave this work to the reader. The theory of unilateral problems and variational inequalities contains more difficult questions than the existence of energy solutions: it studies the problem of regularity of this solution, which is how the smoothness of solutions depends on the smoothness of the load. The interested reader should consult more advanced sources (e.g., [Friedman (1982); Kinderlehrer and Stampacchia (1980)]) for this.

4.15 Exercises

4.1 For all the bodies discussed in § 4.1 (except a stretched bar), write out the functional of total potential energy and the virtual work principle in the case when some part of the object (of its boundary for a 3-D body) is supported by a foundation of Winkler's type (i.e., when there is a contact force of supports whose amplitude is proportional to corresponding displacements).

4.2 By analogy with the material presented in § 4.3, consider the generalized setup for the equilibrium problem for a membrane with mixed boundary conditions. Assume that on some part of the boundary $u = 0$, while on the rest there is a given force $g(s)$. Formulate the corresponding theorem of existence and uniqueness of solution in this setup.

4.3 Consider a beam under bending and stretching. Formulate the generalized setup for this problem, combining the setups for a stretched rod and bent beam. Formulate the corresponding existence-uniqueness theorem.

4.4 (a) Which terms are necessary to add to the equilibrium equation (4.1.10) to include a finite number external point couples and forces acting on the beam into the generalized setup? (b) Is it possible to the consider generalized setup when there is a countable set of point couples and forces?

4.5 For a free plate, consider a case when there are forces given on the edge of the plate. Formulate the form of the potential and the conditions for solvability of the corresponding problem.

4.6 Using the material in § 4.7 as an example, reproduce the form of the Hamilton–Ostrogradskij principle for each type of object we considered.

4.7 Derive equations for solving the minimum problem (4.7.17).

4.8 Show that if $\mathcal{E}$ is not finite dimensional, then the norm $\|u\|_A$ of § 4.11 cannot be equivalent to the initial norm of the space $\mathcal{E}$ because A is compact.

4.9 Demonstrate that the set $\left\{\frac{\sqrt{2}}{\pi}\sin kx\right\}$, $k = 1, 2, \ldots$, is an orthonormal basis of $L^2[0, \pi]$.

4.10 Reformulate the statements of § 4.11 for each of the mechanics problems.

4.11 Suppose that in conditions of Theorem 4.13.2 the minimum point is unique. Prove that any minimizing sequence strongly converges to the minimum point.

4.12 Referring to § 4.14, demonstrate uniqueness of solution of the problem under consideration in $W^{1,2}(\Omega)$, that $w^* + g$ does not depend on the choice of $g \in W^{1,2}(\Omega)$.

Appendix A

Hints for Selected Exercises

Chapter 1

Exercise 1.1. We first show that the Euler equation for the simplest functional can be rewritten in the equivalent form

$$\frac{1}{y'}\left[\frac{d}{dx}(f - f_{y'}y') - f_x\right] = 0.$$

Observe that if $f(x, y, y')$ does not depend explicitly on x, then one integration can be performed to give

$$f - f_{y'}y' = \text{constant}.$$

Indeed, multiplying and dividing the left member of the Euler equation by y', we have

$$\frac{1}{y'}\left[f_y y' - y'\frac{d}{dx}f_{y'}\right] = 0.$$

Adding and subtracting a couple of terms inside the brackets, we obtain

$$\frac{1}{y'}\left[f_x + f_y y' + f_{y'}y'' - f_{y'}y'' - y'\frac{d}{dx}f_{y'} - f_x\right] = 0.$$

But the first three terms inside the brackets add to produce df/dx (total derivative), and the next two terms add to produce $-d(f_{y'}y')/dx$ (product rule).

For the surface of revolution problem, the area functional is

$$\int_a^b 2\pi y\sqrt{1 + (y')^2}\,dx.$$

Note that x does not appear explicitly; using the integrated version of Euler's equation, we get

$$2\pi y\sqrt{1 + (y')^2} - 2\pi y\frac{y'}{\sqrt{1 + (y')^2}}y' = \text{constant} = \alpha.$$

Divide through by 2π, then multiply through by $\sqrt{1+(y')^2}$ and simplify to get $y = \beta\sqrt{1+(y')^2}$ where $\beta = \alpha/2\pi$. Now solve for y' to obtain the separable ODE

$$y' = \sqrt{\frac{y^2}{\beta^2} - 1}.$$

The solution, obtained by direct integration, is

$$\beta \cosh^{-1}\left(\frac{y}{\beta}\right) = x + \gamma,$$

hence

$$y(x) = \beta \cosh\left(\frac{x+\gamma}{\beta}\right)$$

is the general form of the curve sought. The constants β, γ must be determined from the two endpoint conditions. We see that the minimal surface of revolution is a catenoid.

Exercise 1.2. We need to find a smooth curve connecting the points (a, y_0) and (b, y_1), $a < b$. It is clear that for solvability of the problem it is necessary that $y_0 > y_1$.

First show that if f takes the general form

$$f(x, y, y') = p(y)\sqrt{1+(y')^2},$$

where $p(y)$ depends explicitly on y only, then

$$\int \frac{dy}{\sqrt{\frac{p^2(y)}{\alpha^2} - 1}} = x + \beta$$

where α and β are constants. The functional giving the time taken for the motion along a curve $y(x)$ is obtained by putting $p(y) = 1/\sqrt{2gy}$ where g is the acceleration due to gravity.

Using the specific form of p given and introducing a new constant $\gamma = 1/2\alpha^2 g$, we have

$$\int \frac{dy}{\sqrt{\frac{\gamma}{y} - 1}} = x + \beta.$$

The substitution $y = \gamma \sin^2\left(\frac{\theta}{2}\right)$ reduces this to

$$\frac{\gamma}{2}\int (1 - \cos\theta)\, d\theta = x + \beta$$

after the use of a couple of trig identities. Hence

$$x + \beta = \frac{\gamma}{2}(\theta - \sin\theta).$$

The other equation of the cycloid is

$$y = \gamma \sin^2\left(\frac{\theta}{2}\right) = \frac{\gamma}{2}(1 - \cos\theta).$$

Of course, the constants β and γ would be determined by given endpoint conditions.

Exercise 1.3. The Euler equation $f_y - f_{y'x} - f_{y'y}y' - f_{y'y'}y'' = 0$ reduces to

$$f_{y'y'}y'' = 0.$$

This holds if $y'' = 0$ or $f_{y'y'} = 0$. The equation $y'' = 0$ is satisfied by any line of the form $y = c_1x + c_2$. If the equation $f_{y'y'} = 0$ has a real root $y' = \gamma$, then $y = \gamma x + c_3$; this, however, merely gives a family of particular straight lines (all having the same slope γ). In any case, the extremals are all straight lines.

Exercise 1.4. The average kinetic energy is given by

$$\frac{1}{T}\int_0^T \frac{1}{2}mx'^2(t)\,dt.$$

Since the integrand depends explicitly on x' only, the extremal is of the general form $x(t) = c_1t + c_2$. Imposing the end conditions to find the constants c_1 and c_2 we obtain

$$x(t) = \frac{x_1 - x_0}{T}t + x_0.$$

The solution means the motion should be at constant speed. Any acceleration would increase the energy of the motion.

Exercise 1.6. (a) Vanishing of the first variation requires that equation (1.5.4) hold. Let us review for a moment. We know that if we appoint a condition such as $y(a) = c_0$ then, since we need $\phi(a) = 0$ to keep our variations $y(x) + \varphi(x)$ admissible, we need $\varphi(a) = 0$ and equation (1.5.4) yields

$$f_{y'}(b, y(b), y'(b)) = 0.$$

This natural condition makes reference purely to b. Now consider the mixed condition given in the problem. To keep our variations $y(x) + \varphi(x)$ admissible we need $\varphi(a) + \varphi(b) = 0$ or $\varphi(a) = -\varphi(b)$. Equation (1.5.4) yields

$$f_{y'}(b, y(b), y'(b)) + f_{y'}(a, y(a), y'(a)) = 0.$$

This is the supplemental "natural" boundary condition. (b) To keep our variation admissible this time we need

$$\psi(y(a) + \varphi(a), y(b) + \varphi(b)) = 0.$$

As before, we're looking for a relation between $\phi(a)$ and $\phi(b)$ that we can substitute into (1.5.4). Restricting ourselves to infinitesimal variations $\phi(x)$, we use

Taylor's formula in two variables to write, approximately,

$$\psi(y(a)+\varphi(a), y(b)+\varphi(b)) = \psi(y(a), y(b)) + \left(\varphi(a)\frac{\partial}{\partial\alpha} + \varphi(b)\frac{\partial}{\partial\beta}\right)\psi(\alpha,\beta)\bigg|_{\substack{\alpha=y(a)\\ \beta=y(b)}}.$$

The first term on the right-hand side is zero by the condition given in the problem. Therefore we need

$$\varphi(a)\frac{\partial\psi(\alpha,\beta)}{\partial\alpha}\bigg|_{\substack{\alpha=y(a)\\ \beta=y(b)}} + \varphi(b)\frac{\partial\psi(\alpha,\beta)}{\partial\beta}\bigg|_{\substack{\alpha=y(a)\\ \beta=y(b)}} = 0$$

or

$$\varphi(a) = K\varphi(b), \qquad K = -\frac{\dfrac{\partial\psi(\alpha,\beta)}{\partial\beta}\bigg|_{\substack{\alpha=y(a)\\ \beta=y(b)}}}{\dfrac{\partial\psi(\alpha,\beta)}{\partial\alpha}\bigg|_{\substack{\alpha=y(a)\\ \beta=y(b)}}}.$$

Equation (1.5.4) yields

$$f_{y'}(b, y(b), y'(b)) - Kf_{y'}(a, y(a), y'(a)) = 0$$

as the corresponding natural condition. In part (a) we had $\psi(\alpha,\beta) = \alpha+\beta-1$, which gave us $K = -1$.

Exercise 1.7. This is a mixed problem. However, the general solution of the Euler equation is the same as for the brachistochrone problem:

$$x+\beta = \frac{\gamma}{2}(\theta - \sin\theta), \qquad y = \frac{\gamma}{2}(1-\cos\theta).$$

The condition at $x = a$ determines β. The condition at $x = b$ is the free-end condition $f_{y'}\big|_{x=b} = 0$. Here

$$f(x, y, y') = \frac{1}{\sqrt{2gy}}\sqrt{1+(y')^2}$$

(again, the same as for the brachistochrone problem) so that

$$f_{y'} = \frac{y'}{\sqrt{2gy}\sqrt{1+(y')^2}}.$$

Thus the condition at $x = b$ is $y'(b) = 0$; i.e., the required curve must "flatten out" at this endpoint.

Exercise 1.8. Arc length on the cylinder is given by $(ds)^2 = (a\,d\phi)^2 + (dz)^2$. Parameterizing the desired curve as $\phi = \phi(t)$, $z = z(t)$, we seek to minimize the functional

$$\int_a^b [a^2(\phi')^2 + (z')^2]\,dt.$$

Each equation of the system (1.6.4) involves only the derivative of the dependent variable; hence the extremals are straight lines:

$$\phi(t) = c_1 t + c_2, \qquad z(t) = c_3 t + c_4.$$

Eliminating t we find $z(\phi) = \alpha\phi + \beta$, a family of helices on the cylinder.

Exercise 1.9. Repetition of the steps leading to (1.4.3) gives the system

$$\begin{aligned}
&\int_a^b f_y\left(x, \sum_{i=0}^n c_i\varphi_i(x), \sum_{i=0}^n c_i\varphi_i'(x), \sum_{i=0}^n c_i\varphi_i''(x)\right)\varphi_k(x)\,dx \\
&\quad + \int_a^b f_{y'}\left(x, \sum_{i=0}^n c_i\varphi_i(x), \sum_{i=0}^n c_i\varphi_i'(x), \sum_{i=0}^n c_i\varphi_i''(x)\right)\varphi_k'(x)\,dx \\
&\quad + \int_a^b f_{y''}\left(x, \sum_{i=0}^n c_i\varphi_i(x), \sum_{i=0}^n c_i\varphi_i'(x), \sum_{i=0}^n c_i\varphi_i''(x)\right)\varphi_k''(x)\,dx = 0
\end{aligned}$$

for $k = 1, \ldots, n$.

Exercise 1.10. Refer to Chapter 4.

Chapter 2

Exercise 2.2. The result follows from differentiation of the equality

$$\mathbf{\Psi}(t)\cdot\mathbf{\Psi}^{-1}(t) = \mathbf{E}.$$

We have

$$\left(\mathbf{\Psi}(t)\cdot\mathbf{\Psi}^{-1}(t)\right)' = \mathbf{\Psi}'(t)\cdot\mathbf{\Psi}^{-1}(t) + \mathbf{\Psi}(t)\cdot\left(\mathbf{\Psi}^{-1}(t)\right)' = \mathbf{E}' = \mathbf{0},$$

hence

$$\mathbf{\Psi}(t)\cdot\left(\mathbf{\Psi}^{-1}(t)\right)' = -\mathbf{\Psi}'(t)\cdot\mathbf{\Psi}^{-1}(t)$$

and can premultiply both sides by $\mathbf{\Psi}^{-1}(t)$.

Exercise 2.3. Use the linearity of the main part of the increment with respect to the increment of the control function.

Exercise 2.4. Introduce an additional component y_{n+1} of the vector $\mathbf{y}$ by the equations $y_{n+1}'(t) = G(\mathbf{y}(t))$, $y_{n+1}(0) = 0$.

Exercise 2.5. By Pontryagin's maximum principle we get that F take the values $+1$ or -1 for optimal solution. Solve the problems with this F and collect the whole solution using these solutions.

Chapter 3

Exercise 3.1. Assume S is closed in X. Let $\{x_n\} \subset S$ be convergent (in X) so that $x_n \to x$ for some $x \in X$. We want to show that $x \in S$. Let us suppose $x \notin S$ and seek a contradiction. Given any $\varepsilon > 0$ there exists x_k $(\neq x)$ such that $d(x_k, x) < \varepsilon$ (by the assumed convergence), so x is a limit point of S. Therefore S fails to contain all its limit points, and by definition is not closed.

Conversely, assume S contains the limits of all its convergent sequences. Let y be a limit point of S. By virtue of this, construct a convergent sequence $y_n \subset S$ as follows: for each n, take a point $y_n \in S$ such that $d(y_n, y) < 1/n$. Then $y_n \to y$ (in X). By hypothesis then, $y \in S$. This shows that S contains all its limit points, hence S is closed by definition.

Exercise 3.2. (a) Let $B(p,r)$ denote the closed ball centered at point p and having radius r, and let q be a limit point of $B(p,r)$. There is a sequence of points p_k in $B(p,r)$ such that $d(p_k, q) \to 0$ as $k \to \infty$. For each k we have

$$d(q,p) \le d(q,p_k) + d(p_k,p) \le d(q,p_k) + r,$$

hence as $k \to \infty$ we get $d(p,q) \le r$. This proves that $q \in B(p,r)$. (b) True vacuously. (c) Obvious. (d) Let $S = \cap_{i \in I} S_i$ be an intersection of closed sets S_i. If $S = \emptyset$ then it is closed by part (b). Otherwise let q be any limit point of S and choose a sequence $\{p_k\} \subset S$ such that $p_k \to q$. We have $\{p_k\} \subset S_i$ for each i, and each S_i is closed so that we must have $q \in S_i$ for each i. This means that $q \in \cap_{i \in I} S_i$. (e) We communicate the general idea by outlining the proof for a union of two sets. Let $S = A \cup B$ where A, B are closed. Choose a convergent sequence $\{x_n\} \subset S$ and call its limit x. There is a subsequence $\{x_{n_k}\}$ that consists of points belonging to one of the given sets. Without loss of generality suppose $\{x_{n_k}\} \subset A$. But $x_{n_k} \to x$, hence $x \in A$ since A is closed. Therefore $x \in S$.

Exercise 3.3. It is clear that the sequence of centers $\{x_n\}$ is a Cauchy sequence. By completeness, $x_n \to x$ for some $x \in X$. For each n, the sequence $\{x_{n+p}\}_{p=1}^{\infty}$ lies in $B(x_n, r_n)$ and converges to x; since the ball is closed we have $x \in B(x_n, r_n)$. This proves existence of a point in the intersection of all the balls. If y is any other such point, then $d(y,x) \le d(y,x_n) + d(x_n,x) \le 2\varepsilon_n \to 0$ as $n \to \infty$. Hence $y = x$ and we have proved uniqueness.

Exercise 3.4. Let us verify the norm properties for $\|\cdot\|_{X/U}$. Certainly we have $\|x + U\|_{X/U} \ge 0$. Recalling that the zero element of X/U is U, we have

$$\|0_{X/U}\|_{X/U} = \|0_X + U\|_{X/U} = \inf_{u \in U} \|0_X + u\|_X = 0$$

since $0_X \in U$. Conversely, if $\|x+U\|_{X/U} = 0$ then

$$\inf_{u\in U} \|x+u\|_X = 0,$$

hence for every $\varepsilon > 0$ there exists $u \in U$ such that $\|x+u\|_X < \varepsilon$. From this we can infer the existence of a sequence $\{u_k\} \subset U$ such that

$$\lim_{k\to\infty} \|x+u_k\|_X = 0.$$

But this implies $x + u_k \to 0$, or $u_k \to -x$. Since U is closed we have $-x \in U$, hence $x + U = U$. Next,

$$\begin{aligned}\|\alpha(x+U)\|_{X/U} = \|\alpha x + U\|_{X/U} &= \inf_{u\in U} \|\alpha x + u\|_X = |\alpha| \inf_{u\in U} \left\| x + \frac{1}{\alpha}u \right\|_X \\ &= |\alpha| \inf_{u\in U} \|x+u\|_X = |\alpha|\ \|x+U\|_{X/U}\,.\end{aligned}$$

Finally

$$\begin{aligned}\|(x+U)+(y+U)\|_{X/U} = \|(x+y)+U\|_{X/U} &= \inf_{u\in U} \|(x+y)+u\|_X \\ &= \inf_{u,u'\in U} \|(x+y)+u+u'\|_X \\ &= \inf_{u,u'\in U} \|(x+u)+(y+u')\|_X\end{aligned}$$

so that

$$\begin{aligned}\|(x+U)+(y+U)\|_{X/U} &\le \inf_{u,u'\in U} (\|(x+u)\|_X + \|(y+u')\|_X) \\ &= \inf_{u,u'\in U} \|(x+u)\|_X + \inf_{u,u'\in U} \|(y+u')\|_X \\ &= \inf_{u\in U} \|(x+u)\|_X + \inf_{u'\in U} \|(y+u')\|_X \\ &= \|(x+U)\|_{X/U} + \|(y+U)\|_{X/U}\,,\end{aligned}$$

and the triangle inequality holds.

Now suppose X is complete. Choose a Cauchy sequence $\{y_k + U\} \subset X/U$. A "diagonal sequence" argument may be used to extract a subsequence $\{x_k + U\}$ of $\{y_k + U\}$ such that

$$\begin{aligned}\|(x_2+U)-(x_1+U)\|_{X/U} &< 1/2, \\ \|(x_3+U)-(x_2+U)\|_{X/U} &< 1/2^2, \\ &\vdots\end{aligned}$$

i.e., such that

$$\|(x_{k+1}+U)-(x_k+U)\|_{X/U}=\|(x_{k+1}-x_k)+U\|_{X/U}<1/2^k$$

for each k. Then by definition of $\|\cdot\|_{X/U}$ we can assert the existence of an element $u_k \in (x_{k+1}-x_k)+U$ having $\|u_k\|_X < 1/2^k$. Choose a sequence $\{z_k\} \subset X$ such that for each k

$$z_k \in x_k + U, \qquad z_{k+1}-z_k = u_k.$$

(We indicate how this is done; see Bachman [Bachman and Narici (1966)] for a more formal argument. Choose $z_1 \in x_1 + U$. We now wish to choose z_2 so that $z_2 \in x_2 + U$ and $z_2 - z_1 = u_1$. Write

$$u_1 = x_2 - x_1 + v \quad \text{for some } v \in U$$

and also

$$z_1 = x_1 + w \quad \text{for some } w \in U.$$

Then $u_1 + x_1 = x_2 + v$; add w to both sides and let $v + w = w' \in U$ to get

$$z_1 + u_1 = x_2 + w'.$$

Hence define $z_2 = x_2 + w'$. Repeat this procedure to generate $z_3, z_4, \ldots$.) Then

$$\|z_{k+1}-z_k\|_X < 1/2^k.$$

If $m > n$ then

$$\begin{aligned}\|z_m - z_n\|_X &\le \|z_m - z_{m-1}\|_X + \cdots + \|z_{n+1}-z_n\|_X \\ &< \frac{1}{2^{m-1}} + \cdots + \frac{1}{2^n} < \frac{1}{2^{n-1}}\end{aligned}$$

so $\{z_k\}$ is Cauchy in X. Since X is complete, $z_k \to z$ for some $z \in X$. By the way the z_k were defined we have $x_k + U = z_k + U$. Then

$$\begin{aligned}\|(x_k+U)-(z+U)\|_{X/U} &= \|(z_k+U)-(z+U)\|_{X/U} \\ &= \|(z_k-z)+U\|_{X/U} \\ &= \inf_{u\in U}\|(z_k-z)+u\|_X \\ &\le \|z_k-z\|_X \to 0\end{aligned}$$

so that $x_k + U \to z + U$. We have therefore shown that some subsequence of the Cauchy sequence $\{y_k + U\}$ has a limit.

Exercise 3.5. Since X is separable it has a countable dense subset A. The set

$$S = \{[x] \colon x \in A\} \subseteq X/M$$

is evidently countable; let us show that it is also dense in X/M. Because the norm on X/M is given by

$$\| [x] \| = \inf_{m \in M} \|x + m\| ,$$

the distance between any two of its elements $[x]$ and $[y]$ can be expressed as

$$\| [x] - [y] \| = \| [x - y] \| = \inf_{m \in M} \|(x - y) + m\| .$$

So let $[z] \in X/M$ and $\varepsilon > 0$ be given. We can find $w \in A$ such that $\|z - w\| < \varepsilon$. Then the distance between $[z]$ and $[w]$ is given by

$$\begin{aligned} \inf_{m \in M} \|(z - w) + m\| &\leq \inf_{m \in M} (\|z - w\| + \|m\|) \\ &= \|z - w\| + \inf_{m \in M} \|m\| \\ &= \|z - w\| \\ &< \varepsilon. \end{aligned}$$

The element $[w]$ belongs to S and lies within distance ε of $[z]$ in the space X/M.

Exercise 3.6. Let us propose a linear mapping T: to each $[x] \in X/M$ there corresponds the image element $T([x]) = Ax_0$, where x_0 is that representative of $[x]$ which has minimum norm. (The existence of x_0 is guaranteed because M is closed.) We have

$$\|x_0\|_X = \| [x] \|_{X/M} ,$$

so

$$\|T([x])\|_Y = \|Ax_0\|_Y \leq c \|x_0\|_X = c \| [x] \|_{X/M} .$$

Therefore T is bounded.

Exercise 3.7. Let T be defined by $T([x]) = A\bar{x}$, where $\bar{x}$ is the minimum-norm representative of $[x]$. Take a bounded sequence $\{ [x]_n \}$ from X/M so that $\| [x]_n \|_{X/M} < R$ for some finite R. For each n, choose from $[x]_n$ the minimum-norm representative $\bar{x}_n$. We have $T([x]_n) = A\bar{x}_n$ for each n, and the sequence $\{\bar{x}_n\}$ is bounded (in X) because $\|\bar{x}_n\|_X = \| [x]_n \|_{X/M}$. By compactness of A, there is a subsequence $\{\bar{x}_{n_k}\}$ such that $\{A\bar{x}_{n_k}\}$ is a Cauchy sequence in X. Therefore $\{ [x]_n \}$ contains a subsequence $\{ [x]_{n_k} \}$ whose image under T is a Cauchy sequence in X.

Exercise 3.8. (a) Let e_n denote the sequence with nth term 1 and remaining terms 0. Each $e_n \in \ell^2$, and any finite set $\{e_1, \ldots, e_N\}$ is linearly independent. (b) For any positive integer n we have

$$\lim_{p \to \infty} \left(\sum_{k=1}^{n} |x_k|^p \right)^{1/p} = \max_{1 \leq k \leq n} |x_k| \leq \sup_{k \geq 1} |x_k|$$

so that

$$\lim_{n\to\infty}\lim_{p\to\infty}\left(\sum_{k=1}^{n}|x_k|^p\right)^{1/p}=\lim_{p\to\infty}\left(\sum_{k=1}^{\infty}|x_k|^p\right)^{1/p}\le\sup_{k\ge1}|x_k|.$$

But for each $k\ge1$

$$|x_k|\le\lim_{p\to\infty}\left(\sum_{k=1}^{\infty}|x_k|^p\right)^{1/p}$$

so that

$$\sup_{k\ge1}|x_k|\le\lim_{p\to\infty}\left(\sum_{k=1}^{\infty}|x_k|^p\right)^{1/p}.$$

Hence

$$\lim_{p\to\infty}\left(\sum_{k=1}^{\infty}|x_k|^p\right)^{1/p}=\sup_{k\ge1}|x_k|.$$

(c) Assume $q\ge p$. Note that $a\le1$ implies $a^q\le a^p$. If $a_k\le1$ for each k then, we have

$$\sum_{k=1}^{n}(a_k)^q\le\sum_{k=1}^{n}(a_k)^p.$$

Because

$$|x_k|=(|x_k|^p)^{1/p}\le\left(\sum_{j=1}^{n}|x_j|^p\right)^{1/p}=\|\mathbf{x}\|_p, \tag{A.0.1}$$

we have $|x_k|/\|\mathbf{x}\|_p\le1$ for each k, and shall momentarily let $|x_k|/\|\mathbf{x}\|_p$ play the role of a_k above. Now

$$\frac{(\|\mathbf{x}\|_q)^q}{(\|\mathbf{x}\|_p)^q}=\sum_{k=1}^{n}\left(\frac{|x_k|}{\|\mathbf{x}\|_p}\right)^q\le\sum_{k=1}^{n}\left(\frac{|x_k|}{\|\mathbf{x}\|_p}\right)^p=1.$$

Hence $(\|\mathbf{x}\|_q)^q\le(\|\mathbf{x}\|_p)^q$, and the desired inequality follows. (d) To see that $\ell^1\subseteq\ell^p$, observe that

$$\sum_{k=1}^{\infty}|x_k|^p\le\left(\sum_{k=1}^{\infty}|x_k|\right)^p=(\|\mathbf{x}\|_1)^p$$

so $\|\mathbf{x}\|_p\le\|\mathbf{x}\|_1$. If $\mathbf{x}\in\ell^1$ then $\|\mathbf{x}\|_1<\infty$, hence $\|\mathbf{x}\|_p<\infty$ so $\mathbf{x}\in\ell^p$. The inclusion $\ell^p\subseteq\ell^q$ follows from the inequality of part (c). Finally, we may take the supremum of (A.0.1) to obtain $\|\mathbf{x}\|_\infty\le\|\mathbf{x}\|_p$. The inclusion $\ell^p\subseteq\ell^\infty$ follows. (e) Every summable sequence converges to zero, every sequence that converges to zero converges, and every convergent sequence is bounded. (f) Let $p<\infty$ and let $\{\mathbf{x}^n\}$ be a Cauchy sequence in ℓ^p. Each $\mathbf{x}^n = (x_1^n, x_2^n, \ldots, x_k^n, \ldots)$. Let $\varepsilon > 0$ be given and choose N such that whenever $m, n > N$,

$$(\|\mathbf{x}^m - \mathbf{x}^n\|_p)^p = \sum_{k=1}^{\infty} |x_k^m - x_k^n|^p < \varepsilon^p. \tag{A.0.2}$$

Suppose $m \geq n$ and fix $n > N$. By (A.0.1) we have for each k

$$|x_k^m - x_k^n| \leq \|\mathbf{x}^m - \mathbf{x}^n\|_p < \varepsilon;$$

hence, for each k the sequence $\{x_k^m\}$ is a Cauchy sequence in $\mathbb{R}$. By completeness of $\mathbb{R}$ we have $x_k^m \to x_k$, say. Now let $\mathbf{x} = (x_1, x_2, \ldots, x_k, \ldots)$. We will show that $\mathbf{x}^n \to \mathbf{x}$. By (A.0.2) for any finite j we have

$$\sum_{k=1}^{j} |x_k^m - x_k^n|^p < \varepsilon^p.$$

Hence

$$\lim_{m\to\infty} \sum_{k=1}^{j} |x_k^m - x_k^n|^p \leq \lim_{m\to\infty} \varepsilon^p$$

which gives us

$$\sum_{k=1}^{j} |x_k - x_k^n|^p \leq \varepsilon^p.$$

As $j \to \infty$ we therefore have

$$\sum_{k=1}^{\infty} |x_k - x_k^n|^p \leq \varepsilon^p.$$

In other words $\|\mathbf{x} - \mathbf{x}^n\|_p \leq \varepsilon$ whenever $n > N$, hence $\mathbf{x}^n \to \mathbf{x}$. To see that $\mathbf{x} \in \ell^p$ we write

$$\|\mathbf{x}\|_p \leq \|\mathbf{x} - \mathbf{x}^{N+1}\|_p + \|\mathbf{x}^{N+1}\|_p \leq \varepsilon + \|\mathbf{x}^{N+1}\|_p < \infty.$$

Now consider the case $p = \infty$. Let $\{\mathbf{x}^n\}$ be a Cauchy sequence in ℓ^∞. Each $\mathbf{x}^n = (x_k^n)_{k=1}^{\infty}$. Fix $\varepsilon > 0$ and choose N such that whenever $m, n > N$,

$$\sup_k |x_k^m - x_k^n| < \varepsilon.$$

Suppose $m \geq n$ and fix $n > N$. For each k

$$|x_k^m - x_k^n| < \varepsilon, \tag{A.0.3}$$

hence for each k the sequence $\{x_k^m\}$ is a Cauchy sequence of real numbers. By completeness of $\mathbb{R}$ we have $x_k^m \to x_k$, say. Now let $\mathbf{x} = (x_k)_{k=1}^{\infty}$ and show that

$\mathbf{x}^n \to \mathbf{x}$. As $m \to \infty$ (A.0.3) gives

$$|x_k - x_k^n| \le \varepsilon$$

for each k. Hence

$$\sup_k |x_k - x_k^n| \le \varepsilon$$

for $n > N$, proving that $\mathbf{x}^n \to \mathbf{x}$. Since $\|\mathbf{x} - \mathbf{x}^n\|_\infty \le \varepsilon$ for $n > N$ we have

$$\|\mathbf{x}\|_\infty \le \|\mathbf{x} - \mathbf{x}^{N+1}\|_\infty + \|\mathbf{x}^{N+1}\|_\infty \le \varepsilon + \|\mathbf{x}^{N+1}\|_\infty,$$

hence $\mathbf{x} \in \ell^\infty$. (g) Let $\mathbf{x} = (\xi_1, \xi_2, \ldots) \in \ell^p$. Since $\sum_{k=1}^\infty |\xi_k|^p$ converges we can choose n large enough to make $\sum_{k=n+1}^\infty |\xi_k|^p$ as small as desired. Hence we can approximate $\mathbf{x}$ arbitrarily closely by an element $\mathbf{x}_n$ having the form

$$\mathbf{x}_n = (\xi_1, \xi_2, \ldots, \xi_n, 0, 0, 0, \ldots).$$

Furthermore each ξ_i may be approximated by a rational number r_i. The set S consisting of all elements of the form

$$\mathbf{y}_n = (r_1, r_2, \ldots, r_n, 0, 0, 0, \ldots)$$

is countable and dense in ℓ^p. More formally, let $\varepsilon > 0$ be given. Choose n so that $\sum_{k=n+1}^\infty |\xi_k|^p < \varepsilon^p/2$, then choose the r_i so that $|\xi_i - r_i| < \varepsilon/(2n)^{1/p}$ for each $i = 1, \ldots, n$. We have

$$\|\mathbf{x} - \mathbf{y}_n\|^p = \sum_{k=1}^n |\xi_k - r_k|^p + \sum_{k=n+1}^\infty |\xi_k|^p < n\frac{\varepsilon^p}{2n} + \frac{\varepsilon^p}{2} = \varepsilon^p$$

as desired. (h) Fix any countable subset $\{\mathbf{x}^{(n)}\}_{n=1}^\infty$ of ℓ^∞. Denote the components of $\mathbf{x}^{(n)}$ by

$$\mathbf{x}^{(n)} = (\xi_1^{(n)}, \xi_2^{(n)}, \xi_3^{(n)}, \ldots).$$

We now construct $\mathbf{z} \in \ell^\infty$ such that $\|\mathbf{z} - \mathbf{x}^{(n)}\|_\infty \ge 1$ for all n. Denoting

$$\mathbf{z} = (\zeta_1, \zeta_2, \zeta_3, \ldots)$$

we let

$$\zeta_k = \begin{cases} \xi_k^{(k)} + 1, & |\xi_k^{(k)}| \le 1, \\ 0, & |\xi_k^{(k)}| > 1 \end{cases}$$

for each $k = 1, 2, 3, \ldots$. Then

$$\|\mathbf{z} - \mathbf{x}^{(n)}\|_\infty = \sup_{m \ge 1} |\zeta_m - \xi_m^{(n)}| \ge |\zeta_n - \xi_n^{(n)}| \ge 1$$

as desired. (i) Let S be the set of all vectors whose components form rational sequences that converge to 0. This set is evidently countable. We show that it is dense in c_0. Given $\mathbf{x} = (\xi_1, \xi_2, \ldots) \in c_0$ and $\varepsilon > 0$, choose $\mathbf{y} = (r_1, r_2, \ldots) \in S$ such that $|\xi_i - r_i| < \varepsilon$ for all $i = 1, 2, \ldots$. Then $\|\mathbf{x} - \mathbf{y}\|_\infty = \sup_i |\xi_i - r_i| < \varepsilon$.

Exercise 3.9. Let $\{x_n\}$ be a Cauchy sequence in $(\mathbb{R}, d)$. We first show that $\{x_n\}$ is a Cauchy sequence in $(\mathbb{R}, |\cdot|)$. We have

$$|x_n^3 - x_m^3| = \underbrace{|x_n - x_m|}_{\text{factor 1}} \, \underbrace{|x_n^2 + x_n x_m + x_m^2|}_{\text{factor 2}} \to 0 \quad \text{as } m, n \to \infty.$$

This implies that either factor 1 or factor 2 approaches zero, or both. However, if factor 2 approaches zero then $x_n \to 0$ as $n \to \infty$, and this in turn implies that factor 1 approaches zero. So factor 1 must approach zero in any case.

Next, by the known completeness of $(\mathbb{R}, |\cdot|)$, we can name a limit element $x \in \mathbb{R}$ for $\{x_n\}$.

Finally, we show that $x_n \to x$ in $(\mathbb{R}, d)$. This follows from the equality

$$|x_n^3 - x^3| = |x_n - x| \, |x_n^2 + x_n x + x^2|,$$

because the first factor on the right approaches zero and the second factor is bounded (since $\{x_n\}$ is bounded).

Note that here we have no inequality $|x^3 - y^3| < m|x - y|$ for all x, y in $\mathbb{R}$, but the notions of sequence convergence with both metrics are equivalent. This distinguishes the notion of equivalence of metrics from that of equivalence of norms.

Exercise 3.10. Call

$$\alpha = \sup_{\|x\| \neq 0} \frac{\|Ax\|}{\|x\|}.$$

By linearity of A, α is also equal to the other expression given in the exercise. By definition of supremum we have two things:

(1) For every $\varepsilon > 0$ there exists some $x_0 \neq 0$ such that

$$\frac{\|Ax_0\|}{\|x_0\|} > \alpha - \varepsilon.$$

Equivalently, $\|Ax_0\| > (\alpha - \varepsilon) \|x_0\|$. This implies, by the definition of $\|A\|$, that

$$\alpha - \varepsilon < \|A\|.$$

So $\alpha < \|A\| + \varepsilon$, and since $\varepsilon > 0$ is arbitrary we have $\alpha \leq \|A\|$.

(2) For every $x \neq 0$ we have

$$\frac{\|Ax\|}{\|x\|} \leq \alpha.$$

So $\|Ax\| \leq \alpha \|x\|$ for $x \neq 0$; in fact, this obviously holds when $x = 0$ as well so it holds for all x. By definition of $\|A\|$ we have $\|A\| \leq \alpha$.

Combining the inequalities from parts 1 and 2 we obtain $\|A\| = \alpha$.

Exercise 3.11. We can show that the f_i are linearly dependent if and only if the Gram determinant is zero. Our proof can rest on the fact that a linear homogeneous system $Ax = 0$ has a nontrivial solution if and only if $\det A = 0$.

Assume linear dependence. Then $\sum_{i=1}^{n} \alpha_i f_i = 0$ for some α_i not all zero. Taking inner products of this equation with the f_i in succession, we get

$$\begin{aligned} \alpha_1(f_1, f_1) + \cdots + \alpha_n(f_1, f_n) &= 0, \\ &\vdots \\ \alpha_1(f_n, f_1) + \cdots + \alpha_n(f_n, f_n) &= 0, \end{aligned} \tag{A.0.4}$$

or

$$\begin{pmatrix} (f_1, f_1) & \cdots & (f_1, f_n) \\ \vdots & \ddots & \vdots \\ (f_n, f_1) & \cdots & (f_n, f_n) \end{pmatrix} \begin{pmatrix} \alpha_1 \\ \vdots \\ \alpha_n \end{pmatrix} = \begin{pmatrix} 0 \\ \vdots \\ 0 \end{pmatrix}.$$

A nontrivial solution for the vector (α) implies that the Gram determinant vanishes. Conversely, assume the determinant vanishes so that (A.0.4) holds for some nontrivial (α). Rewrite (A.0.4) as

$$\left(f_i, \sum_{j=1}^{n} \alpha_j f_j \right) = 0, \qquad i = 1, \ldots, n,$$

multiply by α_i to get

$$\left(\alpha_i f_i, \sum_{j=1}^{n} \alpha_j f_j \right) = 0, \qquad i = 1, \ldots, n,$$

and then sum over i to obtain

$$\left(\sum_{i=1}^{n} \alpha_i f_i, \sum_{j=1}^{n} \alpha_j f_j \right) = \left\| \sum_{i=1}^{n} \alpha_i f_i \right\|^2 = 0.$$

Hence $\sum_{i=1}^{n} \alpha_i f_i = 0$ for some scalars α_i that are not all zero.

Exercise 3.12. The statement $\|A_n - A\| \to 0$ means that

$$\|(A_n - A)x\| \le c_n \|x\| \quad \text{where } c_n \to 0$$

and each c_n is independent of x. Since $\|x\| \le M$ for all $x \in S$, we have

$$\|A_n x - Ax\| \le c_n M.$$

But $c_n M \to 0$ together with $c_n \to 0$ when $n \to \infty$, thus $A_n x \to Ax$.

Exercise 3.13. We have

$$\left\| \sum_{n=0}^{\infty} c_n g_n \right\|^2 = \left(\sum_{n=0}^{\infty} c_n g_n \, , \, \sum_{k=0}^{\infty} c_k g_k \right) = \sum_{n=0}^{\infty} |c_n|^2 < \infty .$$

Exercise 3.14. Assume $u(t)$ and $v(t)$ are each differentiable at t. Form the difference quotient

$$\frac{(u(t+h), v(t+h)) - (u(t), v(t))}{h} = \frac{1}{h}(u(t+h), v(t+h)) - \frac{1}{h}(u(t), v(t))$$

and on the right-hand side subtract and add the term

$$\frac{1}{h}(u(t), v(t+h))$$

to write the difference quotient as

$$\left(\frac{u(t+h) - u(t)}{h} \, , \, v(t+h) \right) + \left(u(t) \, , \, \frac{v(t+h) - v(t)}{h} \right) .$$

Then let $h \to 0$.

Exercise 3.15. We can use the Cauchy–Schwarz inequality to write

$$\|x_n\| \, \|x\| \geq |(x_n, x)|$$

for each n, hence

$$\liminf_{n\to\infty} \|x_n\| \, \|x\| \geq \liminf_{n\to\infty} |(x_n, x)| = \lim_{n\to\infty} |(x_n, x)| = |(x, x)| = \|x\|^2 .$$

So

$$\|x\| \liminf_{n\to\infty} \|x_n\| \geq \|x\|^2 .$$

For $x \neq 0$ we can divide through by $\|x\|$ to get the desired inequality. It holds trivially when $x = 0$.

Exercise 3.16. Because A is densely defined, for each $x \in V$ there is a sequence $\{x_n\} \subset D(A)$ such that $x_n \to x$. Since this sequence converges it is a Cauchy sequence. Because A is bounded, $\{Ax_n\}$ is a Cauchy sequence in W, hence converges to some $w \in W$. Furthermore, w does not depend on the Cauchy sequence used. (That is, if $x_n \to x$ and $x'_n \to x$, and $Ax_n \to w$, then $Ax'_n \to w$. Indeed for each n we have,

$$0 \leq \left\| Ax_n - Ax'_n \right\| = \left\| Ax_n - Ax + Ax - Ax'_n \right\| \leq \|A\| \left(\|x_n - x\| + \left\| x - x'_n \right\| \right);$$

as $n \to \infty$ we have $\lim_{n\to\infty} \|Ax_n - Ax'_n\| = 0$ and by continuity of the norm we have the conclusion.) Thus we can define an extension A_e by

$$A_e x = \lim_{n\to\infty} Ax_n = w \quad \text{for any } x \in V.$$

Linearity is evident. Since

$$\|A_e x\| = \left\| \lim_{n\to\infty} Ax_n \right\| = \lim_{n\to\infty} \|Ax_n\| \le \lim_{n\to\infty} \|A\| \, \|x_n\| = \|A\| \, \|x\| \, ,$$

A_e is bounded with $\|A_e\| \le \|A\|$. The reverse inequality follows by noting that $Ax = A_e x$ whenever $x \in D(A)$. Finally, we prove uniqueness: if A'_e is another bounded (hence continuous) linear extension of A, then for any sequence $\{x_n\} \subset D(A)$ with $x_n \to x$ we have

$$A'_e x = \lim_{n\to\infty} A'_e x_n = \lim_{n\to\infty} Ax_n = A_e x,$$

which gives $A'_e = A_e$.

Exercise 3.17. Suppose $v_k \rightharpoonup v$ in V where the dimension of V is n. Choose a basis $\{e_k\}$ of V and write

$$v_k = \sum_{j=1}^{n} \alpha_j^{(k)} e_j, \qquad\qquad v = \sum_{j=1}^{n} \alpha_j e_j.$$

For an arbitrary bounded linear functional f on V we have $f(v_k) \to f(v)$ as $k \to \infty$. For $i = 1, \ldots, n$, put f equal to f_i defined for any $x = \sum_{k=1}^{n} \xi_k e_k$ by $f_i(x) = \xi_i$. Then $f_i(v_k) = \alpha_i^{(k)} \to f_i(v) = \alpha_i$ as $k \to \infty$, and we have

$$\lim_{k\to\infty} \|v - v_k\| = \lim_{k\to\infty} \left\| \sum_{j=1}^{n} (\alpha_j^{(k)} - \alpha_j) e_j \right\| \le \lim_{k\to\infty} \sum_{j=1}^{n} |\alpha_j^{(k)} - \alpha_j| \, \|e_j\| = 0.$$

Exercise 3.18. (a) From $x = AA^{-1}x$ we obtain $\|x\| \le \|A\| \, \|A^{-1}\| \, \|x\|$ and the result follows. (b) Using $x = A^{-1}y$ we have $A\varepsilon = r$, hence $\varepsilon = A^{-1}r$. The four inequalities

$$\begin{aligned} \|x\| &\le \|A^{-1}\| \, \|y\| \, , \\ \|r\| &\le \|A\| \, \|\varepsilon\| \, , \\ \|y\| &\le \|A\| \, \|x\| \, , \\ \|\varepsilon\| &\le \|A^{-1}\| \, \|r\| \, , \end{aligned}$$

follow immediately and yield the desired result.

Exercise 3.19. Let B be the unit ball in X. The image of the bounded set B under T is precompact; T^{-1} returns this image into B. But a continuous operator maps precompact sets into precompact sets, hence if T^{-1} were bounded then B would be precompact. Since X is infinite dimensional, this is impossible.

Exercise 3.20. (a) Let $F\colon X \to Y$ be an isometry between metric spaces (X, d_X) and (Y, d_Y). Then, by the definition,

$$d_Y(F(x_2), F(x_1)) = d_X(x_2, x_1) \quad \text{for all } x_1, x_2 \in X.$$

Continuity is evident. To see that F is one-to-one, suppose $F(x_2) = F(x_1)$. Then $d_Y(F(x_2), F(x_1)) = 0 = d_X(x_2, x_1)$, so $x_2 = x_1$ by the metric axioms. (b) First suppose $\|Ax\| = \|x\|$ for all $x \in X$. Replacing x by $x_2 - x_1$ we have $\|Ax_2 - Ax_1\| = \|x_2 - x_1\|$ as required. Conversely suppose that $\|Ax_2 - Ax_1\| = \|x_2 - x_1\|$ for any pair $x_1, x_2 \in X$. Putting $x_1 = 0$ and $x_2 = x$ we have the desired conclusion.

Exercise 3.21. Suppose Parseval's equality holds for all f in H. We fix f and use the equality, equation (3.14.4), and continuity to write

$$\begin{aligned} 0 &= \lim_{n\to\infty} \left(\|f\|^2 - \sum_{k=1}^{n} |(f, g_k)|^2 \right) \\ &= \lim_{n\to\infty} \left\| f - \sum_{k=1}^{n} (f, g_k) g_k \right\|^2 \\ &= \left\| f - \sum_{k=1}^{\infty} (f, g_k) g_k \right\|^2 . \end{aligned}$$

This shows that

$$f = \sum_{k=1}^{\infty} \alpha_k g_k \quad \text{where } \alpha_k = (f, g_k).$$

Exercise 3.22. The inequality

$$\left\| \frac{df}{dx} \right\|_{C(-\infty,\infty)} \le \alpha \, \|f\|_{C^{(1)}(-\infty,\infty)} ,$$

i.e.

$$\sup \left| \frac{df(x)}{dx} \right| \le \alpha \left(\sup |f(x)| + \sup \left| \frac{df(x)}{dx} \right| \right)$$

obviously holds with $\alpha = 1$.

Exercise 3.23. We construct a subset M of the space whose elements cannot be approximated by functions from a countable set. Let α be an arbitrary point of $[0, 1]$. Form M from functions defined as follows:

$$f_\alpha(x) = \begin{cases} 1, & x \ge \alpha, \\ 0, & x < \alpha. \end{cases}$$

The distance from $f_\alpha(x)$ to $f_\beta(x)$ is

$$\|f_\alpha(x) - f_\beta(x)\| = \sup_{x\in[0,1]} |f_\alpha(x) - f_\beta(x)| = 1 \text{ if } \alpha \neq \beta.$$

Take a ball B_α of radius 1/3 about $f_\alpha(x)$. If $\alpha \neq \beta$ then $B_\alpha \cap B_\beta$ is empty.

If a countable subset is dense in the space then each of the B_α must contain at least one element of this subset, but this contradicts Theorem 3.2.2 since the set of balls B_α is of equal power with the continuum.

Exercise 3.24. Let $\{A_n\}$ be a Cauchy sequence in $L(X, Y)$, i.e.,

$$\|A_{n+m} - A_n\| \to 0 \quad \text{as } n \to \infty, \quad m > 0.$$

We must show that there is a continuous linear operator A such that $A_n \to A$. For any $x \in X$, $\{A_n x\}$ is also a Cauchy sequence because

$$\|A_{n+m}x - A_n x\| \leq \|A_{n+m} - A_n\| \, \|x\| \, ;$$

hence there is a $y \in Y$ such that $A_n x \to y$ since Y is a Banach space. For every $x \in X$ this defines a unique $y \in Y$, i.e., defines an operator A such that $y = Ax$. This operator is clearly linear. Since $\{A_n\}$ is a Cauchy sequence, the sequence of norms $\{\|A_n\|\}$ is bounded:

$$\|Ax\| = \lim_{n\to\infty} \|A_n x\| \leq \limsup_{n\to\infty} \|A_n\| \, \|x\| \, .$$

That is, A is continuous.

Exercise 3.25. We can see that the equation $(A+B)x = y$ has a solution for any $y \in Y$ by applying the contraction mapping theorem. Indeed, pre-multiplication by A^{-1} allows us to rewrite this equation as $x = Cx + x_0$ where $C = -A^{-1}B$ and $x_0 = A^{-1}y$. Defining $F(x) = Cx + x_0$, we see that $F(x)$ is a contraction mapping:

$$\|F(x) - F(y)\| = \|Cx - Cy\| \leq \|C\| \, \|x - y\| \, , \quad \|C\| \leq \|A^{-1}\| \, \|B\| < 1.$$

Since the equation $x = F(x)$ has a unique solution $x^* \in X$, so does the original equation.

From $x = A^{-1}Ax$ it follows that $\|x\| \leq \|A^{-1}\| \, \|Ax\|$, hence

$$\|Ax\| \geq \|A^{-1}\|^{-1} \, \|x\| \, .$$

So for any $y \in Y$ we can write

$$\|y\| = \|(A+B)x\| \geq \|Ax\| - \|Bx\| \geq \|A^{-1}\|^{-1} \, \|x\| - \|B\| \, \|x\|$$

and therefore

$$\|x\| \leq (\|A^{-1}\|^{-1} - \|B\|)^{-1} \, \|y\| \, .$$

The desired inequality follows.

Exercise 3.26. First assume that $y = \lambda x$ for some scalar λ. Then

$$|(x, y)| = |(x, \lambda x)| = |\bar{\lambda}||(x, x)| = |\lambda| \, \|x\|^2 = \|x\| \, \|\lambda x\| = \|x\| \, \|y\| ,$$

hence equality holds. Conversely, assume equality holds in (3.9.1). Squaring both sides, we obtain the relation

$$(x, y)\overline{(x, y)} = \|x\|^2 \, \|y\|^2 .$$

Using this it is easily verified that

$$|(y, y)x - (x, y)y|^2 = ((y, y)x - (x, y)y \, , \, (y, y)x - (x, y)y) = 0,$$

hence $(y, y)x - (x, y)y = 0$.

Exercise 3.27. (a) Let us denote $X \backslash S$ by S^c. First suppose that S is open. Let y be an arbitrary point of S. Assume to the contrary that every open ball centered at y contains a point of S^c. In particular, each such ball having radius $1/n$, $n = 1, 2, 3, \ldots$, contains some point $x_n \in S^c$. So there is a sequence $\{x_n\} \subset S^c$ such that $x_n \to y$. But S^c is closed so we must have $y \in S^c$, a contradiction. Conversely, suppose that every point of S is the center of some open ball contained entirely within S. Suppose to the contrary that S is not open. Then S^c is not closed, and there is a convergent sequence $\{z_n\} \subset S^c$ having a limit $y \in S$. This means there are points of $\{z_n\}$ that are arbitrarily close to y, so it is impossible to find a ball centered at y that is contained entirely within S. This contradiction completes the proof. (b) Take an open ball of radius r centered at x, and denote by U the complement of this ball. Now take any sequence $\{x_n\} \subset U$ such that $x_n \to x$. Since $\|x_n - x\| \geq r$ for each n, we have $\|x_0 - x\| \geq r$ by continuity of the norm. This shows that $x_0 \in U$, hence U is closed. So the original ball is open by definition. (c) Let f be continuous and let S be open in Y. The set $f^{-1}(S)$ is open if it is empty, so we suppose it to be nonempty. Choose any $x \in f^{-1}(S)$. Then $f(x) \in S$, and since S is open there is an open ball $B(f(x), \varepsilon)$ contained entirely in S. By continuity there exists a ball $B(x, \delta)$ whose image $f(B(x, \delta))$ is contained in $B(f(x), \varepsilon)$ and therefore in S. So $B(x, \delta)$ is contained in $f^{-1}(S)$. This shows that $f^{-1}(S)$ is open. Next let $f^{-1}(S)$ be open whenever S is open, and pick an arbitrary $x \in X$. The ball $B(f(x), \varepsilon)$ is open so its inverse image is open and contains x. Hence there is a ball $B(x, \delta)$ contained in this inverse image. We have $f(B(x, \delta))$ contained in $B(f(x), \varepsilon)$, so f is continuous at x.

Exercise 3.28. The function

$$f(x) = \begin{cases} 1, & x \text{ rational}, \\ 0, & x \text{ irrational}, \end{cases}$$

can be defined on $\mathbb{R}$. Now for any real number x_0, whether rational or irrational, there are sequences tending to x_0 that consist of purely rational or purely irrational elements (i.e., both the rationals and the irrationals are dense in the reals). For one type of sequence the limit is 1 and for the other type the limit is zero.

Thus at point x_0 there is no limit value and the function is not continuous by definition.

Exercise 3.29. (a) We can write

$$\begin{aligned}
\|Au\|^2_{L^2(0,1)} &= \int_0^1 \left(\int_0^1 k(s,t)u(t)\,dt \right)^2 ds \\
&\le \int_0^1 \left(\int_0^1 |k(s,t)|^2\,dt \right) \left(\int_0^1 u^2(t)\,dt \right) ds \\
&= \left(\int_0^1 \int_0^1 |k(s,t)|^2\,dt\,ds \right) \int_0^1 u^2(t)\,dt \\
&= M^2 \|u\|^2_{L^2(0,1)}
\end{aligned}$$

where

$$M = \left(\int_0^1 \int_0^1 |k(s,t)|^2\,ds\,dt \right)^{1/2}.$$

Therefore

$$\|Au\|_{L^2(0,1)} \le M \|u\|_{L^2(0,1)}$$

and we have $\|A\| \le M$. (b) Since $\|S\mathbf{x}\| = \|\mathbf{x}\|$, we have $\|S\| = 1$.

Exercise 3.30. We have

$$\begin{aligned}
\|Ax - Ay\| &= \max_{t\in[0,1]} \left| \int_0^t x^2(s)\,ds - \int_0^t y^2(s)\,ds \right| \\
&\le \max_{t\in[0,1]} \int_0^t |x(s)+y(s)| \cdot |x(s)-y(s)|\,ds \\
&\le \left(\max_{t\in[0,1]} |x(t)| + \max_{t\in[0,1]} |y(t)| \right) \cdot \max_{t\in[0,1]} |x(t)-y(t)| \cdot \max_{t\in[0,1]} \int_0^t ds \\
&= (\|x\| + \|y\|) \cdot \|x-y\|.
\end{aligned}$$

On any ball of the form $\|x\| \le \frac{1}{2} - \varepsilon$ where $\varepsilon > 0$, we have $\|Ax - Ay\| \le q\,\|x-y\|$ where $q < 1$.

Exercise 3.31. All elements of the form

$$\mathbf{x}_n = \left(1, \frac{1}{2}, \frac{1}{3}, \ldots, \frac{1}{n}, 0, 0, 0, \ldots \right)$$

belong to S. The sequence $\{\mathbf{x}_n\}$ is a Cauchy sequence because for $m \ge 1$ we have

$$\|\mathbf{x}_{n+m} - \mathbf{x}_n\| = \sup_{n+1\le k\le n+m} \frac{1}{k} = \frac{1}{n+1} \to 0 \quad \text{as } n \to \infty.$$

However, the element $\lim_{n\to\infty} \mathbf{x}_n$ does not belong to S.

Exercise 3.32. The Neumann series for $(A-I)^{-1}$ is

$$(A-I)^{-1} = -\sum_{k=0}^{\infty} A^k.$$

So

$$\|(A-I)^{-1}\| \le \sum_{k=0}^{\infty} \|A^k\| \le \sum_{k=0}^{\infty} \|A\|^k = \frac{1}{1-\|A\|}.$$

Exercise 3.33. The reader should verify that the norm axioms are satisfied for the norm in question. Then take a Cauchy sequence $\{(x_k, y_k)\} \subset X \times Y$ so that

$$\begin{aligned}\|(x_m, y_m) - (x_n, y_n)\|_{X\times Y} &= \|(x_m - x_n, y_m - y_n)\|_{X\times Y} \\ &= \max\{\|x_m - x_n\|_X, \|y_m - y_n\|_Y\} \\ &\to 0 \quad \text{as } m, n \to \infty.\end{aligned}$$

This implies that

$$\|x_m - x_n\|_X \to 0 \quad \text{and} \quad \|y_m - y_n\|_Y \to 0 \quad \text{as } m, n \to \infty.$$

So $\{x_k\}$ and $\{y_k\}$ are each Cauchy sequences in their respective spaces X, Y; by completeness of these spaces we have $x_k \to x$ and $y_k \to y$ for some $x \in X$ and $y \in Y$. Finally, we have $(x_k, y_k) \to (x, y)$ in the norm of $X \times Y$:

$$\begin{aligned}\|(x_k, y_k) - (x, y)\| &= \|(x_k - x, y_k - y)\| \\ &= \max\{\|x_k - x\|_X, \|y_k - y\|_Y\} \\ &\to 0 \quad \text{as } k \to \infty.\end{aligned}$$

Exercise 3.34. We have

$$\|y_n - x\| = \left\| \frac{\sum_{i=1}^{n}(x_i - x)}{n} \right\| \le \frac{1}{n}\sum_{i=1}^{n} \kappa_i \quad \text{where } \kappa_i \equiv \|x_i - x\|.$$

Then for any m between 1 and n we can write

$$\begin{aligned}\|y_n - x\| &\le \frac{1}{n}\sum_{i=1}^{m} \kappa_i + \frac{1}{n}\sum_{i=m+1}^{n} \kappa_i \\ &\le \frac{1}{n}\left(m \cdot \max_{1\le i\le m} \kappa_i\right) + \left(\frac{n-m}{n}\right) \cdot \max_{m+1\le i\le n} \kappa_i \\ &\le \frac{1}{n}\left(m \cdot \max_{1\le i\le m} \kappa_i\right) + \max_{i\ge m+1} \kappa_i.\end{aligned}$$

Let $\varepsilon > 0$ be given. Choose and fix m sufficiently large that the second term is less than $\varepsilon/2$. In the first term the quantity in parentheses is then fixed, and we can therefore choose $N > m$ so that the first term is less than $\varepsilon/2$ whenever $n > N$.

Exercise 3.35. Assume $\|\cdot\|_1$ and $\|\cdot\|_2$ have the property that $\|x_n - x\|_1 \to 0$ if and only if $\|x_n - x\|_2 \to 0$. Now suppose to the contrary that there is *no* positive constant C such that $\|x\|_2 \le C\,\|x\|_1$ for all $x \in X$. Then for each positive integer n there exists $x_n \in X$ such that

$$\|x_n\|_2 > n\,\|x_n\|_1 .$$

Define

$$y_n = \frac{1}{\sqrt{n}}\frac{x_n}{\|x_n\|_1}.$$

Then

$$\|y_n\|_1 = \frac{1}{\sqrt{n}} \to 0 \quad \text{as } n \to \infty$$

while

$$\|y_n\|_2 = \frac{1}{\sqrt{n}}\frac{\|x_n\|_2}{\|x_n\|_1} > \frac{1}{\sqrt{n}}\cdot n = \sqrt{n} \to \infty \quad \text{as } n \to \infty.$$

This contradiction shows that the required constant C does exist. Interchange the norms to get the reverse inequality.

Exercise 3.36. We have $|\,\|x_m\| - \|x_n\|\,| \le \|x_m - x_n\| \to 0$ as $m, n \to \infty$, hence the sequence of norms is a Cauchy sequence in $\mathbb{R}$.

Exercise 3.37. Let U be a separable, dense subspace of X. We take a countable dense subset A of U and show that A is also dense in X. Let $x \in X$ and $\varepsilon > 0$ be given. Since U is dense in X there exists $x' \in U$ such that $d(x, x') < \varepsilon/2$. Since A is dense in U there exists $x'' \in A$ such that $d(x', x'') < \varepsilon/2$. So $d(x, x'') < \varepsilon$ as required.

Exercise 3.38. Let X be a Banach space so that any Cauchy sequence in it has a limit. Now let $\sum_{k=1}^{\infty} x_k$ be an absolutely convergent series of elements $x_k \in X$. Denote by s_i the ith partial sum of this series. Now $\{s_i\}$ is a Cauchy sequence in X because for $m > n$ we have

$$\|s_m - s_n\| = \left\| \sum_{k=n+1}^{m} x_k \right\| \le \sum_{k=n+1}^{\infty} \|x_k\| \to 0 \quad \text{as } m, n \to \infty.$$

Therefore $s_i \to s$ for some $s \in X$ by completeness.

Conversely suppose every absolutely convergent series of elements taken from X is convergent. Let $\{x_k\}$ be any Cauchy sequence in X. For every positive integer k we can find $N = N(k)$ such that $\|x_m - x_n\| < 1/2^k$ whenever $m, n > N$; furthermore, we can choose each such N so that $N(k)$ is a strictly increasing function of k. The series $\sum_{k=1}^{\infty}[x_{N(k+1)} - x_{N(k)}]$ converges absolutely:

$$\sum_{k=1}^{\infty} \|x_{N(k+1)} - x_{N(k)}\| < \sum_{k=1}^{\infty} \frac{1}{2^k} = 1.$$

Hence it converges and by definition its sequence of partial sums

$$s_j = \sum_{k=1}^{j}[x_{N(k+1)} - x_{N(k)}] = x_{N(j+1)} - x_{N(1)}$$

converges. Let s be its limit. From the last equality we see that $\{x_{N(j)}\}$ also converges and its limit is $x = s + x_{N(1)}$. But if a subsequence of a Cauchy sequence has a limit the entire sequence converges to it.

Exercise 3.39. It suffices to show that the image of the unit ball, i.e., the set of all vectors $\mathbf{x} \in \ell^2$ having

$$\|\mathbf{x}\|^2 = \sum_{k=1}^{\infty} |x_k|^2 \leq 1,$$

is precompact. We call this image S and show that it is totally bounded (cf., Definition 3.8.2). Let $\varepsilon > 0$ be given. Note that if $\mathbf{z} = A\mathbf{x}$ is any element of S, we have

$$\sum_{n=N+1}^{\infty} |z_n|^2 = \sum_{n=N+1}^{\infty} |2^{-n} x_n|^2 \leq 2^{-2(N+1)} \sum_{n=1}^{\infty} |x_n|^2 \leq 2^{-2(N+1)},$$

hence it is possible to choose $N = N(\varepsilon)$ such that

$$\sum_{n=N+1}^{\infty} |z_n|^2 < \varepsilon^2/2$$

for all $\mathbf{z} \in S$. Now consider the set M of all "reduced" elements of the form $(z_1, \ldots, z_N, 0, 0, 0, \ldots)$ derivable from the elements of S. It is clear that $M \subseteq S$, which is bounded. Besides, the N-tuples of $\mathbf{z}$ belong to a bounded set in the finite dimensional space $\mathbb{R}^N$ in which any bounded set is precompact. Hence there is a finite $\varepsilon^2/2$-net of N-tuples from which for an arbitrary $\mathbf{z}$ we select $(\zeta_1, \ldots, \zeta_N)$ so that

$$\sum_{n=1}^{N} |z_n - \zeta_n|^2 < \varepsilon^2/2.$$

Thus an element $\mathbf{z}^\varepsilon = (\zeta_1, \ldots, \zeta_N, 0, 0, \ldots) \in \ell^2$ is an element of a finite ε-net of S, since

$$\|\mathbf{z} - \mathbf{z}^\varepsilon\|_{\ell^2}^2 = \sum_{n=1}^{N} |z_n - \zeta_n|^2 + \sum_{n=N+1}^{\infty} |z_n|^2 < \varepsilon^2/2 + \varepsilon^2/2 = \varepsilon^2.$$

Exercise 3.40. For $\lambda = 0$ the operator $A - \lambda I$ is the same as A, hence the corresponding resolvent operator is simply A^{-1}. This operator exists; it is the backward-shift operator and its domain is $R(A)$. But $R(A)$ is not dense in ℓ^2 so the conclusion follows.

Exercise 3.41. The ℓ^2-norms of the sequence elements are given by

$$\|\mathbf{x}_k\|_{\ell^2} = \left(\sum_{i=1}^{k} 1^2\right)^{1/2} = k^{1/2}.$$

We see that $\|\mathbf{x}_k\|_{\ell^2} \to \infty$ as $k \to \infty$. But ℓ^2 is a Hilbert space, and in a Hilbert space every weakly convergent sequence is bounded.

Exercise 3.42. It is clear that the sequence $\{\sin kx\}$ converges weakly if and only if the normalized sequence $\{\sqrt{\frac{2}{\pi}}\sin kx\}$ converges weakly. The latter sequence is orthonormal in $L^2(0,\pi)$, and any orthonormal sequence converges weakly to zero. Indeed Bessel's inequality shows that for any orthonormal sequence $\{e_k\}$ and any element $x \in H$ we have

$$\sum_{k=1}^{\infty} |(x, e_k)|^2 < \infty, \qquad \text{hence} \lim_{k\to\infty} (x, e_k) = 0.$$

In the Sobolev space, on the other hand, we have

$$\begin{aligned}\left\|\sqrt{\frac{2}{\pi}}\sin kx\right\|_{W^{1,2}(0,\pi)} &= \left(\int_0^{\pi}\left[\frac{2}{\pi}\sin^2 kx + \frac{2k^2}{\pi}\cos^2 kx\right] dx\right)^{1/2} \\ &= \sqrt{1+k^2} \to \infty \quad \text{as } k \to \infty.\end{aligned}$$

For any subsequence the norms tend to infinity as well. Since any weakly convergent sequence in a Hilbert space is bounded, no subsequence can be weakly convergent.

Exercise 3.43. In the process of introducing Lebesgue integration we obtained the inequality

$$\|F(\mathbf{x})\|_q \le (\operatorname{mes}\Omega)^{\frac{1}{q}-\frac{1}{p}} \|F(\mathbf{x})\|_p, \qquad 1 \le q \le p.$$

So a bound on the norm is $(\operatorname{mes}\Omega)^{\frac{1}{q}-\frac{1}{p}}$. Taking $F = 1$ we see that it is not a simple bound but the norm of the operator.

Exercise 3.44. Since $\{x_n\}$ is an orthonormal sequence, it converges weakly to zero. The image sequence $\{Ax_n\}$ converges strongly to zero by compactness of A.

Exercise 3.45. The subset inclusion $C^{(n)}(\Omega) \subset C(\Omega)$ certainly holds, so the imbedding operator I exists. It is continuous because

$$\|f\|_{C(\Omega)} \le \|f\|_{C^{(k)}(\Omega)},$$

as is seen from the form of the norms on these spaces. We must still show that I is compact.

Take a bounded set $S \subset C^{(n)}(\Omega)$, $n \ge 1$. The image $I(S)$ is uniformly bounded (since it is bounded in the max norm of $C(\Omega)$). Furthermore, S is a bounded subset of $C^{(1)}(\Omega)$. This latter fact, along with the mean value theorem

$$f(\mathbf{y}) - f(\mathbf{x}) = \nabla f(\mathbf{z}) \cdot (\mathbf{y} - \mathbf{x})$$

implies equicontinuity of $I(S)$. (Here $\mathbf{z}$ is an intermediate point on a segment from $\mathbf{x}$ to $\mathbf{y}$.) So $I(S)$ is compact by Arzelà's theorem. Therefore I maps bounded sets into precompact sets as required.

Exercise 3.46. The space P_n with the max norm is complete (since it is a closed subspace of $C(a,b)$) and finite dimensional. Its completion is therefore isomorphic to P_n, and in this sense can be regarded as P_n itself.

Exercise 3.47. We already know that strong convergence implies weak convergence, and this does not depend on the dimension of the space. Let H be an n-dimensional Hilbert space having an orthonormal basis $\{e_1, \ldots, e_n\}$, and suppose $\{x_k\}$ is a sequence of elements in H such that $x_k \rightharpoonup x$. Then

$$\|x_k\|^2 = \sum_{i=1}^{n} |(x_k, e_i)|^2 \to \sum_{i=1}^{n} |(x, e_i)|^2 = \|x\|^2 \quad \text{as } k \to \infty,$$

and we have $x_k \to x$ according to Theorem 3.15.2.

Exercise 3.48. Let M be a closed subspace of a Hilbert space H. Suppose $\{x_n\} \subset M$ converges weakly to $x \in H$. This means that $(x_n, f) \to (x, f)$ for every $f \in H$. Decompose H as $M \oplus M_\perp$. For every $g \in M_\perp$ we have

$$(x, g) = \lim_{n\to\infty} (x_n, g) = 0,$$

so $x \perp M_\perp$. This means that $x \in M$.

Exercise 3.49. (a) Assume S is closed and T is open. Take a sequence $\{x_n\} \subset S \setminus T$ such that $x_n \to x$. Since $\{x_n\} \subset S$, we have $x \in S$. We claim that $x \notin T$. For if not, then x belongs to the open set T and is therefore the center of some small open ball that lies entirely in T — a contradiction. (b) Assume S is open and T is closed. Let $x \in S \setminus T$. Since $x \in S$ we know that x is the center of an open ball that lies entirely in S; we claim that the radius of this ball can be chosen so small that no points of T can belong to it. For if not, then for each n the ball $B(x, 1/n)$ contains a point $x_n \in T$, and the sequence $\{x_n\} \subset T$ is convergent to x. Since T is closed we must have $x \in T$. However, this contradicts the assumption that $x \in S \setminus T$.

Exercise 3.50. For any element f and any $\varepsilon > 0$ we can find an element $f^* \in S$ such that $\|f - f^*\| < \varepsilon/2$. Next, we can approximate f^* with a finite linear sum of system elements up to accuracy $\varepsilon/2$: $\left\|f^* - \sum_k c_k e_k\right\| < \varepsilon/2$. So the same sum approximates f to within accuracy ε.

Exercise 3.51. We can take $\delta = \varepsilon/L$ in the definition of equicontinuity. Since uniform boundedness is given in the problem statement, S satisfies the conditions of Arzelà's theorem.

Exercise 3.52. Suppose S be a compact subset of X. Let $\{y_n\}$ be a convergent sequence in $A(S)$, with $y_n \to y$. We need to show that $y \in A(S)$. The inverse image of $\{y_n\}$ under A is a sequence in S, and contains a convergent subsequence whose limit belongs to S: $x_k \to x \in S$, say. Noting that $\{A(x_k)\}$ is a subsequence of $\{y_n\}$, we have $A(x_k) \to y$. By definition of closed operator it follows that $x \in D(A)$ and $y = Ax$. Since $x \in S$ we have $y \in A(S)$, as desired.

Exercise 3.53. We begin with

$$l|u(x)| \le \left|\int_0^l u(t)\,dt\right| + l\int_0^l |u'(y)|\,dy,$$

square both sides and use the elementary inequality $2|ab| \le a^2 + b^2$ to get

$$l^2|u(x)|^2 \le 2\left|\int_0^l u(t)\,dt\right|^2 + 2l^2\left(\int_0^l |u'(y)|\,dy\right)^2,$$

then integrate this over x:

$$l^2\int_0^l |u(x)|^2\,dx \le 2l\left\{\left|\int_0^l u(t)\,dt\right|^2 + l^2\left(\int_0^l |u'(y)|\,dy\right)^2\right\},$$

so

$$l\int_0^l |u(x)|^2\,dx \le 2\left\{\left|\int_0^l u(t)\,dt\right|^2 + l^2\left(\int_0^l |u'(y)|\,dy\right)^2\right\}.$$

Finally, because of

$$\begin{aligned}\left(\int_0^l |u'(y)|\,dy\right)^2 &= \left(\int_0^l 1\cdot|u'(y)|\,dy\right)^2\\ &\le \int_0^l 1^2\,dy\int_0^l |u'^2(y)|\,dy\\ &= l\int_0^l |u'^2(y)|\,dy\end{aligned}$$

we get

$$l\int_0^l |u(x)|^2\,dx \le 2\left\{\left|\int_0^l u(t)\,dt\right|^2 + l^3\int_0^l |u'^2(y)|\,dy\right\}.$$

Chapter 4

In the following hints, k (with subscripts) denotes Winkler's coefficient, Ω_1, V_1 are subdomains, and γ is a sufficiently smooth curve (may be a part of the boundary).

Exercise 4.1.

(1) *Membrane.* Total potential energy:

$$\frac{1}{2}\int_\Omega \left(\left(\frac{\partial u}{\partial x}\right)^2 + \left(\frac{\partial u}{\partial y}\right)^2 \right) dx\, dy + \frac{1}{2}\int_{\Omega_1} k\,(u(x,y))^2\, dx\, dy + \frac{1}{2}\int_\gamma k_1\,(u(x,y))^2\, ds - \int_\Omega f(x,y)u(x,y)\, dx\, dy.$$

Virtual work principle:

$$\int_\Omega \left(\frac{\partial u}{\partial x}\frac{\partial \varphi}{\partial x} + \frac{\partial u}{\partial y}\frac{\partial \varphi}{\partial y} \right) dx\, dy + \int_{\Omega_1} ku(x,y)\varphi(x,y)\, dx\, dy + \int_\gamma k_1 u(x,y)\varphi(x,y)\, ds = \int_\Omega f(x,y)\varphi(x,y)\, dx\, dy + \int_{\partial\Omega} g(s)\varphi(s)\, ds.$$

(2) *Stretched rod.* Here the notion of Winkler foundation makes no sense, because only longitudinal displacements are taken into account. However, we can suppose that at a point x_0 there is attached a linear spring with coefficient k, acting along the rod (which is analogous to Winkler's foundation). In that case we have the following. Total potential energy:

$$\frac{1}{2}\int_0^l ES(x)u'^2(x)\, dx + \frac{1}{2}\,(ku(x_0))^2 - \int_0^l f(x)u(x)\, dx - Fu(l).$$

Virtual work principle:

$$\int_0^l ES(x)u'(x)v'(x)\, dx + ku(x_0)v(x_0) = \int_0^l f(x)v(x)\, dx + Fv(l).$$

(Consider the case of several springs along the rod as well.)

(3) *Bent beam.* Total potential energy:

$$\frac{1}{2}\int_0^l EI(x)w''^2(x)\, dx + \frac{1}{2}\int_a^b kw^2(x)\, dx + \frac{1}{2}k_1 w^2(x_0)\, dx - \int_0^l f(x)w(x)\, dx - Fw(l).$$

Virtual work principle:

$$\int_0^l EI(x)w''(x)v''(x)\,dx + \int_a^b kw(x)v(x)\,dx + k_1 w(x_0)v(x_0)$$
$$= \int_0^l f(x)v(x)\,dx + Fv(l).$$

Here the region of the foundation is $[a,b]$, $0 \le a < b \le l$. We added a spring with coefficient k_1 at point x_0.

(4) *Plate.* Total potential energy:

$$\frac{D}{2}\int_\Omega \left(w_{xx}^2 + w_{yy}^2 + 2\nu w_{xx}w_{yy} + 2(1-\nu)w_{xy}^2\right) d\Omega$$
$$+\frac{1}{2}\int_{\Omega_1} kw^2\,d\Omega + \frac{1}{2}\int_\gamma k_1 w^2\,ds - \int_\Omega Fw\,d\Omega.$$

Virtual work principle:

$$D\int_\Omega \left(w_{xx}v_{xx} + w_{yy}v_{yy} + \nu\left(w_{xx}v_{yy} + w_{yy}v_{xx}\right) + 2(1-\nu)w_{xy}v_{xy}\right) d\Omega$$
$$+\int_{\Omega_1} kwv\,d\Omega + \int_\gamma k_1 wv\,ds = \int_\Omega Fv\,d\Omega.$$

(5) *3D linearly elastic body.* Total potential energy:

$$\frac{1}{2}\int_V c^{ijkl}e_{kl}(\mathbf{u})e_{ij}(\mathbf{u})\,dV + \frac{1}{2}\int_{\partial V_2} k(\mathbf{u}\cdot\mathbf{n})^2\,dS - \int_V \mathbf{F}\cdot\mathbf{u}\,dV - \int_{\partial V_1} \mathbf{f}\cdot\mathbf{u}\,dS,$$

where $\mathbf{n}$ is the unit outward normal to the boundary. Virtual work principle:

$$\int_V c^{ijkl}e_{kl}(\mathbf{u})e_{ij}(\mathbf{v})\,dV + \int_{\partial V_2} k(\mathbf{u}\cdot\mathbf{n})(\mathbf{v}\cdot\mathbf{n})\,dS = \int_V \mathbf{F}\cdot\mathbf{v}\,dV + \int_{\partial V_1} \mathbf{f}\cdot\mathbf{v}\,dS.$$

Exercise 4.2. For this case the equation of the virtual work principle takes the form

$$\int_\Omega \left(\frac{\partial u}{\partial x}\frac{\partial \varphi}{\partial x} + \frac{\partial u}{\partial y}\frac{\partial \varphi}{\partial y}\right) dx\,dy = \int_\Omega f(x,y)\varphi(x,y)\,dx\,dy + \int_{\partial\Omega_2} g(s)\varphi(s)\,ds.$$

It is valid for all functions $\varphi(x,y) \in C^1(\overline{\Omega})$ such that $\varphi(x,y)|_{\partial\Omega_1} = 0$, when $u = u_0(x,y)$ is a sufficiently smooth solution of the problem under consideration so it satisfies $u(x,y)|_{\partial\Omega_1} = 0$. If $\partial\Omega_1 \cup \partial\Omega_2$ does not cover $\partial\Omega$, this means that on $\Omega \setminus (\partial\Omega_1 \cup \partial\Omega_2)$ there is given zero load and so here $\partial u/\partial n = 0$.

Now the energy inner product takes the same form as for the above considered problems for a membrane $(u,v)_M$, but the energy space $\mathcal{E}_{Mm}$ is the completion of the set of functions $u \in C^1(\overline{\Omega})$ satisfying $u(x,y)|_{\partial\Omega_1} = 0$. On $\mathcal{E}_{Mm}$ the norm induced by the inner product is equivalent to the norm of $W^{1,2}(\Omega)$.

The generalized setup of the problem under consideration is defined by the above equation of the VWP, so $u \in \mathcal{E}_{Mm}$ is a generalized solution if this equation is valid for all $\varphi(x,y) \in \mathcal{E}_{Mm}$.

The minimum problem now takes on the form

$$E_{Mm}(u) = \frac{1}{2}\|u\|_M^2 - \Phi(u),$$

where

$$\Phi(u) = \int_\Omega f(x,y)u(x,y)\,dx\,dy + \int_{\partial\Omega_2} g(s)u(s)\,ds.$$

If

$$f(x,y) \in L^{p_1}(\Omega), \qquad g(s) \in L^{p_2}(\partial\Omega_2), \tag{A.0.5}$$

then $\Phi(u)$ is a linear continuous functional in $\mathcal{E}_{Mm}$. The existence/uniqueness theorem is as follows:

> *Let* (A.0.5) *be valid. In the energy space $\mathcal{E}_{Mm}$ the functional $E_{Mm}(u)$ attains its minimum at $u = u_0$ and the minimizer satisfying the equation of the VWP is unique.*

Exercise 4.3. The total potential energy is now

$$E_{BR}(\mathbf{u}) = \frac{1}{2}\int_0^l ES(x)u'^2(x)\,dx + \frac{1}{2}\int_0^l EI(x)w''^2(x)\,dx$$
$$- \int_0^l f(x)u(x)\,dx - Fu(l) - \int_0^l q(x)w(x)\,dx - Qw(l), \tag{A.0.6}$$

where $q(x)$ is the distributed normal load and Q is the transverse force on the end.

The equation of the VWP is

$$\int_0^l ES(x)u'(x)v'(x)\,dx + \int_0^l EI(x)w''(x)\varphi''(x)\,dx$$
$$= \int_0^l f(x)v(x)\,dx + Fv(l) + \int_0^l q(x)\varphi(x)\,dx + Q\varphi(l). \tag{A.0.7}$$

Now the energy inner product for pairs $\mathbf{u}_i = (u_i, w_i)$ takes the form

$$(\mathbf{u}_1, \mathbf{u}_2)_{BR} = \int_0^l ES(x)u_1'(x)u_2'(x)\,dx + \int_0^l EI(x)w_1''(x)w_2''(x)\,dx.$$

With the boundary conditions $u(0) = 0$ and $w(0) = 0$, $w'(0) = 0$, introduce the energy space $\mathcal{E}_{BR}$. On $\mathcal{E}_{BR}$ its induced norm is equivalent to the norm of

$W^{1,2}(0,l) \times W^{2,2}(0,l)$. The total energy functional now takes the form

$$E_{BR}(\mathbf{u}) = \frac{1}{2} \|u\|^2_{BR} - \Phi_{BR}(u)$$

with

$$\Phi_{BR}(u) = \int_0^l f(x)u(x)\,dx + Fu(l) + \int_0^l q(x)w(x)\,dx + Qw(l).$$

If $f(x) \in L(0,l)$ and $q(x) \in L(0,l)$ the functional $\Phi_{BR}(u)$ is linear and continuous in $\mathcal{E}_{BR}$ and this is enough to state that the total energy functional $E_{BR}(\mathbf{u})$ attains its minimum $\mathbf{u}_0$ in $\mathcal{E}_{BR}$ that is unique. This minimum is a generalized solution to the combined problem under consideration.

Exercise 4.4.

(a) The VWP takes the form

$$\int_0^l EI(x)w''(x)v''(x)\,dx = \int_0^l f(x)v(x)\,dx + \sum_k F_k v(x_k) + \sum_j M_j v'(x_j) + Fv(l),$$

where point force F_k acts at point x_k and point couple M_j acts at point x_j. *Remark:* This is meaningful because the energy space imbeds continuously to the space $C^{(1)}(0,l)$. For membranes and 3-D elastic bodies in the energy setup, point forces are impossible. For a plate we can consider a generalized setup with external point forces acting on the plate.

(b) The generalized setup for countable sets of external point forces and couples is possible when the series $\sum_k F_k$ and $\sum_j M_j$ are absolutely convergent and the the beam ends are clamped, since the corresponding part of the work of external forces $\sum_k F_k v(x_k) + \sum_j M_j v'(x_j)$ is a linear continuous functional in the energy space:

$$\left| \sum_k F_k v(x_k) + \sum_j M_j v'(x_j) \right| \le \max_{[0,l]} |v(x)| \sum_k |F_k| + \max_{[0,l]} |v'(x)| \sum_j |F_j| \le m \|u\|_B.$$

Exercise 4.5. The functional $\Phi(w)$ (the potential) takes the form

$$\Phi(w) = \int_\Omega F(x,y)w(x,y)\,d\Omega + \int_{\partial\Omega} f(s)w(x,y)\,ds + \sum_{k=1}^N F_k w(x_k,y_k).$$

The (self-balance) condition for solvability of the problem is

$$\Phi(ax+by+c) = \int_\Omega F(x,y)(ax+by+c)\,d\Omega + \int_{\partial\Omega} f(s)(ax+by+c)\,ds$$
$$+ \sum_{k=1}^{N} F_k(ax_k+by_k+c) = 0 \quad \text{for all constants } a,b,c.$$

Exercise 4.6. Use the following forms of the kinetic energy functionals. Rod:

$$K = \int_0^l \rho \left(\frac{\partial u}{\partial t}\right)^2 dx.$$

Beam:

$$K = \int_0^l \rho \left(\frac{\partial w}{\partial t}\right)^2 dx.$$

Plate:

$$K = \int_\Omega \rho \left(\frac{\partial w}{\partial t}\right)^2 dx.$$

Exercise 4.7. It is necessary to solve the following simultaneous algebraic equations with respect to $a_1, \ldots, a_n$:

$$\sum_{k=1}^{n} a_k(\varphi_k, \varphi_1)_M = (u_0^*, \varphi_1)_M,$$
$$\sum_{k=1}^{n} a_k(\varphi_k, \varphi_2)_M = (u_0^*, \varphi_2)_M,$$
$$\vdots$$
$$\sum_{k=1}^{n} a_k(\varphi_k, \varphi_n)_M = (u_0^*, \varphi_n)_M.$$

Exercise 4.8. In case of infinite dimensional space $\mathcal{E}$ the inequality $\|u\|_A \geq m\|u\|_{\mathcal{E}}$ with constant $m > 0$ independent of u is impossible. Indeed, take an orthonormal sequence $\{e_n\}$ in $\mathcal{E}$, so $\|e_n\|_{\mathcal{E}} = 1$. This sequence converges to zero weakly and thus, because A is compact, we get $\|Ae_n\|_{\mathcal{E}} \to 0$. Then $\|e_n\|_A^2 = (Ae_n, e_n)_{\mathcal{E}} \to 0$ as well.

Exercise 4.9. Use the fact that this set is the set of eigenfunctions of the eigenvalue problem

$$u'' + \lambda^2 u = 0, \qquad u(0) = 0 = u(\pi).$$

What is the energy space for this problem where the set is an orthogonal basis?

Exercise 4.10. We recall only that for each of our problems the operator A is introduced by the following equalities (and the Riesz representation theorem). Beam:

$$(Aw, v)_B = \int_0^l \rho w(x)v(x)\,dx.$$

Plate:

$$(Aw, v)_P = \int_\Omega \rho w(x,y)v(x,y)\,d\Omega.$$

3-D elastic body:

$$(A\mathbf{u}, \mathbf{v})_E = \int_V \rho \mathbf{u}\cdot\mathbf{v}\,dV.$$

These operators have all the properties needed in Theorem 4.11.1, and so the theorem can be formulated for each of the problems without change.

Exercise 4.11. Suppose there is a minimizing sequence $\{x_n\}$ that does not strongly converge to x_0. This means that there is $\varepsilon > 0$ and a subsequence $\{x_{n_k}\}$ such that $\|x_0 - x_{n_k}\|_H > \varepsilon$. But $\{x_{n_k}\}$ is a minimizing sequence as well, and so it contains a subsequence that strongly converges to a minimizer (by the theorem). By uniqueness this minimizer is x_0, which contradicts the above inequality.

Exercise 4.12. Suppose that for g_1 and g_2 we get solutions $w_1^* + g_1$ and $w_2^* + g_2$. Then $(g_2 - g_1)|_{\partial\Omega} = 0$. Consider the "difference" of the corresponding equations. We come to the same problem for $w_3 = w_2 - w_1$ with $f = 0$ and the function $(g_1 - g_2)$ taken as g. This problem, by the theorem, has a unique solution w_3^*. By the structure of the equation of the problem it is evident that $w_3^* = g_1 - g_2$, and so $w_1^* + g_1 = w_2^* + g_2$.

References

Adams, R.A. *Sobolev Spaces.* Academic Press, New York, 1975.

Antman, S.S. *Nonlinear Problems of Elasticity.* Springer–Verlag, New York, 1996.

Bachman, G., and Narici, L. *Functional Analysis.* Academic Press, New York, 1966.

Brechtken-Manderschied, U. *Introduction to the Calculus of Variations.* Chapman and Hall, London, 1991.

Ciarlet, P.G. *Mathematical Elasticity.* North Holland, 1988–2000.

Cloud, M.J., and Drachman, B.C. *Inequalities with Applications to Engineering.* Springer–Verlag, New York, 1998.

Courant, R., and Hilbert, D. *Methods of Mathematical Physics, Volume 1.* Wiley, New York, 1989.

Ewing, G.M. *Calculus of Variations with Applications.* Dover Publications, New York, 1985.

Fichera, G. Existence theorems in elasticity (XIII.15), and Boundary value problems of elasticity with unilateral constraints (YII.8, XIII.15, XIII.6), in Handbuch der Physik YIa/2, C. Truesdell, ed., Springer–Verlag, 1972.

Fox, C. *An Introduction to the Calculus of Variations.* Dover Publications, New York, 1988.

Friedman, A. *Variational Principles and Free-Boundary Problems.* Wiley, New York, 1982.

Gelfand, I.M., and Fomin, S.V. *Calculus of Variations.* Prentice–Hall, Englewood Cliffs, NJ, 1963.

Griffel, D.H. *Applied Functional Analysis.* Dover Publications, New York, 2002.

Hanson, G.W., and Yakovlev, A.B. *Operator Theory for Electromagnetics.* Springer–Verlag, New York, 2001.

Hardy, G.H., Littlewood, J.E., and Pólya, G. *Inequalities.* Cambridge University Press, 1952.

Kinderlehrer, D., and Stampacchia, G. *An Introduction to Variational Inequalities and their Applications.* Academic Press, 1980.

Koenig, H.A. *Modern Computational Methods.* Taylor & Francis, Philadelphia, 1998.

Kolmogorov, A.N., and Fomin, S.V. *Elements of the Theory of Functions and Functional Analysis.* Dover Publications, New York, 1999.

Lanzcos, C. *The Variational Principles of Mechanics.* Dover Publications, New York, 1986.

Lebedev, L.P., and Cloud, M.J. *Tensor Analysis.* World Scientific, Singapore, 2003.

Lebedev, L.P., Vorovich, I.I., and Gladwell, G.M.L. *Functional Analysis: Applications in Mechanics and Inverse Problems.* Kluwer Academic Publishers, 1996.

Lebedev, L.P., and Vorovich, I.I. *Functional Analysis in Mechanics.* Springer–Verlag, New York, 2002.

Mikhlin, S.G. *The Problem of Minimum of a Quadratic Functional.* Holden–Day, San Francisco, 1965.

Miklavcic, M. *Applied Functional Analysis and Partial Differential Equations.* World Scientific, Singapore, 1998.

Mura, T., and Koya, T. *Variational Methods in Mechanics.* Oxford University Press, New York, 1992.

Nachbin, L. *Introduction to Functional Analysis: Banach Spaces and Differential Calculus.* Marcel Dekker, New York, 1981.

Petrov, I. *Variational Methods in Optimum Control Theory.* Academic Press, New York, 1968.

Pinch, E. *Optimal Control and the Calculus of Variations.* Oxford University Press, Oxford, UK, 1993.

Pugachev, V.S., and Sinitsyn, I.N. *Lectures on Functional Analysis and Applications.* World Scientific, Singapore, 1999.

Reddy, J.N. *Energy Principles and Variational Methods in Applied Mechanics.* Wiley, New York, 2002.

Riesz, F. Über lineare Funktionalgleichungen. Acta Math. 41 71–98, 1918.

Sagan, H. *Introduction to the Calculus of Variations.* Dover Publications, New York, 1993.

Schauder, J. Über lineare, volstetige Funktionaloperationen. Stud. Math. 2, 1–6, 1930.

Sobolev, S.L. *Some Applications of Functional Analysis to Mathematical Physics.* LGU, 1951.

Vorovich, I.I. *Nonlinear Theory of Shallow Shells.* Springer–Verlag, New York, 1999.

Weinstock, R. *Calculus of Variations.* Dover Publications, New York, 1974.

Yosida, K. *Functional Analysis.* Springer–Verlag, New York, 1965.

Index